AF361936

Machine Learning with Metaheuristic Algorithms for Sustainable Water Resources Management

Machine Learning with Metaheuristic Algorithms for Sustainable Water Resources Management

Editor

Ozgur Kisi

MDPI • Basel • Beijing • Wuhan • Barcelona • Belgrade • Manchester • Tokyo • Cluj • Tianjin

Editor
Ozgur Kisi
Civil Engineering
Ilia State University
Tbilisi
Georgia

Editorial Office
MDPI
St. Alban-Anlage 66
4052 Basel, Switzerland

This is a reprint of articles from the Special Issue published online in the open access journal *Sustainability* (ISSN 2071-1050) (available at: www.mdpi.com/journal/sustainability/special_issues/ Machine_Learning_with_Metaheuristic_Algorithms).

For citation purposes, cite each article independently as indicated on the article page online and as indicated below:

LastName, A.A.; LastName, B.B.; LastName, C.C. Article Title. *Journal Name* **Year**, *Volume Number*, Page Range.

ISBN 978-3-0365-1720-9 (Hbk)
ISBN 978-3-0365-1719-3 (PDF)

Contents

About the Editor

Ozgur Kisi

Dr. Kisi has been working as a professor at Ilia State University. He received his Ph.D. in Institute of Science and Technology (Hydraulics Division) at the Istanbul Technical University, Turkey (2003). His research fields are developing novel algorithms and methods towards the innovative solution of hydrologic forecasting and modelling; suspended sediment modelling; forecasting, estimating, spatial and temporal analysis of hydro-climatic variables, such as precipitation, streamflow, suspended sediment, evaporation, evapotranspiration, groundwater, lake level, and water quality parameters; hydro-informatics. He is an active participant in numerous national research projects and supervisor of several M.Sc. and Ph.D. works. Author of several peer reviewed scientific publications. He is serving as an Editorial Board Member of several reputed journals. He has authored more than 500 research articles, 12 chapters, and 30 discussions. He is the recipient of the 2006 International Tison Award.

Preface to "Machine Learning with Metaheuristic Algorithms for Sustainable Water Resources Management"

Management of available water resources needs well planning and prognostication of hydrological parameters (parameters of hydrological cycle such as rainfall, runoff, solar radiation, groundwater, evaporation, and evapotranspiration) is necessary to do this. The prediction of such phenomena is a highly non-linear issue and necessitates the usage of new mathematical methods, called machine learning (ML). There are plenty of new ML methods and their application to water resource areas is very common. ML methods have several advantages, and the main ones are being able to model complex non-linear phenomena using related data without learning any physical relationship and having a fast processing time to reach a solution. However, they also have some disadvantages, and the main one is not having physical basis and not being easily applied to other climatic regions without calibration and training with new data. With the advancement in technology over recent decades (e.g., satellite data), attaining hydrological data is much easier and this provides opportunity to use data-driven ML tools to solve related problems. Related literature indicates that the ML methods are generally tested with point prediction with a high level of uncertainty. Determination of model uncertainty is an important issue in modelling water resources with ML methods and uncertainty analysis should be considered for an efficient decision making. The use of a high number of data (quantity of input variable and data) is very essential in developing ML models, otherwise correctly tuning hyperparameters will be very hard especially for the advanced algorithms.

This book involves several studies mainly covering the application of new ML methods or algorithms in modeling hydrological and water resources phenomena, such as streamflow, stage-discharge relationship, flood routing, and ground water level. Modelling streamflow as a main component of hydrological cycle is an important issue in water resource management. Forecasting streamflow is essential for planning and management of water resources, including early flood warning and flood mitigation, planning and operating reservoirs, hydro-electricity production, water supply for industry or domestic use, and managing droughts.

The observation of water level (stage) and river discharge is very important in water resource planning and management. Adequate estimation of the stage-discharge relationship is essential in designing hydraulic structures such as dams, canals, bridges, and culverts. In some cases (e.g., compound or dynamic streams), measuring streamflow may be difficult and not feasible. In such cases, stage-discharge rating curves (RCs) are used. However, simple regression-based RCs cannot produce discharge calculations and, therefore, ML methods have been preferred in developing stage-discharge relationship for a long time.

Flood routing is also essential in water resource management. Floods are catastrophic events and they may cause a great deal of damage, such as loss of life, damaging infrastructure, and other economic loses, etc. The Muskingum method (MM) is widely used for flood routing because of its easy application and accurate calculation. In recent decades, ML algorithms have been successfully used in improving MM for flood routing.

Modeling groundwater (GW) as one of the main components of the hydrological cycle is very important in water resource management. GW is very important source for water supply for industry, domestic use, or irrigation purposes. Accurate prediction of GW level is essential for sustainable management of water resources. ML methods and metaheuristic algorithms have proven less costly, time-consuming, and data-intensive compared to mathematical models that use GW dynamics successfully in modelling GW.

The main aim of this book is to present various implementations of ML methods and metaheuristic algorithms to improve modelling and prediction hydrological and water resources phenomena having vital importance in water resource management. I hope that all readers of this book will benefit from learning about the state-of-the-art ML methods and their applications in hydrological phenomena, such as stage-discharge relationship, flood routing, and groundwater level.

Ozgur Kisi
Editor

Article

Machine Learning Improvement of Streamflow Simulation by Utilizing Remote Sensing Data and Potential Application in Guiding Reservoir Operation

Shaokun He [1], Lei Gu [2], Jing Tian [1], Lele Deng [1], Jiabo Yin [1,3,*], Zhen Liao [1], Ziyue Zeng [4], Youjiang Shen [1] and Yu Hui [5]

1   State Key Laboratory of Water Resources and Hydropower Engineering Science, Wuhan University, Wuhan 430072, China; he_shaokun@whu.edu.cn (S.H.); jingtian@whu.edu.cn (J.T.); leledeng@whu.edu.cn (L.D.); zyliao@whu.edu.cn (Z.L.); yjshen@whu.edu.cn (Y.S.)
2   School of Civil and Hydraulic Engineering, Huazhong University of Science and Technology, Wuhan 430074, China; shisan@hust.edu.cn
3   Hubei Provincial Key Lab of Water System Science for Sponge City Construction, Wuhan University, Wuhan 430072, China
4   Changjiang River Scientific Research Institute, Wuhan 430015, China; zengzy@mail.crsri.cn
5   Changjiang Institute of Survey, Planning, Design and Research, Wuhan 430015, China; whuhy@whu.edu.cn
*   Correspondence: jboyn@whu.edu.cn

Abstract: Hydro-meteorological datasets are key components for understanding physical hydrological processes, but the scarcity of observational data hinders their potential application in poorly gauged regions. Satellite-retrieved and atmospheric reanalysis products exhibit considerable advantages in filling the spatial gaps in in-situ gauging networks and are thus forced to drive the physically lumped hydrological models for long-term streamflow simulation in data-sparse regions. As machine learning (ML)-based techniques can capture the relationship between different elements, they may have potential in further exploring meteorological predictors and hydrological responses. To examine the application prospects of a physically constrained ML algorithm using earth observation data, we used a short-series hydrological observation of the Hanjiang River basin in China as a case study. In this study, the prevalent modèle du Génie Rural à 9 paramètres Journalier (GR4J-9) hydrological model was used to initially simulate streamflow, and then, the simulated series and remote sensing data were used to train the long short-term memory (LSTM) method. The results demonstrated that the advanced GR4J9–LSTM model chain effectively improves the performance of the streamflow simulation by using more remote sensing data related to the hydrological response variables. Additionally, we derived a reservoir operation model by feeding the LSTM-based simulation outputs, which further revealed the potential application of our proposed technique.

Keywords: ungauged basin; machine learning; streamflow simulation; satellite precipitation; atmospheric reanalysis

Citation: He, S.; Gu, L.; Tian, J.; Deng, L.; Yin, J.; Liao, Z.; Zeng, Z.; Shen, Y.; Hui, Y. Machine Learning Improvement of Streamflow Simulation by Utilizing Remote Sensing Data and Potential Application in Guiding Reservoir Operation. *Sustainability* **2021**, *13*, 3645. https://doi.org/10.3390/su13073645

Academic Editor: Ozgur Kisi

Received: 30 January 2021
Accepted: 23 March 2021
Published: 25 March 2021

Publisher's Note: MDPI stays neutral with regard to jurisdictional claims in published maps and institutional affiliations.

1. Introduction

The availability of reliable hydro-meteorological material is an initial yet crucial part of water resource planning and management. Using hydrological simulation (or forecasting) as an example can have significant repercussions on socio-economic growth and development prospects from rainfall-runoff observations [1–5]. However, the scarcity of hydro-meteorological monitoring and inaccessibility issues pose obstacles to conducting effective integrated evaluation research, developing modeling frameworks, and recommending policies for resilience, especially for developing countries. Obtaining access to actual long-series hydro-meteorological processes is worthy of further investigation.

In recent decades, both satellite telemetry and data inversion techniques have been mined deeply, which compensate for the deficiencies of meteorological stations and provide an attractive prospect for ungauged areas [6]. For example, the quantitative precipitation output produced by remote sensing covers a wide range of observations with high spatio-temporal resolution. On the premise of controlling these open-source datasets (e.g., pilot balloon, unmanned aerial vehicle, and satellite), numerous studies have developed data assimilation techniques to further reconstruct long time-series historical climatic processes. This achieved huge success in data-scarce areas [7–9]. Guan et al. [9] evaluated six widely used satellite-derived rainfall products against gauge observations from the Chinese Meteorology Administration and investigated their effect on four different hydrological models over the upper Yellow River Basin in China. Bastola and François [10] constructed two key chronological records of rainfall and potential evapotranspiration for flow simulation modeling in the Lake Chad Basin, and then analyzed the error propagated through a distributed hydrological model. However, traditional hydrological models are more suitable for simulating natural runoff with a consistent assumption of the underlying surface [11]. They would fail in real-life situations where engineering measurements (e.g., dam construction, agricultural irrigation, and inter-basin water diversion) often alter streamflow regime, further resulting in serious overestimates or underestimates in the streamflow variability [12].

As a state-of-the-art model-free approach, machine learning (ML) techniques have started to play an important role in the hydrological time-series process [13]. Since ML can fit the complex high-dimensional relationship, it can be developed as a reliable high-precision model (using satellite and reanalysis data as the input and hydrological streamflow as the response variable), even if the black-box feature makes the physical process ambiguous. Among these ML models, artificial neural networks (ANNs), support vector machines (SVMs), and classification and regression trees (CARTs) are the most prevalent tools. For example, Sadler et al. [14] efficiently ran a sophisticated database by the ML model to predict flood hazards in Dongjiang River, China. Previous studies have demonstrated that these models are competent for short-period simulations but fail in different flow regimes with local optimal solutions and gradient disappearance [15–18]. With feedback from both time-delayed input and output, the emerging recurrent neural networks (RNNs) can retain both short-and long-term information, and thus, RNNs are preferred for complicated dynamic timing and sequential issues [19,20]. Although RNNs still have some limitations (i.e., time-consuming, gradient vanishing, and exploding trouble), they are preferred for dynamic hydrological processes [21,22]. Cheng et al. [23] systematically analyzed an ANN and long short-term memory (LSTM, a modified version of RNN) in long lead-time streamflow forecasting and reported that LSTM could prevail and assist in strategic decisions for water resource management. Fu et al. [24] explored the advantages of LSTM in processing steady streamflow data in a dry period and its ability to capture data features in the rapidly fluctuant streamflow data in a wet period. However, all these cases considered a small-scale watershed area and had relatively complete hydro-meteorological data; in other words, inputs and output flow were highly correlated. Further work is expected to verify the effectiveness of LSTM for a large watershed with remote sensing data.

To this end, we selected China's Hanjiang River Basin for the experiment. The objective of this study was to propose a novel ML-based framework to simulate hydrologic streamflow using remote sensing and to test its potential application value. The remainder of the paper is structured as follows: Section 2 introduces the hydrological characteristics of the study area and the remote sensing data. Section 3 details the hydrological simulation techniques and water management policy. In Section 4, the results of hydrological variables as well as operating performance are presented and discussed. Finally, we end with the conclusion.

2. Study Area and Data

2.1. Study Area

The Hanjiang River, illustrated in Figure 1, was chosen as the case study. The river is the largest tributary of the Yangtze River. It lies between 30.28° and 34.5° N and 106.42° and 114.55° E, with a mainstream length of 1577 km and a total drainage area of 159,000 km^2. It originates from the southern Qin Mountain, flows through the Shanxi and Hubei provinces and converges into the Yangtze River in Wuhan. Characterized by a subtropical monsoon climate and annual precipitation of between 700 and 1100 mm, this basin has abundant water resources; 75% of the total annual precipitation occurs in the flood season (June to September). During the flood season, the sudden rainstorms in early summer and persistent rainfalls in autumn typically induce large-scale flooding [25]. For water resource regulation, a key water conservancy project named the Danjiangkou Reservoir was built in the middle of the Hanjiang River Basin. It not only serves as a source for the Middle Route of the South-to-North Water Diversion Project (MSWDP) but also plays an important role in China's One Belt, One Road construction. Therefore, the sub-basin over Danjiangkou Reservoir is a useful candidate for conducting our proposed approach (in Section 3).

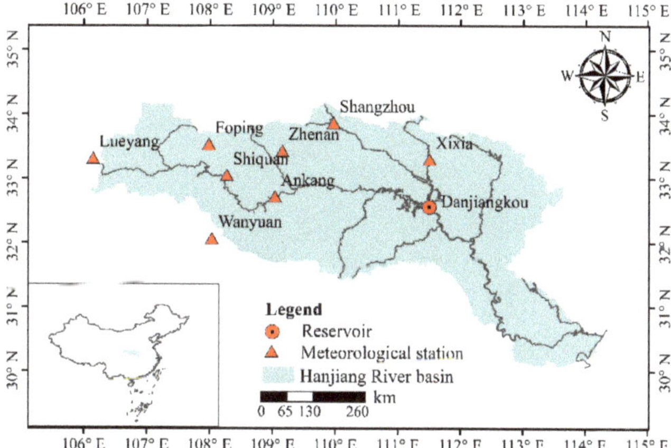

Figure 1. Geographic information of the Danjiangkou Reservoir in the Hanjiang River Basin.

The Danjiangkou Reservoir has eased chronic water shortages in several of China's provinces and urban cities, including the capital, Beijing. An official water diversion diagram developed by the Ministry of Water Resources of China is used for guidance of the water diversion. As shown in Figure 2, it defines a pre-set water diversion value. For example, if the reservoir water level at the initial time of the water diversion is in Region 3 (in Figure 2), the ideal water diversion flow should be 300 m^3/s. Apart from water diversion, the Danjiangkou Reservoir also works as a hydropower source. Water supply and hydropower consist of the main positive purposes of the Danjiangkou Reservoir; these two objectives compete with each other, as part of the reservoir water release is redirected for water diversion instead of power generation. The basic reservoir parameters are listed in Table 1.

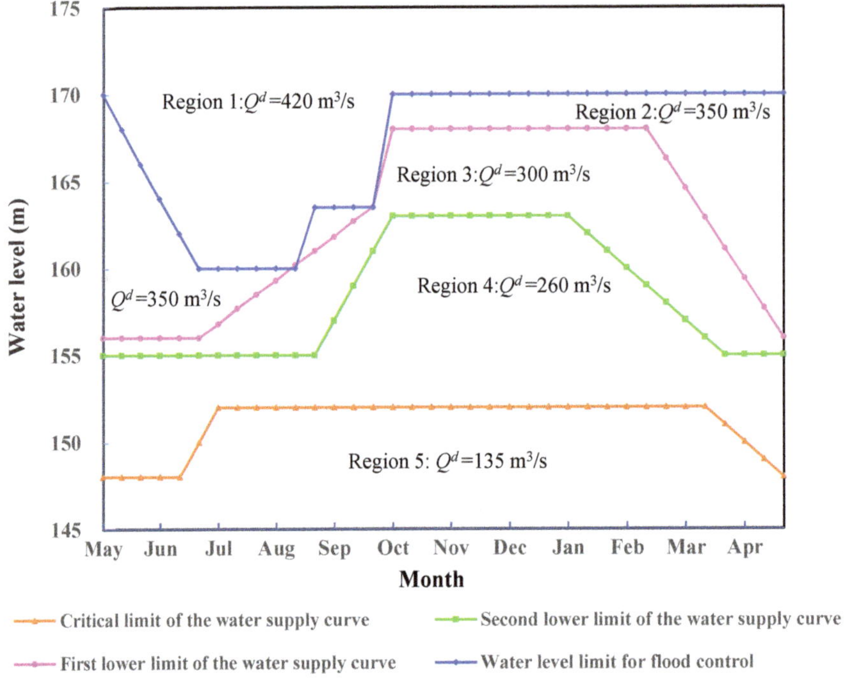

Figure 2. The operation rule curves of the Danjiangkou Reservoir for water supply.

Table 1. Characteristic parameters of the Danjiangkou Reservoir.

Characteristic	Unit	Value
Flood limited water level (FLWL)	m	160.0/163.5
Normal pool level	m	170.0
Crest elevation	m	176.6
Storage capacity for flood control	billion m³	11.44/11.00
Total storage capacity	billion m³	33.91
Guaranteed hydropower capacity	MW	247
Installed hydropower capacity	MW	900

Note: FLWL has two different values for the summer and autumn flood seasons, respectively.

2.2. Data Collection

Three kinds of datasets (i.e., satellite-based observation, atmospheric reanalysis, and short-series streamflow) were collected and used in this study. The GPM Core Observatory is equipped with the first space-borne Ku/Ka-band dual-frequency radar and a multi-channel microwave imager, which improves the monitoring ability of light and solid precipitation. Since the first release of Integrated multi-satellite retrievals for GPM (IMERG) products in 2015, it has undergone many improvements, and the latest version (V06B) has been retrospectively processed, including TRMM-era data since June 2000. Due to the infusion of the Global Precipitation Climatology Centre (GPCC) rain gauge data, the final operation of IMERG provides a more accurate estimation and was therefore adopted in this study.

ERA5 was used as another meteorological product, which is a global atmospheric reanalysis dataset developed by ECMWF. ERA5 data are generated by the combination of model simulations and observations using physics laws, which are based on data assimilation by the Integrated Forecasting System (IFS Cy31r2). This assimilation system includes a four-dimensional variational (4D-Var) analysis method and considers the exact time of observation and model evolution in the assimilation window to estimate the deviation between observations and select high-quality data from poor data. The hourly output resolution is $0.25° \times 0.25°$, which offers a more sophisticated simulation of weather processes. The hourly near-surface air temperature, dew point temperature (T_{dew}), and wind speed from the ERA5 dataset were considered. All the sub-daily satellite/reanalysis data covering 2002–2019 were aggregated into a daily scale.

As the boundary of the upper and mid-lower reaches of the Hanjiang River Basin, the short-series inflow of the Danjiangkou Reservoir was selected to calibrate the parameters of the hydrological model. Observed streamflow data spanning 2003–2007 were obtained from the Yangtze River Water Conservancy Commission.

3. Methodology

A flowchart of the framework module is presented in Figure 3, which is further elaborated in the following sections. It is worth mentioning that our proposed framework can be used in hydrological simulation rather than forecasting.

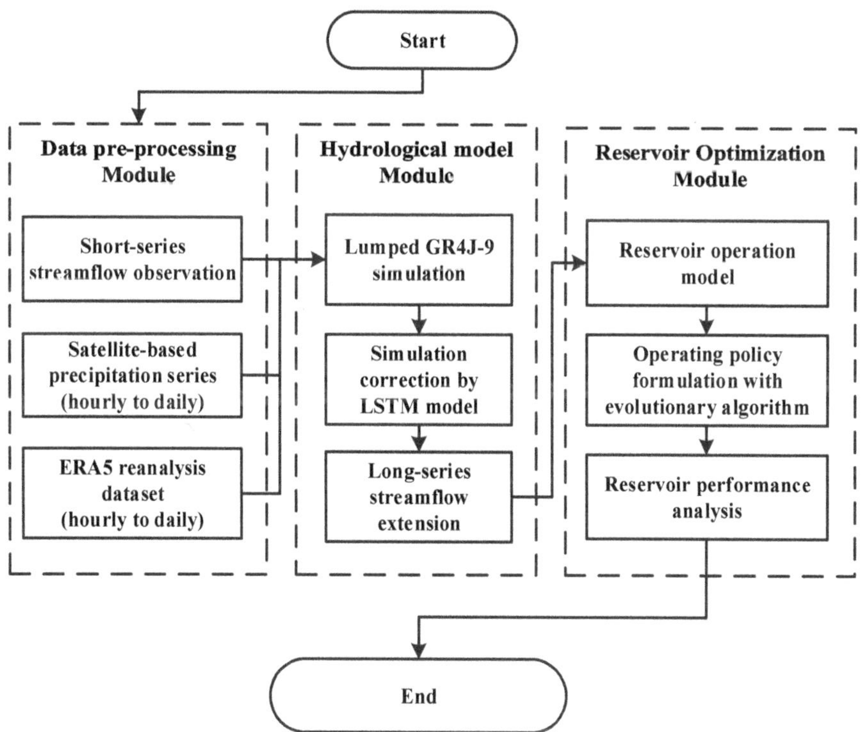

Figure 3. Flowchart of streamflow retrieval and its potential application value.

3.1. Hydrological Model

The modèle du Génie Rural à 9 paramètres Journalier (GR4J-9) hydrological model was used to initially simulate the hydrology of the upper Hanjiang River watershed. The GR4J-9 model is a daily lumped nine-parameter rainfall-runoff model that integrates a traditional GR4J (five-parameter version) hydrological model with the CemaNeige snowfall accumulation and snowmelt module (which occupies four parameters). The GR4J-9 model belongs to the family of soil moisture accounting models that route runoff through two interconnected reservoirs (i.e., production and routing reservoirs) and two-unit hydrographs. It has several main parameters: the maximum capacity of the production reservoir, the groundwater exchange coefficient, the 1-day maximum retention capacity of the routing reservoir, and the time base of the unit hydrograph. This model has been tested in a large sample of catchments and has shown competitive performance compared to more complex models with more parameters [26,27]. Yang et al. [28] showed that the performance of GR4J is more stable than other models (i.e., WASMOD, HBV, and XAJ) in a changing climate. They also set the fixed coefficient of percolation leakage as a free parameter to better fit the study area and calibrated it for the objective watershed. The potential evaporation in the GR4J-9 model is obtained from the temperature-based Oudin method [29].

The observed daily precipitation (P), maximum and minimum temperature (T_{max} and T_{min}, respectively) and flow discharge were fed to calibrate and validate the GR4J-9 model for the experimental watershed. We optimized the parameters of the hydrological model using the Shuffled Complex Evolution (SCE-UA) method developed at the University of Arizona [30]. The SCE-UA method integrates the advantages of several effective global optimization concepts and employs both deterministic search strategies and random schemes to achieve an effective search ability.

3.2. Long Short-Term Memory (LSTM) for Streamflow Simulation

3.2.1. LSTM Model

An LSTM model is a specific kind of RNN designed to overcome the drawbacks caused by a vanishing gradient or exploding in the process of training the RNN using backpropagation through time (BPTT) [31,32]. It sets up a dedicated memory cell that stores information over long periods, potentially making it an ideal candidate for modeling dynamic systems such as watersheds. An unfolded computational graph, as depicted in Figure 4, can reveal the working principle of the LSTM method. One LSTM unit is composed of an input gate, a forget gate, a memory cell and an output gate. The input gate decides which new value regulated by the memory cell will be updated in the cell state, and the forget gate controls the information to remove or retain in the cell state. A general memory block of an LSTM structure can be described by the following equations:

$$C(t+1) = \sigma[w_f X(t+1) + W_f H(t) + b_f] \otimes C(t) + \sigma[w_i X(t+1) + W_i H(t) + b_i] \otimes \tanh[w_c X(t+1) + W_c H(t) + b_c] \tag{1}$$

$$H(t+1) = \sigma[w_o X(t+1) + W_o H(t) + b_o] \otimes \tanh[C(t+1)] \tag{2}$$

where $C(t+1)$ and $C(t)$ are the cell state at time $t+1$ and t, respectively; $X(t+1)$ and $H(t+1)$ are the network input and the recurrent input at time $t+1$, respectively. At the initial time step, both the cell and hidden states are initialized as a vector of zeros. w and W are the weights of the link between gates and layers, respectively; b_i, b_f, b_c, and b_o are learnable bias parameters for each gate; $\sigma[\cdot]$ is the sigmoid function and $\tanh[\cdot]$ is the hyperbolic tangent function; both are activation functions with objective values ranging from 0 to 1.

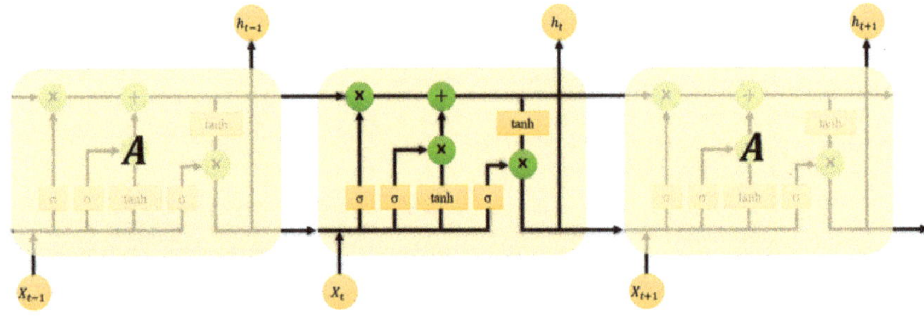

Figure 4. Architecture diagram of the long short-term memory (LSTM) model.

In this study, the two hyperparameters (the initial learning rate and the number of hidden nodes) diversely affecting LSTM model performance needed to be determined. A larger number of hidden neurons lead to a fully trained model, which may overfit the data; conversely, a small number of hidden neurons may cause randomness with high bias. A more detailed process of hyper-parameter tuning is described in Section 4.2. For simplicity, we chose a three-layer LSTM network as the fully connected structure, which consists of one input layer, one hidden layer, and one output layer. Preliminary investigations found that LSTMs with one single hidden layer are capable of simulating streamflow in Hanjiang River Basin [33]. The BPTT algorithm [34] was used to train the LSTM model and the adaptive moment estimation (ADAM) algorithm [35] was employed as the learning rate method. Finally, the mean square error was treated as the loss index. The general model implementation can be accessed from the Statistics and Machine Learning Toolbox of the MATLAB software (website: https://ww2.mathworks.cn/products/statistics.html# machine-learning, accessed on 22 March 2021) [36].

3.2.2. Input Variable Selection (IVS)

The appropriate determination of inputs substantially influences the development of the LSTM model [37]. For streamflow simulation, a dataset of candidate inputs typically covers observed predictors (e.g., initial basin conditions or climate elements) as well as lagged streamflow observations. A wide array of potentially hydrological components can be fed into the model; however, many may only add redundancy or a high level of noise into the model. Furthermore, some candidate inputs (e.g., long-lagged temporal data) may add little or no value to the rainfall–runoff model. To this end, the idea of input variable selection (IVS) [38] is introduced.

A tree-based IVS method developed by Galelli and Castelletti [39] was implemented to identify the optimal input combination. The extremely random tree (extra tree) method is a non-parametric tree regression approach that partitions the input space into mutually exclusive regions on a predefined principle of splitting nodes [40]. In this particular structure, the extra-tree can rank the importance of the input variables by scoring each input variable by evaluation of the relative variance reduction. It adopts a goodness-of-fit criterion of the coefficient of determination (R^2) to systematically select the most significant and non-redundant input space, which was found to consistently indicate the optimum LSTM structure in water resource modeling applications [41].

We used basin-averaged daily mean air temperature, precipitation, wind speed, relative humidity (RH), and the simulated daily discharge (corresponding to the observed reservoir inflow) from the simulations of the GR4J-9 model as the candidate inputs of the LSTM, and used observed daily discharge as the response output. Considering a typical e-folding time scale (recession time) of streamflow, we set the time lag at 4 days.

3.3. Simulation Performance Assessment

To ensure that the trained simulation model did not contain known or detectable defects and could be used on any unseen data, a comprehensive assessment was performed considering different aspects of the modeled simulation flow. Specifically, the Kling–Gupta efficiency (*KGE*) index was selected to describe the statistical accuracy of the hydrological models, and the objective function was to maximize the *KGE* value during the calibration period.

$$KGE = 1 - \sqrt{(r-1)^2 + (\alpha - 1)^2 + (\beta - 1)^2} \tag{3}$$

where r refers to Pearson's linear correlation coefficient between the observation and the simulations, and α (β) indicates the ratio of standard deviations (mean value) of the observed and simulated streamflow. *KGE* varies $(-\infty, 1]$; a value closer to 1 represents a better simulation.

As shown by previous studies [42–44], another metric (mean relative absolute error (*MRAE*)) can be coupled with *KGE* to evaluate the overall deterministic performance.

$$MRAE = \frac{1}{T}\sum_{t=1}^{T}\frac{|y_o^t - y_s^t|}{y_o^t} \tag{4}$$

where *MRAE* varies $[0, \infty)$ with a perfect fit at *MRAE* = 0.

3.4. Policy Optimization for Reservoir Operation

3.4.1. Operation Model

As mentioned in Section 2.1, water supply and power generation are the two main yet conflicting objectives of the Danjiangkou Reservoir. Their mathematical formulations can be expressed by Equations (5) and (6).

$$W = \sum_{t=1}^{T} Q_t^d \cdot \Delta t \tag{5}$$

$$E = \sum_{t=1}^{T} N_t \cdot \Delta t, N_t = k \cdot Q_t^P \cdot H_t \tag{6}$$

where W and E are water supply yield (m^3) and power generation (kW·h) per year, respectively; N_t is the power output at time t (kW); Q_t^d and Q_t^P are water diversion flow and release discharge for power generation at time t (m^3/s), respectively; k is hydropower generation efficiency; H_t is the average water head at time t (m); Δt is the time step (s); and T is the total number of operational periods.

The reservoir operation model obeys some physical constraints, which were outlined by He et al. [45]. The mathematical equations of these constraints are omitted for the sake of brevity.

3.4.2. Operating Strategy of Reservoir Release

The optimal reservoir operation determines the reservoir release sequence Q_t^{out} during the whole operating period for the maximization of W and E. As the optimization strategy of reservoir release involves a high-dimensional and non-linear property, Gaussian radial bias functions (RBFs) are taken as the operating policy, since they are flexible to make decisions with strong universal approximation [46,47]. In the RBFs method, Q_t^{out} can be expressed in Equations (7) and (8).

$$Q_t^{out} = \sum_{u=1}^{U} \omega_u \varphi_u(X_t), \ t \in [1, T] \tag{7}$$

$$\varphi_u(X_t) = \exp\left[-\sum_{m=1}^{M}\frac{((X_t)_m - c_{m,u})^2}{b_u}\right] \quad c_{m,u} \in [-1,1], b_{m,u} \in (0,1) \tag{8}$$

where U is the total number of RBFs $\varphi(\cdot)$; ω_u is the weight of the uth RBF, the sum of all weights is 1, e.g., $\sum_{u=1}^{U} \omega_u = 1$. M is the number of input variables of X_t; and $c_{m,u}$ and b_u are the mth-dimensional center and radius of the uth RBF, respectively. For an individual reservoir, X_t usually consists of three variables, namely time at time t, current reservoir storage (V_t) and reservoir inflow information (Q_t^{in}) [48]; thus, M was set to 3. Moreover, each RBF can be regarded as one pattern of decision-making in reservoir operation based on X_t and, ultimately, decision-making is determined by the combination of four patterns (i.e., U is 4) as suggested by Yang et al. [48].

Consequently, there were 20 parameters to be calibrated for the sum of the RBFs. We optimized the parameter combination based on the parameterization–simulation–optimization (PSO) framework using the non-dominated sorting genetic algorithm II

(NSGA-II). To converge the Pareto front, the evolutionary NSGA-II algorithm adopts the fast non-dominated sorting and crowding distance strategies. The experimental setup of NSGA-II was: population size = 100, generation number = 1000, crossover probability = 0.9, and mutation probability = 0.1.

4. Results and Discussion

4.1. Initial Simulation of the GR4J-9 Model

We first calibrated and validated the GR4J-9 model using observed reservoir inflow during 2003–2007. With the first year serving as the warm-up period, the *KGE* values of the calibration (2004–2006, three years) and validation periods (2007, one year) were 0.76 and 0.54, respectively; the *MRAE* values were 0.56 and 0.73, respectively. As recommended by previous studies, the model performance is judged to be satisfactory for flow simulations if the daily *KGE* is greater than 0.5 and the *MRAE* is less than 0.85 for watershed-scale models [2,49].

Figure 5 depicts the simulation results for reservoir inflow. It shows that the GR4J-9 model could fit the inflow hydrograph of the calibration well, especially for a low inflow regime. However, it achieved a relatively low *KGE* value in the validation period, with a serious underestimation. In detail, the GR4J-9 model cannot capture the peak discharge of 28,900 m^3/s, and instead, gives a lower value of 12,461 m^3/s. Similar results were also found at other different times, which were caused by several aspects. From the view of model inputs, the basin-averaged daily precipitation still has a relatively low resolution compared to actual precipitation observations. From the view of model structure, GR4J-9 has a simple structure with nine model parameters. Although it has superior performance compared to distributed models in the ungauged basin (the latter need more sub-basin observations to calibrate parameters), it inevitably leads to a model error where the simple structure assumption fails to cater to the actual hydrological condition. Therefore, it is necessary to develop a state-of-the-art technique to further improve streamflow accuracy.

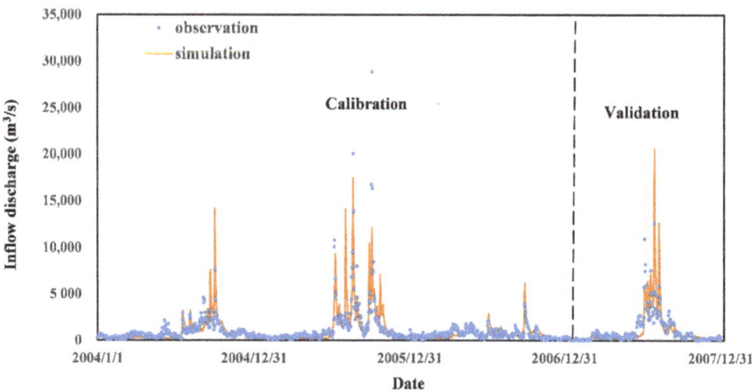

Figure 5. Simulation result of reservoir inflow by the modèle du Génie Rural à 9 paramètres Journalier (GR4J-9) model.

4.2. LSTM Performance

As stated in Section 3.2.2, we fed the GR4J-9 model output into the advanced LSTM model, which also included wind speed and RH. To derive RH data for LSTM inputs, we used the daily dew point temperature (T_{dew}) and daily mean temperature (T_{mean}) from ERA5. The actual vapor pressure (e) and saturated vapor pressure (e_{sa}) were derived using the Clausius–Clapeyron equation. The determined input variables were normalized to eliminate the influence of magnitude, thereby improving the accuracy and efficiency of network learning.

As anticipated, the tree-based IVS algorithm could score and rank input variables in terms of their relevance to the output. To ensure a reliable ranking result, the experiment was cross-validated multiple times with different shuffled datasets, and the inputs were sorted in decreasing order. The score results of the IVS run are presented in Table 2, which illustrates the importance of each selected variable. This tree-based method can make sense for the ex-post physical interpretation of the cause–effect relationships captured by the model. For the upper Hanjiang River Basin, the simulated flow at time t by the GR4J-9 model (Q_t^{sim}), precipitation (P_t), antecedent simulated flow and precipitation with 1- or 2-time lag (Q_{t-1}^{sim}, Q_{t-2}^{sim} and P_{t-1}, respectively) were the top five most important variables (about 68% of the ensemble total score), followed by relative humidity and wind speed at time t (i.e., RH_t and WD_t, respectively). Except for the simulated streamflow element, which was highly related to the observed inflow, we found that P_{t-1} and P_t ranked in the top positions, with relative scores of 13% and 6%, respectively. This may be due to the hydraulic characteristics of this large catchment, which is drained by base flow with a long period of concentration. The basin-averaged RH and wind speed are less important, but not negligible.

Table 2. The top 11 input ranking results for the upper Hanjiang River Basin dataset.

Variable	Q_t^{sim}	Q_{t-1}^{sim}	P_{t-1}	Q_{t-2}^{sim}	P_t	RH_t	WD_t	P_{t-2}	Q_{t-3}^{sim}	Q_{t-4}^{sim}	P_{t-3}
Score (%)	23.84	16.14	13.54	8.38	6.55	5.27	5.03	4.06	2.31	2.01	1.34
Ranking	1	2	3	4	5	6	7	8	9	11	12

Finally, antecedent simulated flow with 1–4 time lags (Q_t^{sim}, Q_{t-1}^{sim}, Q_{t-2}^{sim}, Q_{t-3}^{sim}, and Q_{t-4}^{sim}, respectively), antecedent simulated flow with 1–3 time lags (P_t, P_{t-1}, P_{t-2} and P_{t-3}, respectively), RH_t and WD_t consisted of 11 inputs for the LSTM machine learning model. With a hidden layer of 64 neurons and an initial learning rate of 0.1 (identified by the trial-and-error method), the LSTM-based model could substantially improve the accuracy of the streamflow simulation compared to the GR4J-9 benchmark model. The daily streamflow trajectories are shown in Figure 6a. The model achieved a high *KGE* value of 0.87 and *MRAE* of 0.56 for the daily discharge in the calibration period; additionally, the *KGE* was 0.68 and the *MRAE* value was 0.71 in the validation period. The results demonstrated that this method can competently ameliorate the hydrological data scarcity. Compared to the benchmark GR4J-9 model, the LSTM model uses related hydrological variables (i.e., RH and wind speed) as driving inputs to improve streamflow accuracy. Compared to physically distributed hydrological models, which have limited applications in basins with short-series datasets due to the complex characteristics [50], the LSTM model can sufficiently use remote sensing data and has the features of easy-to-use and highly efficient. However, the LSTM model displayed serious overestimation behavior in the validation period. This is due to the LSTM model being overly reliant on the calibrated data without exception. Figure 6a,b show that the streamflow simulations corrected by the LSTM model are closer to the peak discharge in the calibration. This overfitting performance was unfortunately carried into the validation period, which caused the LSTM model to provide a relatively high value for streamflow discharge under low flow regimes. In general, the LSTM model has room for improvement, but this does not hinder its application value.

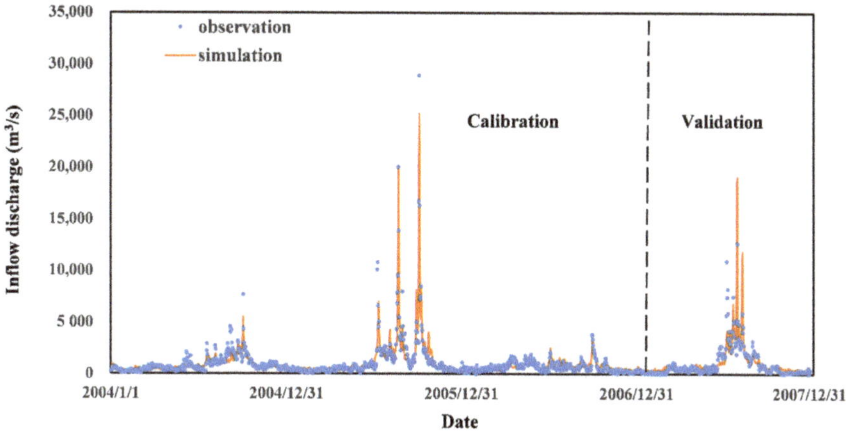

(**a**) Streamflow hydrograph of the whole period (2004–2007)

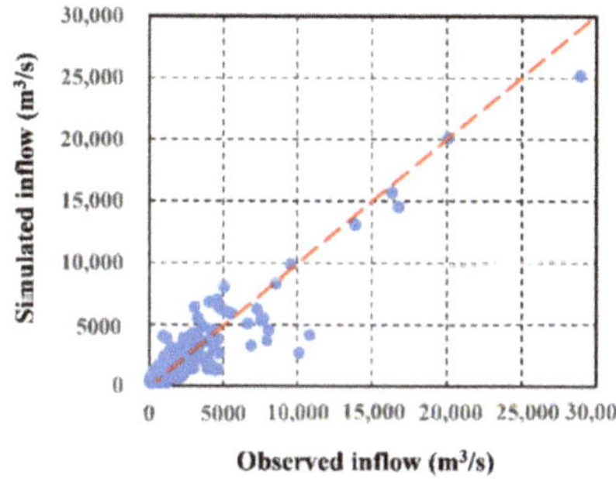

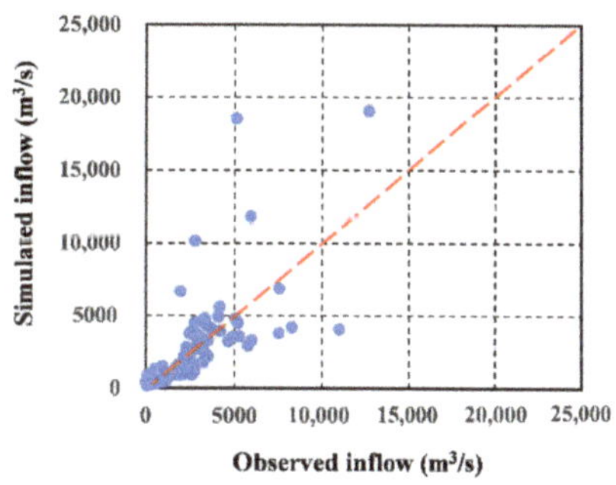

(**b**) the calibration period of 2004–2006

(**c**) the validation period of 2007

Figure 6. Simulation result of reservoir inflow by the long and short-term model (LSTM) model. (**a**) Inflow hydrograph in the whole period; (**b**) Scatter plot of the calibration period; (**c**) Scatter plot of the validation period.

4.3. Potential Application in Reservoir Management

We acquired a long-series daily streamflow simulation from the period of 2008–2019 by feeding the remote sensing data into the calibrated LSTM model. We could formulate more scientific strategies for basins with hydrological data scarcity. Using the medium- and long-term management of the Danjiangkou Reservoir operation as an example, we aimed to improve hydropower benefits and water supply yield and balance them as much as possible. Before performing the NSGA-II reservoir optimization method, daily scale simulated streamflow was converted into 10-day average runoff.

The optimization results of W and E by NSGA-II are presented in Figure 7, and the operation results merely based on short-series observation (2003–2007) are also provided for further comparison. Both of the Pareto fronts under different scenarios are widely and evenly distributed between the two conflicting objectives. Considering the pursuit of maximum economic profit (official electricity price: 0.21 RMB/kWh; water price for the MSWTP project: 0.13 RMB/m³), the final two optimal operating rules were chosen for promising economic prospect, namely Solution I under long-series simulated scenario of 2003–2019 and Solution II under short-series observed scenario of 2003–2007 (in Figure 7).

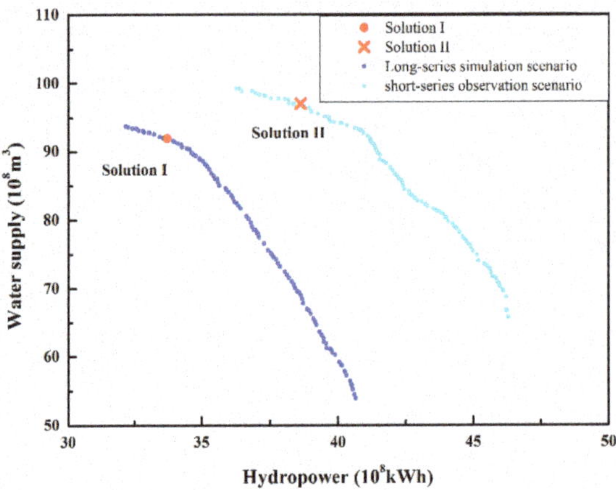

Figure 7. Two sets of Pareto fronts in different streamflow scenarios.

With these two optimal operating rules guiding reservoir operation during the period of 2003–2019, their objective results are summarized in Table 3. It can be inferred from water supply, hydropower, and economic profit (W, E, and H, respectively) that Solution I prefers a higher value of W than Solution II, no matter whether in the observation period or in the whole period when decision-makers want to pursue more economic profit. As shown in Table 3, while the performance of Solution I (obtained through LSTM-based streamflow simulation) is similar to that of Solution II, it provides decision-makers with another viable operating way that requires more real observational data to verify.

Table 3. Objective results of two optimal rules in different periods.

Operating Rule	Different Season	Observation Period (2003–2007)			Whole Period (2003–2019)		
		W (10⁸ m³)	E (10⁸ kWh)	H (10⁸ RMB)	W (10⁸ m³)	E (10⁸ kWh)	H (10⁸ RMB)
Solution I	flood	34.25	14.68	7.54	32.16	13.67	7.05
	non-flood	64.72	22.76	13.19	59.83	20.08	12.00
	annual	98.97	37.44	20.73	91.99	33.75	19.05
Solution II	wet	33.77	15.20	7.58	31.24	13.62	6.92
	dry	63.48	23.62	13.21	58.21	20.85	11.95
	annual	97.25	38.82	20.79	89.45	34.47	18.86

Note: W = water supply, E = hydropower, and H = economic profit.

5. Conclusions

Hydro-meteorological data scarcity impairs hydrological simulation, manifesting a pressing need to develop an alternative scheme in this field. Satellite-based and atmospheric reanalysis estimation may provide feasible access to reproduce the hydrological recycle process and may have potential value. To this end, this paper proposed a novel method to integrate open-source remote sensing data and ML-based inversion techniques for hydrology. Furthermore, we applied it in reservoir management. According to the results, we reached the following conclusions:

(1) Driven by the synthetic data generated by a lumped hydrological GR4J model, satellite-based data and ERA5 could overcome the limitation of historical observation scarcity. With a *KGE* value of 0.54 in the validation period for the upper

Hanjiang River Basin, the traditional lumped hydrological model can be applied to the ungauged basins but there is still room for improvement.

(2) Compared to the traditional GR4J model, the ML-based data-driven model showed its superior performance in capturing the long-series time sequence using a sophisticated network structure. Along with inheriting the simulation output of the traditional hydrological model, the LSTM model can further mine the value of remote sensing data related to hydrological variables. Compared to traditional distributed hydrological models, its model structure is simple and can be highly efficient.

(3) The LSTM-based streamflow simulation scenario can provide the basis for another scientific operation way for reservoir managers, which requires future validation of the potential value of the LSTM-based method.

Despite the outstanding performance of the developed methodology, some work remains for further exploration. First, hydrological models have different behaviors depending on the flow regimes. However, in this study, the hydro-meteorological data with a one-day timescale were fed to drive one set of hydrological models, yet the separation of flood seasons and non-flood seasons was neglected. Besides, a simpler LSTM model should be taken into consideration and compared with our proposed hybrid model for hydrological performance. Secondly, the methodology was merely applied for hydrological simulation rather than hydrological forecasts. Some products such as the Global Ensemble Forecast System (GEFS) Reforecast can be included to improve this methodology. In the future, we will explore the ML-based method with remote sensing data to verify its generalizability in more ungauged basins.

Author Contributions: Conceptualization and software, S.H., L.G. and J.T.; data curation, J.Y. and Z.L.; formal analysis, S.H., Z.Z., and Y.S.; writing—Original draft preparation, S.H.; writing—Review and editing, J.Y., L.D. and Y.H. All authors have read and agreed to the published version of the manuscript.

Funding: This work was funded by the National Natural Science Foundation of China (52009091; 51879192), the Natural Science Foundation of Hubei Province (2020CFB239 and 2020CFB132) and the China Postdoctoral Science Foundation (2020M682478). This work was partly supported by the Fundamental Research Funds for the Central Universities (No. 2042020kf0003) and the Post-Doctoral Innovative Talent Support Program of China (BX20200257). This study was also funded by the Ministry of Foreign Affairs of Denmark, administered by the Danida Fellowship Centre (file number: 18-MCC01-DTU).

Data Availability Statement: The data presented in this study are available on request from the corresponding author.

Acknowledgments: The authors would like to express their gratitude to anonymous reviewers for their insightful and constructive comments.

Conflicts of Interest: The authors declare no conflict of interest.

References

1. Ferreira, R.G.; da Silva, D.D.; Elesbon, A.A.A.; Fernandes-Filho, E.I.; Veloso, G.V.; Fraga, M.D.S.; Ferreira, L.B. Machine learning models for streamflow regionalization in a tropical watershed. *J. Environ. Manag.* **2021**, *280*, 111713. [CrossRef]
2. Gu, L.; Chen, J.; Yin, J.; Xu, C.Y.; Zhou, J. Responses of Precipitation and Runoff to Climate Warming and Implications for Future Drought Changes in China. *Earths Future* **2020**, *8*, 8. [CrossRef]
3. Zhou, Y.; Guo, S.; Chang, F.-J. Explore an evolutionary recurrent ANFIS for modelling multi-step-ahead flood forecasts. *J. Hydrol.* **2019**, *570*, 343–355. [CrossRef]
4. Suwal, N.; Kuriqi, A.; Huang, X.F.; Delgado, J.; Mlynski, D.; Walega, A. Environmental Flows Assessment in Nepal: The Case of Kaligandaki River. *Sustainability* **2020**, *12*, 8766. [CrossRef]
5. Yin, J.; Guo, S.; Gentine, P.; Sullivan, S.C.; Gu, L.; He, S.; Chen, J.; Liu, P. Does the Hook Structure Constrain Future Flood Intensification Under Anthropogenic Climate Warming? *Water Resour. Res.* **2021**, *57*. [CrossRef]
6. Shen, Y.; Liu, D.; Jiang, L.; Yin, J.; Nielsen, K.; Bauer-Gottwein, P.; Guo, S.; Wang, J. On the Contribution of Satellite Altimetry-Derived Water Surface Elevation to Hydrodynamic Model Calibration in the Han River. *Remote Sens.* **2020**, *12*, 4087. [CrossRef]
7. Bastola, S.; Misra, V. Evaluation of dynamically downscaled reanalysis precipitation data for hydrological application. *Hydrol. Process.* **2014**, *28*, 1989–2002. [CrossRef]

8. Weedon, G.P.; Balsamo, G.; Bellouin, N.; Gomes, S.; Best, M.J.; Viterbo, P. The WFDEI meteorological forcing data set: WATCH Forcing Data methodology applied to ERA-Interim reanalysis data. *Water Resour. Res.* **2014**, *50*, 7505–7514. [CrossRef]

9. Guan, X.X.; Zhang, J.Y.; Yang, Q.L.; Tang, X.P.; Liu, C.S.; Jin, J.L.; Liu, Y.; Bao, Z.X.; Wang, G.Q. Evaluation of Precipitation Products by Using Multiple Hydrological Models over the Upper Yellow River Basin, China. *Remote Sens.* **2020**, *12*, 4023. [CrossRef]

10. Bastola, S.; François, D. Temporal extension of meteorological records for hydrological modelling of Lake Chad Basin (Africa) using satellite rainfall data and reanalysis datasets. *Meteorol. Appl.* **2012**, *19*, 54–70. [CrossRef]

11. Mazzoleni, M.; Alfonso, L.; Solomatine, D. Influence of spatial distribution of sensors and observation accuracy on the assimilation of distributed streamflow data in hydrological modelling. *Hydrol. Sci. J.* **2017**, *62*, 389–407. [CrossRef]

12. Kuriqi, A.; Ali, R.; Pham, Q.B.; Gambini, J.M.; Gupta, V.; Malik, A.; Linh, N.T.T.; Joshi, Y.; Anh, D.T.; Nam, V.T.; et al. Seasonality shift and streamflow flow variability trends in central India. *Acta Geophys.* **2020**, *68*, 1461–1475. [CrossRef]

13. Chang, L.C.; Chang, F.J.; Hsu, H.C. Real-Time Reservoir Operation for Flood Control Using Artificial Intelligent Techniques. *Int. J. Nonlin. Sci. Num.* **2010**, *11*, 887–902. [CrossRef]

14. Sadler, J.M.; Goodall, J.L.; Morsy, M.M.; Spencer, K. Modeling urban coastal flood severity from crowd-sourced flood reports using Poisson regression and Random Forest. *J. Hydrol.* **2018**, *559*, 43–55. [CrossRef]

15. Kopeć, A.; Trybała, P.; Głąbicki, D.; Buczyńska, A.; Owczarz, K.; Bugajska, N.; Kozińska, P.; Chojwa, M.; Gattner, A. Application of Remote Sensing, GIS and Machine Learning with Geographically Weighted Regression in Assessing the Impact of Hard Coal Mining on the Natural Environment. *Sustainability* **2020**, *12*, 9338. [CrossRef]

16. Manfreda, S.; Samela, C. A digital elevation model based method for a rapid estimation of flood inundation depth. *J. Flood Risk Manag.* **2019**, *12*. [CrossRef]

17. Solomatine, D.P.; Shrestha, D.L. A novel method to estimate model uncertainty using machine learning techniques. *Water Resour. Res.* **2009**, *45*. [CrossRef]

18. Adnan, R.M.; Zounemat-Kermani, M.; Kuriqi, A.; Kisi, O. Machine Learning Method in Prediction Streamflow Considering Periodicity Component. In *Intelligent Data Analytics for Decision-Support Systems in Hazard Mitigation: Theory and Practice of Hazard Mitigation*; Deo, R.C., Samui, P., Kisi, O., Yaseen, Z.M., Eds.; Springer: Singapore, 2021; pp. 383–403. [CrossRef]

19. Zhang, D.; Peng, Q.; Lin, J.; Wang, D.; Liu, X.; Zhuang, J. Simulating Reservoir Operation Using a Recurrent Neural Network Algorithm. *Water* **2019**, *11*, 865. [CrossRef]

20. Misra, S.; Sarkar, S.; Mitra, P. Statistical downscaling of precipitation using long short-term memory recurrent neural networks. *Theor. Appl. Climatol.* **2017**, *134*, 1179–1196. [CrossRef]

21. Nourani, V.; Baghanam, A.H.; Adamowski, J.; Kisi, O. Applications of hybrid wavelet-Artificial Intelligence models in hydrology: A review. *J. Hydrol.* **2014**, *514*, 358–377. [CrossRef]

22. Zhang, H.B.; Singh, V.P.; Bin Wang, B.; Yu, Y.H. CEREF: A hybrid data-driven model for forecasting annual streamflow from a socio-hydrological system. *J. Hydrol.* **2016**, *540*, 246–256. [CrossRef]

23. Cheng, M.; Fang, F.; Kinouchi, T.; Navon, I.M.; Pain, C.C. Long lead-time daily and monthly streamflow forecasting using machine learning methods. *J. Hydrol.* **2020**, *590*, 125376. [CrossRef]

24. Fu, M.; Fan, T.; Ding, Z.a.; Salih, S.Q.; Al-Ansari, N.; Yaseen, Z.M. Deep Learning Data-Intelligence Model Based on Adjusted Forecasting Window Scale: Application in Daily Streamflow Simulation. *IEEE Access* **2020**, *8*, 32632–32651. [CrossRef]

25. He, S.K.; Guo, S.L.; Yang, G.; Chen, K.B.; Liu, D.D.; Zhou, Y.L. Optimizing Operation Rules of Cascade Reservoirs for Adapting Climate Change. *Water Resour. Manag.* **2020**, *34*, 101–120. [CrossRef]

26. Kunnath-Poovakka, A.; Eldho, T.I. A comparative study of conceptual rainfall-runoff models GR4J, AWBM and Sacramento at catchments in the upper Godavari river basin, India. *J. Earth Syst. Sci.* **2019**, *128*, 33. [CrossRef]

27. Edijatno; Nascimento, N.D.; Yang, X.L.; Makhlouf, Z.; Michel, C. GR3J: A daily watershed model with three free parameters. *Hydrol. Sci. J.* **1999**, *44*, 263–277.

28. Yang, W.S.; Chen, H.; Xu, C.Y.; Huo, R.; Chen, J.; Guo, S.L. Temporal and spatial transferabilities of hydrological models under different climates and underlying surface conditions. *J. Hydrol.* **2020**, *591*, 125276. [CrossRef]

29. Oudin, L.; Hervieu, F.; Michel, C.; Perrin, C.; Andréassian, V.; Anctil, F.; Loumagne, C. Which potential evapotranspiration input for a lumped rainfall–runoff model? *J. Hydrol.* **2005**, *303*, 290–306. [CrossRef]

30. Duan, Q.; Sorooshian, S.; Gupta, V. Effective and efficient global optimization for conceptual rainfall-runoff models. *Water Resour. Res.* **1992**, *28*, 1015–1031. [CrossRef]

31. Hochreiter, S.; Schmidhuber, J. Long short-term memory. *Neural Comput.* **1997**, *9*, 1735–1780. [CrossRef] [PubMed]

32. Bengio, Y.; Simard, P.; Frasconi, P. Learning Long-Term Dependencies with Gradient Descent Is Difficult. *IEEE Trans. Neural Netw.* **1994**, *5*, 157–166. [CrossRef]

33. Hu, Q.; Cao, S.; Yang, H.; Wang, Y.; Li, L.; Wang, L. Daily runoff predication using LSTM at the Ankang Station, Hanjing River. *Prog. Geogr.* **2020**, *39*, 636–642. [CrossRef]

34. Werbos, P.J. Backpropagation through Time-What It Does and How to Do It. *Proc. IEEE* **1990**, *78*, 1550–1560. [CrossRef]

35. Chang, Z.H.; Zhang, Y.; Chen, W.B. Electricity price prediction based on hybrid model of adam optimized LSTM neural network and wavelet transform. *Energy* **2019**, *187*, 115804. [CrossRef]

36. Zhou, Y.L. Real-time probabilistic forecasting of river water quality under data missing situation: Deep learning plus post-processing techniques. *J. Hydrol.* **2020**, *589*, 125164. [CrossRef]

37. Kratzert, F.; Klotz, D.; Shalev, G.; Klambauer, G.; Hochreiter, S.; Nearing, G. Towards learning universal, regional, and local hydrological behaviors via machine learning applied to large-sample datasets. *Hydrol. Earth Syst. Sci.* **2019**, *23*, 5089–5110. [CrossRef]
38. Humphrey, G.B.; Gibbs, M.S.; Dandy, G.C.; Maier, H.R. A hybrid approach to monthly streamflow forecasting: Integrating hydrological model outputs into a Bayesian artificial neural network. *J. Hydrol.* **2016**, *540*, 623–640. [CrossRef]
39. Galelli, S.; Castelletti, A. Tree-based iterative input variable selection for hydrological modeling. *Water Resour. Res.* **2013**, *49*, 4295–4310. [CrossRef]
40. Jaxa-Rozen, M.; Kwakkel, J. Tree-based ensemble methods for sensitivity analysis of environmental models: A performance comparison with Sobol and Morris techniques. *Environ. Model. Softw.* **2018**, *107*, 245–266. [CrossRef]
41. Li, Y.T.; Bao, T.F.; Gong, J.; Shu, X.S.; Zhang, K. The Prediction of Dam Displacement Time Series Using STL, Extra-Trees, and Stacked LSTM Neural Network. *IEEE Access* **2020**, *8*, 94440–94452. [CrossRef]
42. Gupta, H.V.; Kling, H.; Yilmaz, K.K.; Martinez, G.F. Decomposition of the mean squared error and NSE performance criteria: Implications for improving hydrological modelling. *J. Hydrol.* **2009**, *377*, 80–91. [CrossRef]
43. He, S.K.; Guo, S.L.; Liu, Z.J.; Yin, J.B.; Chen, K.B.; Wu, X.S. Uncertainty analysis of hydrological multi-model ensembles based on CBP-BMA method. *Hydrol. Res.* **2018**, *49*, 1636–1651. [CrossRef]
44. Moriasi, D.N.; Arnold, J.G.; Van Liew, M.W.; Bingner, R.L.; Harmel, R.D.; Veith, T.L. Model evaluation guidelines for systematic quantification of accuracy in watershed simulations. *Trans. ASABE* **2007**, *50*, 885–900. [CrossRef]
45. He, S.K.; Guo, S.L.; Chen, K.B.; Deng, L.L.; Liao, Z.; Xiong, F.; Yin, J.B. Optimal impoundment operation for cascade reservoirs coupling parallel dynamic programming with importance sampling and successive approximation. *Adv. Water Resour.* **2019**, *131*. [CrossRef]
46. Giuliani, M.; Castelletti, A. Is robustness really robust? How different definitions of robustness impact decision-making under climate change. *Clim. Chang.* **2016**, *135*, 409–424. [CrossRef]
47. Giudici, F.; Castelletti, A.; Giuliani, M.; Maier, H.R. An active learning approach for identifying the smallest subset of informative scenarios for robust planning under deep uncertainty. *Environ. Model. Softw.* **2020**, *127*, 104681. [CrossRef]
48. Yang, G.; Guo, S.L.; Liu, P.; Li, L.P.; Xu, C.Y. Multiobjective reservoir operating rules based on cascade reservoir input variable selection method. *Water Resour. Res.* **2017**, *53*, 3446–3463. [CrossRef]
49. Yin, J.B.; Guo, S.L.; Gu, L.; He, S.K.; Ba, H.H.; Tian, J.; Li, Q.X.; Chen, J. Projected changes of bivariate flood quantiles and estimation uncertainty based on multi-model ensembles over China. *J. Hydrol.* **2020**, *585*, 124760. [CrossRef]
50. Yang, S.; Yang, D.; Chen, J.; Santisirisomboon, J.; Lu, W.; Zhao, B. A physical process and machine learning combined hydrological model for daily streamflow simulations of large watersheds with limited observation data. *J. Hydrol.* **2020**, *590*, 125206. [CrossRef]

sustainability

MDPI

Article

Novel Ensemble Forecasting of Streamflow Using Locally Weighted Learning Algorithm

Rana Muhammad Adnan [1], Abolfazl Jaafari [2,*], Aadhityaa Mohanavelu [3], Ozgur Kisi [4,*] and Ahmed Elbeltagi [5]

1   State Key Laboratory of Hydrology-Water Resources and Hydraulic Engineering, Hohai University, Nanjing 210098, China; rana@hhu.edu.cn
2   Forest Research Division, Research Institute of Forests and Rangelands, Agricultural Research, Education and Extension Organization (AREEO), Tehran 1496813111, Iran
3   Department of Civil Engineering, Amrita School of Engineering, Amrita Vishwa Vidyapeetham, Amritanagar, Coimbatore 641 112, India; aadhityaa65@gmail.com
4   Civil Engineering Department, Ilia State University, 0162 Tbilisi, Georgia
5   Agricultural Engineering Department, Faculty of Agriculture, Mansoura University, Mansoura 35516, Egypt; ahmedelbeltagy81@mans.edu.eg
*   Correspondence: jaafari@rifr-ac.ir (A.J.); ozgur.kisi@iliauni.edu.ge (O.K.)

Abstract: The development of advanced computational models for improving the accuracy of streamflow forecasting could save time and cost for sustainable water resource management. In this study, a locally weighted learning (LWL) algorithm is combined with the Additive Regression (AR), Bagging (BG), Dagging (DG), Random Subspace (RS), and Rotation Forest (RF) ensemble techniques for the streamflow forecasting in the Jhelum Catchment, Pakistan. To build the models, we grouped the initial parameters into four different scenarios (M1–M4) of input data with a five-fold cross-validation (I–V) approach. To evaluate the accuracy of the developed ensemble models, previous lagged values of streamflow were used as inputs whereas the cross-validation technique and periodicity input were used to examine prediction accuracy on the basis of root correlation coefficient (R), root mean squared error (RMSE), mean absolute error (MAE), relative absolute error (RAE), and root relative squared error (RRSE). The results showed that the incorporation of periodicity (i.e., MN) as an additional input variable considerably improved both the training performance and predictive performance of the models. A comparison between the results obtained from the input combinations III and IV revealed a significant performance improvement. The cross-validation revealed that the dataset M3 provided more accurate results compared to the other datasets. While all the ensemble models successfully outperformed the standalone LWL model, the ensemble LWL-AR model was identified as the best model. Our study demonstrated that the ensemble modeling approach is a robust and promising alternative to the single forecasting of streamflow that should be further investigated with different datasets from other regions around the world.

Keywords: ensemble modeling; additive regression; bagging; dagging; random subspace; rotation forest

Citation: Adnan, R.M.; Jaafari, A.; Mohanavelu, A.; Kisi, O.; Elbeltagi, A. Novel Ensemble Forecasting of Streamflow Using Locally Weighted Learning Algorithm. *Sustainability* **2021**, *13*, 5877. https://doi.org/10.3390/su13115877

Academic Editor: Giuseppe Barbaro

Received: 21 April 2021
Accepted: 20 May 2021
Published: 24 May 2021

1. Introduction

To understand the current state, potential, and prospects of water availability, systematic studies on all aspects of basin hydrology (e.g., precipitation, surface, and sub-surface water) and investigation of all indicators are required [1–3]. Streamflow is one such indicator that has a direct influence on local drinking water supply and the quantity of water available for irrigation, hydro-electricity generation, and other needs [4]. Indeed, projections have shown that 20% of the river discharge is controlled by human interventions [5]. Changes in land use and land cover over time, glaciers, snowfields, topographic boundaries, dams, and reservoir management are some of the key factors influencing streamflow trends [6]. Streamflow data are a very valuable asset if available over a long period of years. Advantages of streamflow forecasting include early flood warning and mitigation, reservoir planning and management, quantification of available water resources for water supply projects, etc. [7]. Accurate forecasting of streamflow is crucial for the efficient management of water reservoir systems, such as dams, under competing demand for water for irrigation, domestic use, and hydro-power generation activities while at the same time maintaining an adequate environment in the river (or stream) system [8]. In addition, both short-term and long-term streamflow forecasting is necessary pertaining to the optimization of the hydrological components of water resource systems mainly during flood or drought periods [9]. Early prediction of streamflow could provide an imminent warning to disaster management organizations to prepare in response to floods quite early thus preventing the costly socio-economic losses incurred from such extreme events [10].

Since streamflow is a derivative of a complex physical system, the predictions of streamflow using physical-based models generally have significant, inherent uncertainty caused by inaccurate or simple representation of hydrological processes, incomplete or incorrect antecedent conditions, bias or errors in the input variables, or uncertainty in the model parameters. In addition, the requirement for big data (n number of parameters) to simulate the hydrological process also restricts the application of physical models [11]. The application of statistics-based time series models such as the autoregressive integrated moving average (ARIMA) model and its derivatives such as periodic or seasonal ARIMA models and more complex multivariate models such as transfer function-noise (TFN) models have been particularly widely used in forecasting monthly streamflow [12]. However, these models are mostly built upon the assumption that the process follows a normal distribution, however the streamflow process is generally non-linear and stochastic in nature [13]. Machine learning (ML) models, which have been widely used in recent decades to model many real-world problems [14–20], have the unique ability to identify the complex non-linear relationships between the predictors (inputs) and targets (outputs) without the need for the physical characterization of the system or the requirement of making any underlying assumptions. Many hybrid ensemble ML models with the integration of different data preprocessing techniques such as wavelet transformations, empirical mode decomposition, etc. have very high efficiency in accurately forecasting the future streamflow using only antecedent streamflow time series data as input [12,21,22]. Examples of the most recent works on streamflow forecasting can be found in Adnan, Liang, Heddam, Zounemat-Kermani, Kisi and Li [2], Ferreira, et al. [23], Piazzi, et al. [24], Saraiva, et al. [25], and Tyralis, et al. [26].

Although several ML-derived models have been suggested and used to forecast streamflow, there is no model that can forecast streamflow without any biases or with utmost certainty based on the time series of antecedent streamflow values. While literature shows evidence that some single and hybrid ML models, such as OSELM, BGWO-RELM, MLR—KNN, RMGM-BP, RBF-ANN, and MARS-DE, are very effective in forecasting streamflow in river basins across the world, none of these models have been proven to forecast streamflow without any biases or with utmost certainty based on the time series of antecedent streamflow values [2,9,12,24,27]. Hence, the development and application of novel and sophisticated machine learning algorithms for streamflow forecasting are critical to overcoming such limitations in favor of improving the overall forecasting accuracy and model performance. Locally Weighted Learning (LWL) is one such novel machine learning algorithm that has proven efficient for modeling environmental problems. Recently LWL-based ensemble models have been successfully used to model groundwater potential [28] and forest fire susceptibility [29]. One unique advantage of LWL is that for each point of interest a local model is created based on neighboring data of the query point instead of building a whole global model for the entire functional space. Based on this strategy, data points closer to the query point receive a higher weight that can control overprediction. In this study, we combine the LWL algorithm with five ensemble learning techniques, that is, Additive Regression (AR), Bagging (BG), Dagging (DG), Random Subspace (RS), and

Rotation Forest (RF), to develop five ensemble models for a novel ensemble forecasting of streamflow. We apply the models to the lagged streamflow time-series input derived from the antecedent streamflow data. To the best of our knowledge, the LWL technique has not yet been investigated for streamflow forecasting and this study is the first to use and compare different versions of the LWL-based ensemble models for this purpose.

2. Case Study

For this study, the Jhelum Catchment located in the western Himalayas in the north part of Pakistan was selected. This catchment originates from India and drains the southern slope of the Greater Himalayas and the northern slope of the Pir Punjal Mountains. The upstream side of the basin located in India is occupied with great glaciers. Due to climate change in recent years, this transboundary river in Pakistan side is greatly affected by glacier melt. Pakistan has a key reservoir (i.e., Mangla Reservoir) downstream of this basin. This reservoir is the second biggest reservoir in Pakistan with an installed capacity of 1000 MW and fulfills 6% of the electricity generation demand of the country. Therefore, precise estimation of this key catchment is very crucial for the economy and sustainability of water resources in Pakistan. This catchment mainly consists of two main sub-basins, that is, the Naran and Neelum basins. The catchment covers a drainage area of 33,342 km^2 up to Mangla Dam with an elevation variation of 200 m to 6248 m. For accurate estimation of streamflow in this basin, the key hydraulic station, that is, Kohala station, at the main river Jhelum streamline after the confluence of both key tributaries (Neelum and Naran) was selected as shown in Figure 1. For model development, the monthly streamflow data of the selected station were obtained from the Water and Power Development Authority (WAPDA) of Pakistan for the duration of 1965 to 2012. For a robust data analysis with the models, a cross-validation scheme was applied. Therefore, data were divided into four equal datasets where each dataset was used for model testing whereas the other three datasets were set aside for model training.

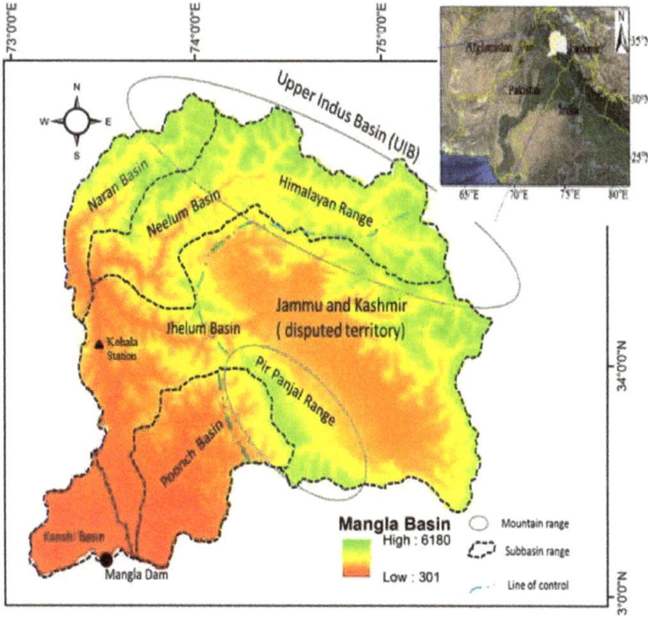

Figure 1. Location map of study area.

3. Methods

3.1. Locally Weighted Learning (LWL) Algorithm

The locally weighted learning (LWL) algorithm is motivated by the classification of example-based approaches [30]. In this algorithm, the regression model is not processed unless the output value of the new vector is presented. This is required to correctly execute all learning at the prediction moment. LWL is an advanced type of M5 method in a way that suits both linear and non-linear regression in space for the unique fields of example [31]. Based on the weighted results, distance according to the questionnaire was used to allocate the weights to the training datasets and a regression equation is produced. There is a wide range of methods of distance-based weighting that can be used on the basis of the problem preference in LWL [32]. The statistical model for basic linear regression and the linear model of the multiple regression are presented, respectively, in Equations (1) and (2):

$$y_i = \beta_0 + \beta_{1xi} + \varepsilon_i \; i = 1, \, 2, \, 3 \ldots .n \tag{1}$$

$$y_i = \beta_0 + \beta_{1xi1} + \beta_{2xi2} + \cdots + \beta_{kxik} + \varepsilon_i \tag{2}$$

where y is the response (dependent variable), x is the predictor (independent variable), y_i and ε_i represent random variables, and xi is constant. The linear existence of the model is due to β-parameters. The LWL objective function of squared error is expressed as follows:

$$\text{Minimize } F = \frac{1}{2N} \sum_{K=1}^{N} w_k \left(\propto_{K0} + \sum_{n=1}^{M} \propto_{kn} x_{kn} + \varepsilon_k - y_k \right)^2 \tag{3}$$

where F is the function of objective, w is the weight function matrix, M is total variables number, ε_k is the random error, and $\propto_{K0} \ldots \ldots \propto_{kn}$ are regression coefficients.

3.2. Bagging

Bagging or "Bootstrap Aggregating" is a method composed of two major steps for getting more stable, robust, and precise models [33,34]. Bagging is one of the stable ensemble learning techniques used for resampling the training dataset. The first phase consists of bootstrapping the raw data samples that make up the various sets of training data. From these training datasets, multiple models are created. Prediction is generated from the continuous training processes for datasets and multiple models. The underlying notion of the Bagging technique is straightforward. Instead of generating predictions from a standalone model that is appropriate for the actual data, the relationship between the input-output variables is defined by multiple models generated. Then using the weighted average in the Bagged algorithm, various models are coupled to form a single output [35,36]. This strategy can effectively reduce the possible uncertainties in the modeling process. Previous works prove that Bagging is a favorable choice for ensemble modeling of many environmental problems [29].

3.3. Additive Regression

Additive Regression was first developed by Stone [37] as a nonparametric method to approximate a multivariate function by using multiple unary functions. For the dependent variable Y and the independent variables $X_1, X_2, \ldots , X_p$, the nonparametric additive model can be given by:

$$E(Y|X_1. \, X_2. \ldots X_p = \propto + \sum_{i=1}^{P} f_i(X_i) \tag{4}$$

where $f_i(X_i)$ is a unary nonparametric function. To satisfy the identifiable conditions, it is generally required that $f_i(X_i) = 0$, $i = 1, 2, \ldots , p$. Compared to traditional linear models, the nonparametric regression model does not pre-suppose the relationship between variables and the form of the regression function. Further, it is an adaptable and robust data-driven model that can yield a better approximation for nonlinear nonhomogeneous problems [38]. Given these advantages, many researchers applied this technique to study the linear and nonlinear relationships in environmental problems [39].

3.4. Random Subspace (RS)

Random Subspace (RS) was developed by Ho [40] as a new ensemble learning technique for resolving real-world problems. The numerous classifiers of this technique are combined and trained on an altered feature space to generate multiple training subsets for the classifiers, which are the training bases. RS applies multiple samples on function space, as opposed to the example space as in other ensemble models, as stated by Havlíček, et al. [41]. This strategy takes advantage of bootstrapping and grouping. The RS inputs are the training set (x), the base-classifier (w), and the subspaces number (L) [42]. It is strongly recommended by Pham, et al. [43] that this approach be used to prevent over-fitting issues and to cope with the most unnecessary datasets.

3.5. Dagging

Ting and Witten [44] pioneered the Dagging algorithm as a resampling ensemble technique that uses most votes to combine various classifiers to get improved prediction accuracy for the base classifier. Dagging generates multiple different samples instead of producing the bootstrap samples to acquire the base classifier. In recent years, it has been considered a promising machine learning algorithm for classification problems. In the real world, the Dagging ensemble technique has been applied to solve different classification problems. The development of an M dataset can occur with a specific training dataset containing N samples which may come from the existing training datasets [45,46]. There are many n ($n < N$) samples in any dataset that are distinct from each other. In the particular training datasets, the variables are not replaced and can be chosen as a part of the dataset specified where the size of sample datasets is expanded. According to that, a base classifier is installed on any sample dataset. Ultimately, depending on the training dataset, many classifiers can be acquired. The capability of Dagging has been frequently proven for obtaining improved predictive modeling of different classification problems [29,47].

3.6. Rotation Forest

Rotation Forest (RF) is an ensemble learning technique that independently trains L decision trees using, for each tree, a different set of extracted features. Suppose the $x = (x_1, \ldots x_n)$ T represents an example defined by n characteristics (attributes) and let X be an $N \times n$ matrix including examples of the training process. We assume that the actual class labels of all instances of training are also given. Let go of $D = \{D_1, \ldots D_L\}$ is the set of classifiers for L and F is the set of characteristics. The purpose of Rotation Forest is to create precise and diverse classifiers. As in Bagging, bootstrap samples are taken as the training collection for the individual classifiers. The key heuristic is to introduce extraction of features and to recreate a complete feature set for each classifier in the ensemble afterward [48]. The feature collection is randomly divided into K subsets to do this. The principal component analysis (PCA) is run on each subset separately, and a new set of n linear extracted features is constructed by pooling all main components. The data are translated into the new space of the function linearly. With this data collection, classifier D_i is educated. Multiple splits of the collection of features will contribute to various extracted features, thereby leading to the diversity of the bootstrap sampling implemented.

4. Ensemble Forecasting

Ensemble forecasting of the monthly streamflow was performed using the LWL algorithm that was used as the base model and was combined with the Additive Regression (AR), Bagging (BG), Dagging (DG), Random Subspace (RS), and Rotation Forest (RF) ensemble techniques. This combination resulted in five ensemble models, namely the ensemble LWL-AR, LWL-BG, LWL-DG, LWL-RS, and LWL-RF models. In each model, the ensemble learning technique performs resampling of the training dataset to train the base LWL algorithm. Table 1 details the summary of statistical characteristics of the data used in this study. To build the models, we grouped the initial input parameters into four different scenarios of input data. They include:

(i)   Qt-1
(ii) Qt-1, Qt-2
(iii) Qt-1, Qt-2, Qt-3
(iv) Qt-1, Qt-2, Qt-3, MN

where Qt-1 is the streamflow at 1 previous month and vice versa and MN is the month number of the streamflow.

Table 1. An overview of the statistical characteristics of the data used.

Statistics	Whole Dataset (m^3/s) 1965 to 2012	M1 Dataset (m^3/s) 2001 to 2012	M2 Dataset (m^3/s) 1989 to 2000	M3 Dataset (m^3/s) 1977 to 1988	M4 Dataset (m^3/s) 1965 to 1976
Mean	772.9	794.0	783.7	835.8	678.0
Min.	110.7	112.3	134.9	127.0	110.7
Max.	2824	2824	2426	2773	2014
Skewness	0.886	0.931	0.716	0.845	0.888
Std. dev.	609.2	645.1	600.6	651.7	514.1
Variance	371,069	416,106	360,780	424,712	264,330

In a cross-validation approach, data were divided into four equal sets such that three sets were used for model training and the remaining set was used for validation [49–51]. We used several performance metrics to measure the performance of the models during both training and validation phases. These metrics include: correlation coefficient (R) (Equation (5), root mean square error (RMSE) (Equation (6), mean absolute error (MAE) (Equation (7), relative absolute error (RAE) (Equation (8), and root-relative square error (RRSE) (Equation (9). A full description of these metrics can be found in the corresponding literature [2,24,52–55].

$$R = \frac{\sum (P_i - \overline{P})(T_i - \overline{T})}{\sqrt{\sum (P_i - \overline{P})^2 \sum (T_i - \overline{T})^2}} \tag{5}$$

$$RMSE = \sqrt{\frac{\sum_{i=1}^{n}(P_{ij} - T_j)^2}{N}} \tag{6}$$

$$MAE = \frac{\sum_{i=1}^{n}|P_{ij} - T_j|}{N} \tag{7}$$

$$RAE = |\frac{P_{ij} - T_j}{T_j}| \times 100 \tag{8}$$

$$RRSE = \sqrt{\frac{\sum_{i=1}^{n}(P_{ij} - T_j)^2}{\sum_{i=1}^{n}(T_j - \overline{T}_j)^2}} \tag{9}$$

where P is the value predicted, T is the target value, $\overline{P}$ and $\overline{T}$ are the mean predicted and target values.

We developed the models using the open-source Weka software on an HP Laptop with an Intel(R) Core (TM) i3-3110M CPU @ 2.40GHz, 4 GB of RAM, an x64-based processor, and the Microsoft Windows 8.1 operating system. The optimum value for each model parameter was identified via a trial-and-error process. To do so, we arbitrarily entered different values until the best model performance was achieved [36,56]. Table 2 details the optimum parameter setting of each model.

Table 2. Optimum parameter setting of the models.

Parameter	Model					
	LWL	**AR**	**BG**	**DG**	**RS**	**RF**
Debug	False	False	False	False	False	False
Search algorithm	Linear NN search	-	-	-	-	-
Weighting kernel	0	-	-	-	-	-
Number of iterations	-	14	12	10	10	11
Shrinkage	-	0.1	-	-	-	-
Bag size percent	-	-	100	-	-	-
Seed	-	-	1	1	1	1
Number of folds	-	-	-	10	-	-
Verbose	-	-	-	False	-	-
Number of boosting iterations	-	30	-	-	-	-
Subspace size	-	-	-	-	0.5	-
Max group	-	-	-	-	-	3
Min group	-	-	-	-	-	3
Number of groups	-	-	-	-	-	False
Projection filter	-	-	-	-	-	PCA
Removed percentage	-	-	-	-	-	50

5. Results

Table 3 shows the results of the single LWL model with different input combinations and datasets. Given the mean values of each metric obtained from each input combination and dataset, the model with input combination IV performed the best and achieved RMSE = 244.6 m^3/s, MAE = 175 m^3/s, RAE = 34.47 m^3/s, RRSE = 40.90 m^3/s, and R = 0.834 in the training phase and RMSE = 274.8 m^3/s, MAE = 199.2 m^3/s, RAE = 38.70 m^3/s, RRSE = 44.08 m^3/s, and R = 0.809 in the testing phase. Importing periodicity (i.e., MN) as an additional input variable into the model considerably improved both the training performance and prediction performance. A comparison between the results obtained from the input combinations III and IV revealed a significant performance improvement, that is, RMSE, MAE, RAE, and RRSE decreased up to 10.12, 14.59, 15.41, and 10.69% in the training phase and 6.17, 9.41, 8.40, and 9.56% in the testing phase, respectively. In terms of the R metric, the results showed 5.3 and 6.1% training and testing improvements when we used input combination IV. Further, the results revealed that the best and worst predictive performance (i.e., testing performance) was obtained with the datasets M3 and M2, respectively.

Table 3. Results of the single LWL model.

Metric	Data Set	Training				Testing			
		Input Combination				Input Combination			
		I	II	III	IV	I	II	III	IV
RMSE	M1	358.6	300.3	295.5	255.9	365.8	308.6	311.4	**295.4**
	M2	358.7	303.7	275.5	242.1	397.0	370.2	369.5	**328.0**
	M3	**358.8**	**283.8**	**271.5**	**244.0**	**382.1**	**303**	**292.9**	**274.8**
	M4	362.3	306.5	300.4	252.3	397.9	342.2	312.4	**277.7**
	Mean	359.6	298.6	285.7	248.6	385.7	331.0	321.6	**294.0**
MAE	M1	282.6	227.7	226.0	183.5	271.0	231.3	241.4	**207.7**
	M2	279.9	227.0	210.1	178.8	306.9	263.9	265.0	**222.2**
	M3	**274.4**	**213.8**	**204.9**	**175.0**	**291.5**	**228.8**	**219.9**	**199.2**
	M4	281.5	230.8	229.0	183.5	309.8	257.2	240.9	**200.0**
	Mean	279.6	224.8	217.5	180.2	294.8	245.3	241.8	**207.3**
RAE	M1	52.24	42.09	41.78	35.92	57.57	49.12	51.27	**44.12**
	M2	55.67	44.14	41.39	35.57	56.68	48.74	48.95	**41.03**
	M3	**53.35**	**42.51**	**40.75**	**34.47**	**56.01**	**43.95**	**42.25**	**38.70**
	M4	55.47	45.47	44.53	33.67	57.56	47.79	44.75	**38.16**
	Mean	54.18	43.55	42.11	34.91	56.96	47.40	46.81	**40.50**
RRSE	M1	58.80	47.42	46.64	39.94	65.81	57.8	58.32	**55.32**
	M2	60.51	49.62	46.19	40.85	63.60	56.34	56.24	**49.91**
	M3	**58.63**	**47.88**	**45.80**	**40.90**	**60.42**	**50.43**	**48.74**	**44.08**
	M4	60.74	51.38	49.09	41.72	61.62	52.99	48.38	**46.24**
	Mean	59.67	49.08	46.93	40.85	62.86	54.39	52.92	**48.89**
R	M1	0.659	0.776	0.783	0.841	0.594	0.672	0.676	**0.746**
	M2	0.642	0.750	0.792	0.834	0.612	0.687	0.694	**0.759**
	M3	**0.658**	**0.773**	**0.792**	**0.834**	**0.629**	**0.746**	**0.762**	**0.809**
	M4	0.634	0.736	0.759	0.826	0.619	0.723	0.757	**0.789**
	Mean	0.648	0.759	0.782	0.834	0.614	0.707	0.722	**0.776**

The best performance is shown in bold.

The results of the five ensemble models, that is, LWL-AR, LWL-BG, LWL-DG, LWL-RS, and LWL-RF, are summed up in Tables 4–8. Similar to the single LWL model, the performance of the ensembles models was predominantly influenced by the input combination and dataset. For example, RMSE of the testing phase ranged from 223.9 m^3/s (M3-IV) to 407.8 m^3/s (M2-I) for LWL-AR, from 255.3 m^3/s (M3-IV) to 345.2 m^3/s (M2-III) for LWL-BG, from 233.5 m^3/s (M3-IV) to 390.6 m^3/s (M2-I) for LWL-DG, from 242.8 m^3/s (M3-IV) to 397.2 m^3/s (M4-I) for LWL-RS, and from 229.4 m^3/s (M3-IV) to 397 m^3/s (M2-I) for LWL-RF. Given these values and also the values of other performance metrics, it is evident that the best performance of all models was achieved by the dataset M3 and the input combination IV (i.e., M3-IV).

Table 4. Results of the ensemble LWL-AR model.

Metric	Dataset	Training				Testing			
		Input Combination				Input Combination			
		I	II	III	IV	I	II	III	IV
RMSE	M1	321.0	184.5	162.4	143.7	327.5	292.8	293.3	**261.9**
	M2	310.0	183.8	170.3	128.3	407.8	334.4	315.2	**273.5**
	M3	**306.9**	**174.0**	**152.1**	**138.4**	**373.6**	**264.4**	**258.8**	**223.9**
	M4	314.2	193.3	169.1	139.1	377.6	294.6	284.2	**242.9**

Table 4. *Cont.*

| Metric | Dataset | Training | | | | Testing | | | |
| | | Input Combination | | | | Input Combination | | | |
		I	II	III	IV	I	II	III	IV
	Mean	313.0	183.9	163.5	137.4	371.6	296.6	287.9	**250.6**
	M1	248.1	135.8	115.8	97.47	247.2	199.8	195.9	**171.1**
	M2	241.6	130.2	120	88.47	310.9	224.2	209.4	**168.6**
MAE	**M3**	**232.2**	**125.1**	**104.8**	**95.72**	**292.1**	**191.7**	**183.8**	**150.7**
	M4	242.6	136.6	117.8	95.72	295.1	200.4	198.6	**156.0**
	Mean	241.1	131.9	114.6	94.30	286.3	204.0	196.9	**161.6**
	M1	45.87	25.11	21.43	17.70	52.51	42.44	41.61	**33.90**
	M2	48.04	25.31	23.32	17.60	57.43	41.41	38.67	**31.13**
RAE	**M3**	**45.15**	**24.87**	**20.83**	**18.95**	**56.11**	**36.82**	**35.31**	**28.95**
	M4	47.79	26.92	23.20	18.60	54.83	37.23	36.89	**31.80**
	Mean	46.71	25.55	22.20	18.21	55.22	39.48	38.12	**31.45**
	M1	50.69	29.13	25.64	21.86	61.33	54.83	54.93	**45.49**
	M2	52.29	30.03	27.83	21.64	62.06	50.90	47.96	**41.62**
RRSE	**M3**	**50.15**	**29.36**	**25.65**	**23.48**	**62.18**	**44.00**	**43.08**	**23.26**
	M4	52.67	32.4	28.34	23.31	58.47	45.63	44.02	**40.57**
	Mean	51.45	30.23	26.87	22.57	61.01	48.84	47.50	**37.74**
	M1	0.743	0.916	0.935	0.953	0.621	0.740	0.733	**0.823**
	M2	0.728	0.910	0.924	0.953	0.612	0.743	0.773	**0.828**
R	**M3**	**0.750**	**0.914**	**0.935**	**0.945**	**0.616**	**0.808**	**0.821**	**0.867**
	M4	0.723	0.903	0.922	0.947	0.658	0.794	0.806	**0.835**
	Mean	0.736	0.911	0.929	0.950	0.627	0.771	0.783	**0.838**

The best performance is shown in bold.

Table 5. Results of the ensemble LWL-BG model.

| Metric | Dataset | Training | | | | Testing | | | |
| | | Input Combination | | | | Input Combination | | | |
		I	II	III	IV	I	II	III	IV
	M1	363.6	290.0	272.9	240.9	345.5	294.0	274.3	**261.2**
	M2	352.3	285.8	266.5	237.9	398.1	340.6	345.2	**306.5**
RMSE	**M3**	**336.6**	**262.0**	**250.7**	**229.4**	**359.4**	**292.8**	**276.1**	**255.3**
	M4	342.8	289.5	272.6	250.5	376.4	319.1	294.7	**258.8**
	Mean	348.8	281.8	265.7	239.7	369.9	311.6	297.6	**270.5**
	M1	284.4	224.2	212.6	177.4	269.0	226.6	218.5	**187.7**
	M2	273.6	217.4	202.1	174.8	310.9	243.5	247.3	**208.0**
MAE	**M3**	**265.9**	**202.0**	**191.0**	**169.8**	**279.4**	**223.6**	**209.5**	**191.7**
	M4	270.6	220.8	205.1	183.0	300.5	248.6	225.3	**182.8**
	Mean	273.6	216.1	202.7	176.3	290.0	235.6	225.2	**192.6**
	M1	52.58	41.45	39.31	32.90	57.13	48.14	46.40	**38.83**
	M2	53.19	42.27	39.83	34.45	57.42	44.97	45.69	**38.43**
RAE	**M3**	**52.88**	**40.16**	**37.98**	**33.77**	**53.68**	**42.95**	**40.24**	**34.88**
	M4	53.32	43.50	39.88	35.57	55.83	46.19	41.86	**36.83**
	Mean	52.99	41.85	39.25	34.17	56.02	45.56	43.55	**37.24**
	M1	57.42	45.79	43.10	38.04	64.71	55.05	51.37	**48.46**
	M2	57.56	46.71	44.67	39.88	60.58	51.83	52.54	**46.65**
RRSE	**M3**	**56.77**	**44.19**	**42.29**	**38.70**	**59.82**	**48.73**	**45.96**	**40.45**
	M4	57.47	48.53	44.54	40.94	58.30	49.42	45.64	**42.48**
	Mean	57.31	46.31	43.65	39.39	60.85	51.26	48.88	**44.51**
	M1	0.672	0.794	0.817	0.859	0.590	0.694	0.743	**0.781**
	M2	0.669	0.783	0.803	0.845	0.627	0.736	0.731	**0.796**
R	**M3**	**0.679**	**0.808**	**0.824**	**0.854**	**0.646**	**0.762**	**0.789**	**0.845**
	M4	0.671	0.766	0.803	0.834	0.661	0.760	0.799	**0.821**
	Mean	0.673	0.788	0.812	0.848	0.631	0.738	0.766	**0.811**

The best performance is shown in bold.

Table 6. Results of the ensemble LWL-DG model.

Metric	Dataset	Training				Testing			
		Input Combination				**Input Combination**			
		I	**II**	**III**	**IV**	**I**	**II**	**III**	**IV**
RMSE	M1	369.2	310.1	279.1	241.0	320.3	274.1	259.0	**249.4**
	M2	355.4	296.3	270.9	264.8	390.6	335.8	326.4	**298.8**
	M3	**338.3**	**285.6**	**262.6**	**234.5**	**349.6**	**288.2**	**253.9**	**233.5**
	M4	346.6	299.3	271.7	239.5	337.6	324.9	286.0	**247.1**
	Mean	352.4	297.8	271.1	245.0	349.5	305.8	281.3	**257.2**
MAE	M1	286.0	236.6	211.7	171.3	248.0	217.4	206.0	**191.1**
	M2	275.2	225.8	201.7	194.0	299.3	246.8	229.3	**209.4**
	M3	**262.2**	**220.1**	**200.1**	**166.9**	**274.3**	**227.2**	**197.8**	**181.3**
	M4	267.5	225.2	207.3	181.4	298.8	252.2	220.8	**179.4**
	Mean	272.7	226.9	205.2	178.4	280.1	235.9	213.5	**190.3**
RAE	M1	52.87	43.75	39.14	31.67	52.67	46.18	43.76	**39.85**
	M2	53.51	43.89	39.21	37.71	55.28	45.59	42.36	**38.68**
	M3	**52.15**	**43.77**	**39.80**	**32.89**	**52.69**	**43.66**	**37.99**	**33.34**
	M4	52.70	44.37	40.84	36.07	55.52	46.85	41.02	**36.72**
	Mean	52.81	43.95	39.75	34.59	54.04	45.57	41.28	**37.15**
RRSE	M1	58.30	48.97	44.07	38.06	59.98	51.32	48.50	**44.56**
	M2	58.07	48.42	44.27	43.28	59.44	51.10	49.67	**45.48**
	M3	**57.06**	**48.18**	**44.30**	**39.30**	**58.18**	**47.96**	**42.26**	**38.26**
	M4	58.09	50.18	45.54	40.40	58.64	50.32	44.29	**41.52**
	Mean	57.88	48.94	44.55	40.26	59.06	50.18	46.18	**42.46**
R	M1	0.663	0.766	0.814	0.867	0.623	0.724	0.753	**0.803**
	M2	0.663	0.771	0.806	0.815	0.643	0.753	0.778	**0.797**
	M3	**0.676**	**0.774**	**0.815**	**0.848**	**0.663**	**0.774**	**0.824**	**0.847**
	M4	0.663	0.753	0.797	0.841	0.659	0.764	0.821	**0.828**
	Mean	0.666	0.766	0.808	0.843	0.647	0.754	0.794	**0.819**

The best performance is shown in bold.

Table 7. Results of the ensemble LWL-RS model.

Metric	Dataset	Training				Testing			
		Input Combination				**Input Combination**			
		I	**II**	**III**	**IV**	**I**	**II**	**III**	**IV**
RMSE	M1	371.3	329.9	287.0	270.4	351.4	302.1	274.1	**248.7**
	M2	358.7	319.8	301.1	282.0	397.0	362.1	371.6	**345.5**
	M3	**359.8**	**317.2**	**279.3**	**268.2**	**382.1**	**326.8**	**302.7**	**242.8**
	M4	362.3	296.7	319.8	280.4	397.9	344.3	319.2	**302.5**
	Mean	363.0	315.9	296.8	275.3	382.1	333.8	316.9	**284.9**
MAE	M1	282.6	261.1	225.2	200.3	271.0	248.1	221.8	**192.7**
	M2	279.9	238.0	231.8	207.2	306.9	295.5	274.1	**243.0**
	M3	**274.4**	**245.1**	**215.7**	**205.9**	**291.5**	**263.8**	**238.5**	**191.4**
	M4	281.5	248.6	238.0	219.3	309.8	276.7	240.7	**228.3**
	Mean	279.6	248.2	227.7	208.2	294.8	271.0	243.8	**213.9**
RAE	M1	52.25	46.76	41.64	37.09	57.56	52.68	47.11	**40.66**
	M2	55.68	49.13	45.06	40.29	56.68	53.43	50.63	**44.88**
	M3	**53.35**	**50.64**	**42.51**	**40.57**	**56.01**	**51.27**	**45.81**	**37.03**
	M4	55.47	48.78	49.13	43.62	57.56	54.41	44.73	**42.41**
	Mean	54.19	48.83	44.59	40.39	56.95	52.95	47.07	**41.25**
RRSE	M1	58.63	49.88	45.32	42.70	65.81	61.71	51.32	**45.47**
	M2	60.51	55.97	49.20	46.08	60.42	58.54	56.56	**52.58**
	M3	**58.80**	**54.89**	**46.82**	**44.96**	**63.60**	**57.68**	**50.39**	**41.39**
	M4	60.74	54.83	55.97	47.30	61.62	57.72	49.43	**46.86**
	Mean	59.67	53.89	49.33	45.26	62.86	58.91	51.93	**46.58**
R	M1	0.659	0.676	0.806	0.837	0.594	0.637	0.736	**0.796**
	M2	0.642	0.714	0.769	0.814	0.629	0.659	0.702	**0.773**
	M3	**0.659**	**0.682**	**0.790**	**0.821**	**0.612**	**0.676**	**0.750**	**0.848**
	M4	0.634	0.679	0.714	0.792	0.619	0.671	0.769	**0.815**
	Mean	0.649	0.688	0.770	0.816	0.614	0.661	0.739	**0.808**

The best performance is shown in bold.

Table 8. Results of the ensemble LWL-RF model.

Metric	Dataset	Training				Testing			
		Input Combination				Input Combination			
		I	II	III	IV	I	II	III	IV
RMSE	M1	371.3	259.9	261.7	225.4	351.4	289.8	278.3	**232.4**
	M2	359.8	271.7	269.4	229.1	397.0	307.4	336.3	**300.2**
	M3	**358.7**	**242.9**	**253.0**	**213.8**	**382.1**	**265.6**	**266.3**	229.4
	M4	362.3	271.0	311.5	230.6	397.9	297.3	311.4	**266.4**
	Mean	363.0	261.4	273.9	224.7	382.1	290.0	298.1	**257.1**
MAE	M1	282.6	196.2	200.0	167.4	271.0	218.0	212.8	**195.7**
	M2	274.4	203.4	205.2	165.0	306.9	223.1	237.4	**201.7**
	M3	**279.9**	**184.2**	**193.0**	**160.0**	**291.5**	**195.5**	**195.6**	173.0
	M4	281.5	204.1	237.8	167.3	309.8	227.8	237.8	**175.7**
	Mean	279.6	197.0	209.0	164.9	294.8	216.1	220.9	**186.5**
RAE	M1	52.25	36.27	36.97	30.96	57.57	46.30	45.19	**37.31**
	M2	53.35	39.55	39.90	32.07	56.68	41.20	43.86	**37.25**
	M3	**55.68**	**36.62**	**38.38**	**31.36**	**56.01**	**37.56**	**37.57**	33.23
	M4	55.47	40.22	44.18	32.96	57.56	42.32	44.18	**36.37**
	Mean	54.19	38.17	39.86	31.84	56.96	41.85	42.70	**36.04**
RRSE	M1	58.63	41.03	41.33	35.59	65.81	54.27	52.11	**42.96**
	M2	58.80	44.41	44.03	37.44	60.42	46.78	51.17	**45.68**
	M3	**60.51**	**40.98**	**42.68**	**36.06**	**63.60**	**44.21**	**44.32**	38.68
	M4	60.74	45.42	48.24	38.66	61.62	46.04	48.24	**41.26**
	Mean	59.67	42.96	44.07	36.94	62.86	47.83	48.96	**42.15**
R	M1	0.659	0.834	0.830	0.88	0.594	0.714	0.753	**0.821**
	M2	0.659	0.805	0.806	0.869	0.629	0.787	0.750	**0.806**
	M3	**0.642**	**0.835**	**0.819**	**0.882**	**0.612**	**0.808**	**0.805**	0.858
	M4	0.634	0.796	0.771	0.856	0.619	0.799	0.771	**0.846**
	Mean	0.649	0.818	0.807	0.872	0.614	0.777	0.770	**0.833**

The best performance is shown in bold.

A comparison between the results obtained from the single LWL model and its ensembles clearly indicates that the ensemble learning techniques considerably improved the training and testing performances of the base LWL algorithm. The ensemble models achieved greater training performance than the single LWL model by about 44.7, 44.7, 47.8, 44.7, and 13.9% in terms of the RMSE, MAE, RAE, RRSE, and R metrics, respectively. In the case of the testing performance, LWL-AR showed 53.3, 54.5, 55, 53.8, and 22.4% improvements. Similarly, testing performance improvements in the corresponding metrics are 8, 7.1, 8, 9, and 4.5% by applying LWL-BG, 12.5, 8.2, 8.3, 13.2, and 5.5% by applying LWL-DG, 3.1, 3.2, 1.9, 4.7, and 4.1% by applying LWL-RS, 12.6, 10, 11, 13.8, and 7.3% by applying LWL-RF, respectively.

A comparison of the models' outcomes also reveals that the ensemble LWL-AR model performed better than the other models in both training and testing phases of the monthly streamflow modeling. The LWL-DG and LWL-RF models showed similar performance and ranked as the second-best models, followed by the LWL-RS model that was identified as the least effective ensemble model.

To further compare the models' performance, we used time variation, scatter plots, and Taylor and violin diagrams to visualize the results obtained from the best input combination (i.e., M3-IV). Figure 2 shows that LWL-AR predictions are much closer to the observed values compared to the other models. Figure 3 reveals that the ensemble LWL-AR model performed better compared to other models in catching the extreme streamflow values (minimum and maximum), which is an important indicator in water resource management and for the evaluation of extreme events such as drought and flood.

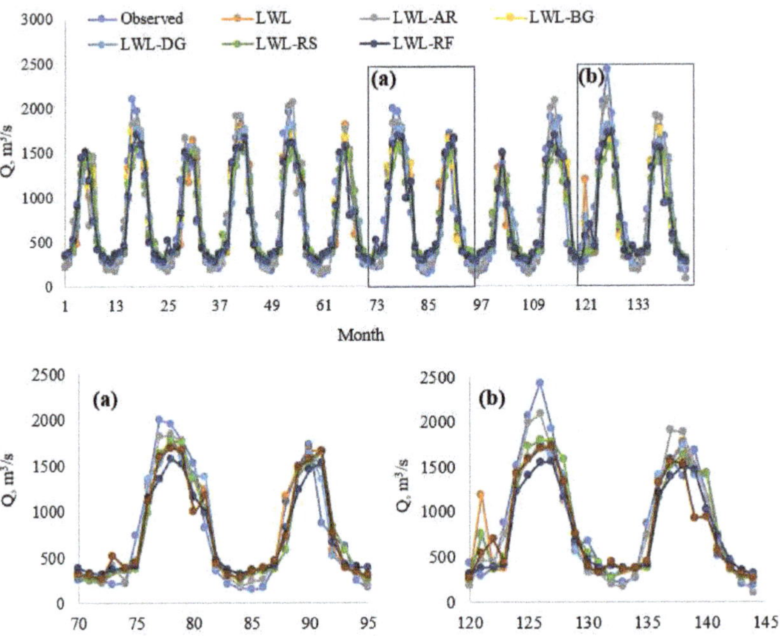

Figure 2. Time variation graphs of the observed and predicted streamflow by different LWL-based ensemble models in the testing phase using the best input combination (M3-IV).

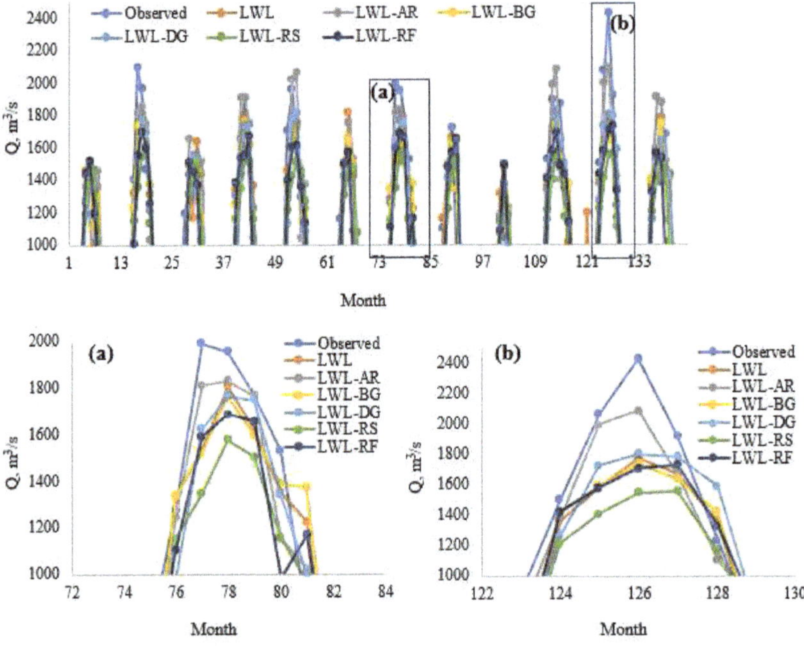

Figure 3. Comparison of different LWL-based ensemble models for forecasting peak streamflow in the testing phase using the best input combination (M3-IV).

Figure 4 compares the single LWL model with its ensemble models in low streamflow (i.e., lower than 500 m^3/s) prediction and clearly demonstrates the superiority of LWL-AR in catching the minimums of streamflow. Figure 5 shows the scatter plots of the observed and predicted monthly streamflow for the best input combination (i.e., M3-IV). While the single LWL model resulted in a highly scattered prediction with R^2 = 0.809, the LWL-AR ensemble model produced a fit line equation ($y = 0.9401x + 56.669$) close to the exact line ($y = x$) with the highest R^2 value (0.867) compared to the other models.

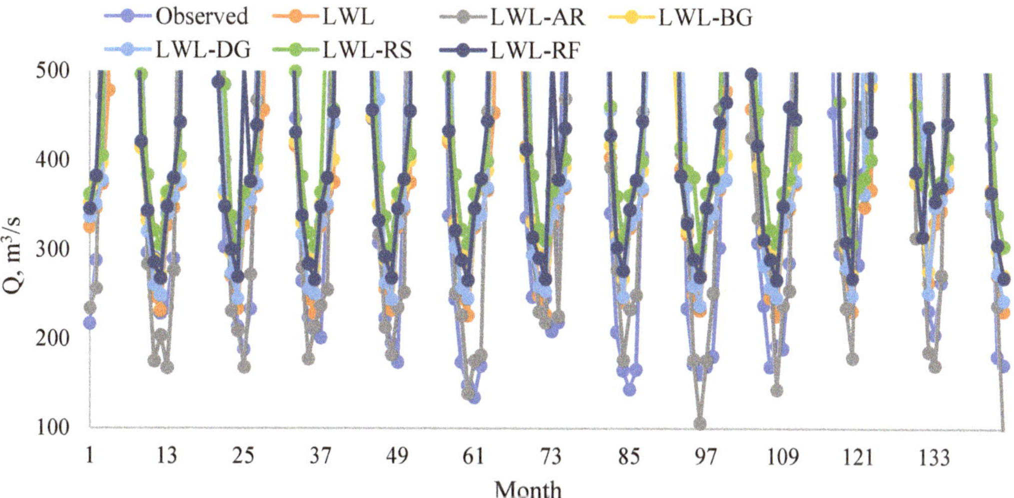

Figure 4. Comparison of different LWL-based ensemble models for forecasting low streamflow in the testing phase using the best input combination (M3-IV). Notice that only the streamflow values lower than 500 m^3/s are shown.

Figure 6 shows the Taylor diagram of the models and indicates how well the models match each other in terms of their standard deviation and correlation difference. Among the different models, LWL-AR achieved a closer standard deviation to the observed data with the lowest square error and highest correlation, which is followed by the LWL-BG and LWL-DG models. Figure 7 shows the violin graph of the models and indicates that LWL-AR achieved a data distribution similar to the observed data, which is followed by the LWL-DG model.

Overall, our case study demonstrated that the ensemble models successfully outperformed the single LWL model and provided promising accuracy for streamflow forecasting. Due to the non-linear nature of many environmental processes and phenomena (e.g., streamflow), hybrid ensemble models that benefit from the advantages of multiple methods/models can better capture the complexity of these phenomena and often yield more accurate results than single simple models.

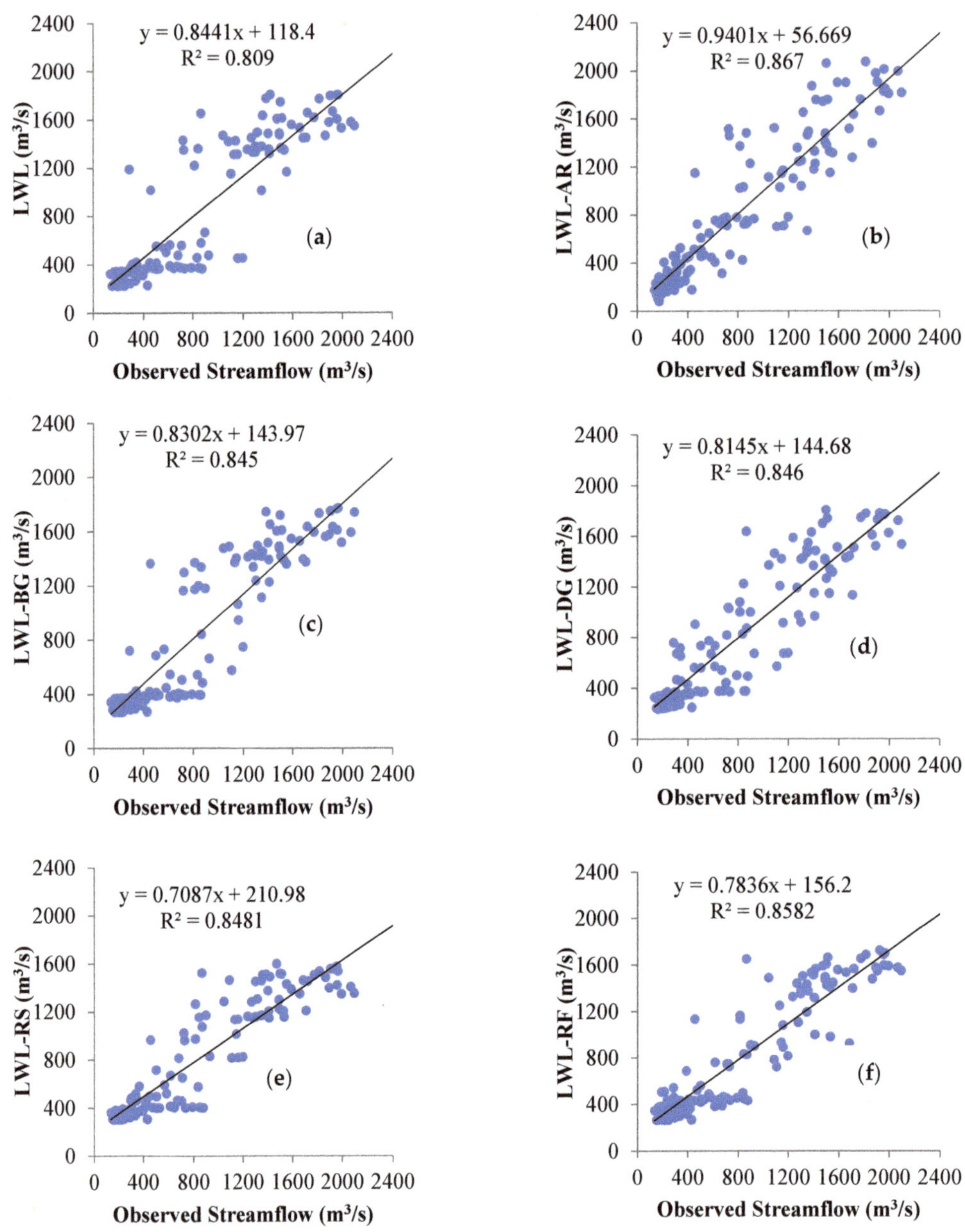

Figure 5. Scatterplots of the observed and predicted streamflow by (**a**) LWL, (**b**) LWL-AR, (**c**) LWL-BG, (**d**) LWL-DG, (**e**) LWL-RS, (**f**) LWL-RF ensemble models in the testing phase using the best input combination (M3-IV).

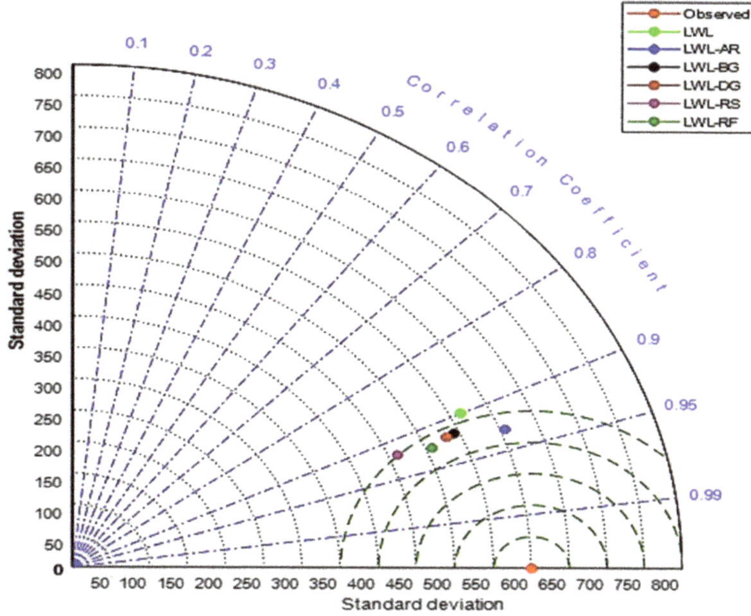

Figure 6. Taylor diagram of different LWL-based ensemble models in the testing phase using the best input combination (M3-IV).

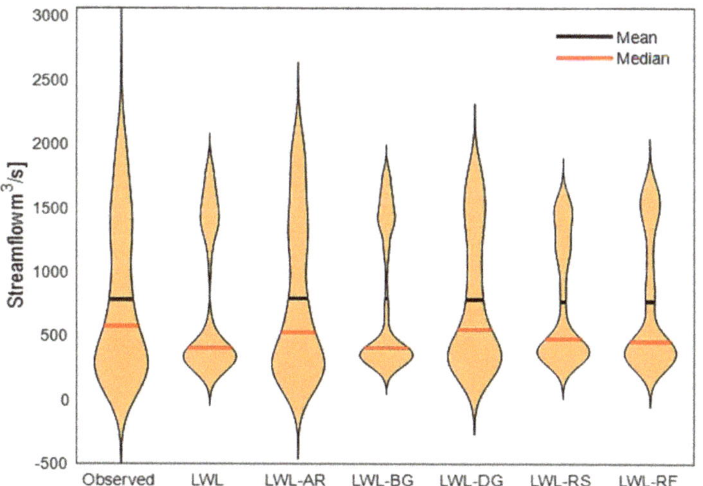

Figure 7. Violin diagrams of different LWL-based ensemble models in the testing using the best input combination (M3-IV).

6. Discussion

In all ensemble models, considering periodicity (i.e., MN) as an additional input variable substantially improved both the training performance and predictive performance. During the testing phase, for the LWL-AR model, the improvements in RMSE, MAE, RAE, RRSE, and R were up to 13, 17.9, 17.5, 20.5, and 7%, respectively. For the LWL-BG model, the metrics improved up to 9.1, 14.5, 14.5, 8.9, and 5.9%, respectively. For the LWL-DG model, the metrics improved up to 8.6, 10.9, 10, 8.1, and 3.1%, respectively. For the LWL-RS model, the metrics improved up to 10.1, 12.3, 12.4, 10.3, and 9.3%, respectively. For the LWL-RF model, the metrics improved up to 13.8, 15.6, 15.6, 13.9, and 8.2%, respectively. These results are in agreement with the previous studies that reported on the improvement of predictive accuracy using the periodicity variable. For example, Kişi [57] demonstrated the improved performance of the three types of ANN models using the periodicity variable for the prediction of monthly streamflow of the Canakdere and Goksudere rivers, Turkey. Adnan, et al. [58] used the periodicity variable to improve the predictive capability of the FFNN, RBNN, GRNN, and ANFIS models for the prediction of the monthly streamflow of the Gilgit River, Pakistan. In a recent study, Adnan, Zounemat-Kermani, Kuriqi and Kisi [53] achieved an improved performance of the long short-term memory (LSTM), extreme learning machines (ELM), and random forest (RF) models for the monthly streamflow of the Kohala and Garhi Habibullah stations in Pakistan. They showed that the inclusion of the periodicity component (MN) decreased the RMSE of the optimal LSTM, ELM, and RF models by 11.9%, 6.9%, and 1% for the Garhihabibullah Station and by 20.8%, 20.5%, and 3.7% for the Kohala Station, respectively.

A comparison of the models' outcomes revealed that the ensemble LWL-AR model performed better than the other models in both training and testing phases of the monthly streamflow modeling. The LWL-DG and LWL-RF models showed similar performance and ranked as the second-best models, followed by the LWL-RS model that was identified as the least effective ensemble model. The results of other modeling studies support our findings that the application of the ensemble learning techniques can considerably improve the capability of the base models for modeling different environmental problems [26,29,47,59]. Overall, our case study demonstrated that the ensemble models successfully outperformed the single LWL model and provided promising accuracy for streamflow forecasting. Due to the non-linear nature of many environmental processes and phenomena (e.g., streamflow), hybrid ensemble models that benefit from the advantages of multiple methods/models can better capture the complexity of these phenomena and often yield more accurate results than single simple models.

7. Conclusions

This study investigated the capability of five ensemble models, that is, LWL-AR, LWL-BG, LWL-DG, LWL-RS, and LWL-RF, for monthly streamflow forecasting. The results were validated using several performance metrics and compared to those of a single LWL model. Based on the results obtained, we conclude that:

- The ensemble models are predominantly superior to the single LWL model for monthly streamflow forecasting.
- Among the ensemble methods, the LWL-AR model surpasses the other models in both training and testing performances.
- The most accurate models are developed when the periodicity variable (MN, month number) is incorporated into the modeling process.
- Ensemble forecasting is a robust and promising alternative to the single forecasting of streamflow.

Although the developed ensemble models were verified using a regional-scale dataset from Pakistan, they are sufficiently general to be applied in any other region around the world with minor adjustments in the variables relative to local conditions. Future research can extend this ensemble forecasting approach by using other ensemble learning techniques (e.g., AdaBoost, MultiBoost, LogitBoost, Decorate, etc.) and, perhaps even more interesting, by testing various types of state-of-the-art machine learning methods as the base classifier. The idea of coupling machine learning methods with ensemble learning techniques with the aim of enhancing the computational performance and improving the predictive accuracy can be extended beyond forecasting monthly streamflow to solve many other complex geo-hydrology problems. In this study, previous streamflow values and periodicity information were considered as inputs to the ensemble models. In future works, streamflow forecasting considering the flood mitigation capacity of Mangla Dam can be investigated using ensemble models. Furthermore, by taking into account the landforms (the digital terrain model) and the dimensions of the river basin as inputs, the implemented methods may provide more accurate forecasting results.

Author Contributions: Conceptualization: R.M.A., O.K., and A.J.; formal analysis: A.J., A.E., O.K., and R.M.A.; validation: R.M.A., A.J., A.M., A.E., and O.K.; supervision: A.J. and O.K.; writing original draft: R.M.A., A.J., A.M., A.E., and O.K.; visualization: R.M.A., A.J., and A.E.; investigation: R.M.A., A.J., and A.M. All authors have read and agreed to the published version of the manuscript.

Funding: This research was supported by the National Key R&D Program of China. (2016YFC0402706).

Institutional Review Board Statement: Not applicable.

Informed Consent Statement: Not applicable.

Data Availability Statement: The data presented in this study will be available on interested request from the corresponding author.

Conflicts of Interest: There is no conflict of interest in this study.

References

1. Zhang, W.; Hu, Y.; Liu, J.; Wang, H.; Wei, J.; Sun, P.; Wu, L.; Zheng, H. Progress of ethylene action mechanism and its application on plant type formation in crops. *Saudi J. Biol. Sci.* **2020**, *27*, 1667–1673. [CrossRef] [PubMed]
2. Adnan, R.M.; Liang, Z.; Heddam, S.; Zounemat-Kermani, M.; Kisi, O.; Li, B. Least square support vector machine and multivariate adaptive regression splines for streamflow prediction in mountainous basin using hydro-meteorological data as inputs. *J. Hydrol.* **2020**, *586*, 124371. [CrossRef]
3. Yuan, X.; Chen, C.; Lei, X.; Yuan, Y.; Adnan, R.M. Monthly runoff forecasting based on LSTM–ALO model. *Stoch. Environ. Res. Risk Assess.* **2018**, *32*, 2199–2212. [CrossRef]
4. Liu, J.; Liu, Y.; Wang, X. An environmental assessment model of construction and demolition waste based on system dynamics: A case study in Guangzhou. *Environ. Sci. Pollut. Res.* **2019**, *27*, 37237–37259. [CrossRef]
5. Mehran, A.; AghaKouchak, A.; Nakhjiri, N.; Stewardson, M.J.; Peel, M.C.; Phillips, T.J.; Wada, Y.; Ravalico, J.K. Compounding Impacts of Human-Induced Water Stress and Climate Change on Water Availability. *Sci. Rep.* **2017**, *7*, 6282. [CrossRef]
6. Zhang, C.; Zhang, B.; Li, W.; Liu, M. Response of streamflow to climate change and human activity in Xitiaoxi river basin in China. *Hydrol. Process.* **2014**, *28*, 43–50. [CrossRef]
7. Adnan, R.M.; Liang, Z.; Parmar, K.S.; Soni, K.; Kisi, O. Modeling monthly streamflow in mountainous basin by MARS, GMDH-NN and DENFIS using hydroclimatic data. *Neural Comput. Appl.* **2021**, *33*, 2853–2871. [CrossRef]
8. Gibbs, M.S.; Dandy, G.C.; Maier, H.R. Assessment of the ability to meet environmental water requirements in the Upper South East of South Australia. *Stoch. Environ. Res. Risk Assess.* **2013**, *28*, 39–56. [CrossRef]
9. Kişi, Ö. Streamflow Forecasting Using Different Artificial Neural Network Algorithms. *J. Hydrol. Eng.* **2007**, *12*, 532–539. [CrossRef]
10. Yossef, N.C.; Winsemius, H.; Weerts, A.; Van Beek, R.; Bierkens, M.F.P. Skill of a global seasonal streamflow forecasting system, relative roles of initial conditions and meteorological forcing. *Water Resour. Res.* **2013**, *49*, 4687–4699. [CrossRef]
11. Aqil, M.; Kita, I.; Yano, A.; Nishiyama, S. A comparative study of artificial neural networks and neuro-fuzzy in continuous modeling of the daily and hourly behaviour of runoff. *J. Hydrol.* **2007**, *337*, 22–34. [CrossRef]
12. Abudu, S.; Cui, C.-L.; King, J.P.; Abudukadeer, K. Comparison of performance of statistical models in forecasting monthly streamflow of Kizil River, China. *Water Sci. Eng.* **2010**, *3*, 269–281.
13. Wang, W. *Stochasticity, Nonlinearity and Forecasting of Streamflow Processes*; IOS Press: Amsterdam, The Netherlands, 2006.
14. Rajaee, T. Wavelet and Neuro-fuzzy Conjunction Approach for Suspended Sediment Prediction. *CLEAN Soil Air Water* **2010**, *38*, 275–286. [CrossRef]
15. Mehdizadeh, S.; Fathian, F.; Safari, M.J.S.; Adamowski, J.F. Comparative assessment of time series and artificial intelligence models to estimate monthly streamflow: A local and external data analysis approach. *J. Hydrol.* **2019**, *579*, 124225. [CrossRef]
16. Adnan, R.M.; Petroselli, A.; Heddam, S.; Santos, C.A.G.; Kisi, O. Short term rainfall-runoff modelling using several machine learning methods and a conceptual event-based model. *Stoch. Environ. Res. Risk Assess.* **2021**, *35*, 597–616. [CrossRef]
17. Rahgoshay, M.; Feiznia, S.; Arian, M.; Hashemi, S.A.A. Simulation of daily suspended sediment load using an improved model of support vector machine and genetic algorithms and particle swarm. *Arab. J. Geosci.* **2019**, *12*. [CrossRef]
18. Kim, C.M.; Parnichkun, M. Prediction of settled water turbidity and optimal coagulant dosage in drinking water treatment plant using a hybrid model of k-means clustering and adaptive neuro-fuzzy inference system. *Appl. Water Sci.* **2017**, *7*, 3885–3902. [CrossRef]
19. Affes, Z.; Kaffel, R.H. Forecast Bankruptcy Using a Blend of Clustering and MARS Model—Case of US Banks. *SSRN Electron. J.* **2016**, *281*, 27–64. [CrossRef]

20. Adnan, R.M.; Liang, Z.; Trajkovic, S.; Zounemat-Kermani, M.; Li, B.; Kisi, O. Daily streamflow prediction using optimally pruned extreme learning machine. *J. Hydrol.* **2019**, *577*, 123981. [CrossRef]

21. Zhang, X.; Peng, Y.; Zhang, C.; Wang, B. Are hybrid models integrated with data preprocessing techniques suitable for monthly streamflow forecasting? Some experiment evidences. *J. Hydrol.* **2015**, *530*, 137–152. [CrossRef]

22. Tongal, H.; Booij, M.J. Simulation and forecasting of streamflows using machine learning models coupled with base flow separation. *J. Hydrol.* **2018**, *564*, 266–282. [CrossRef]

23. Ferreira, R.G.; da Silva, D.D.; Elesbon, A.A.A.; Fernandes-Filho, E.I.; Veloso, G.V.; Fraga, M.D.S.; Ferreira, L.B. Machine learning models for streamflow regionalization in a tropical watershed. *J. Environ. Manag.* **2021**, *280*, 111713. [CrossRef]

24. Piazzi, G.; Thirel, G.; Perrin, C.; Delaigue, O. Sequential Data Assimilation for Streamflow Forecasting: Assessing the Sensitivity to Uncertainties and Updated Variables of a Conceptual Hydrological Model at Basin Scale. *Water Resour. Res.* **2021**, *57*, 57. [CrossRef]

25. Saraiva, S.V.; Carvalho, F.D.O.; Santos, C.A.G.; Barreto, L.C.; Freire, P.K.D.M.M. Daily streamflow forecasting in Sobradinho Reservoir using machine learning models coupled with wavelet transform and bootstrapping. *Appl. Soft Comput.* **2021**, *102*, 107081. [CrossRef]

26. Tyralis, H.; Papacharalampous, G.; Langousis, A. Super ensemble learning for daily streamflow forecasting: Large-scale demonstration and comparison with multiple machine learning algorithms. *Neural Comput. Appl.* **2021**, *33*, 3053–3068. [CrossRef]

27. Zhang, K.; Ruben, G.B.; Li, X.; Li, Z.; Yu, Z.; Xia, J.; Dong, Z. A comprehensive assessment framework for quantifying climatic and anthropogenic contributions to streamflow changes: A case study in a typical semi-arid North China basin. *Environ. Model. Softw.* **2020**, *128*, 104704. [CrossRef]

28. Yen, H.P.H.; Pham, B.T.; Van Phong, T.; Ha, D.H.; Costache, R.; Van Le, H.; Nguyen, H.D.; Amiri, M.; Van Tao, N.; Prakash, I. Locally weighted learning based hybrid intelligence models for groundwater potential mapping and modeling: A case study at Gia Lai province, Vietnam. *Geosci. Front.* **2021**, *12*, 101154. [CrossRef]

29. Tuyen, T.T.; Jaafari, A.; Yen, H.P.H.; Nguyen-Thoi, T.; Van Phong, T.; Nguyen, H.D.; Van Le, H.; Phuong, T.T.M.; Nguyen, S.H.; Prakash, I.; et al. Mapping forest fire susceptibility using spatially explicit ensemble models based on the locally weighted learning algorithm. *Ecol. Inform.* **2021**, *63*, 101292. [CrossRef]

30. Atkeson, C.G.; Moore, A.W.; Schaal, S. Locally Weighted Learning. *Artif. Intell. Rev.* **1997**, *11*, 11–73. [CrossRef]

31. Ahmadianfar, I.; Jamei, M.; Chu, X. A novel Hybrid Wavelet-Locally Weighted Linear Regression (W-LWLR) Model for Electrical Conductivity (EC) Prediction in Surface Water. *J. Contam. Hydrol.* **2020**, *232*, 103641. [CrossRef] [PubMed]

32. Kisi, O.; Ozkan, C. A New Approach for Modeling Sediment-Discharge Relationship: Local Weighted Linear Regression. *Water Resour. Manag.* **2016**, *31*, 1–23. [CrossRef]

33. Chen, T.; Ren, J. Bagging for Gaussian process regression. *Neurocomputing* **2009**, *72*, 1605–1610. [CrossRef]

34. Zhou, Y.; Tian, L.; Zhu, C.; Jin, X.; Sun, Y. Video Coding Optimization for Virtual Reality 360-Degree Source. *IEEE J. Sel. Top. Signal Process.* **2020**, *14*, 118–129. [CrossRef]

35. Azhari, M.; Abarda, A.; Alaoui, A.; Ettaki, B.; Zerouaoui, J. Detection of Pulsar Candidates using Bagging Method. *Procedia Comput. Sci.* **2020**, *170*, 1096–1101. [CrossRef]

36. Xue, X.; Zhang, K.; Tan, K.C.; Feng, L.; Wang, J.; Chen, G.; Zhao, X.; Zhang, L.; Yao, J. Affine Transformation-Enhanced Multifactorial Optimization for Heterogeneous Problems. *IEEE Trans. Cybern.* **2020**, 1–15. [CrossRef] [PubMed]

37. Stone, C.J. Additive Regression and Other Nonparametric Models. *Ann. Stat.* **1985**, *13*, 689–705. [CrossRef]

38. Piegorsch, W.W.; Xiong, H.; Bhattacharya, R.N.; Lin, L. Benchmark Dose Analysis via Nonparametric Regression Modeling. *Risk Anal.* **2013**, *34*, 135–151. [CrossRef]

39. Zhang, M.; Yang, Z.; Liu, L.; Zhou, D. Impact of renewable energy investment on carbon emissions in China—An empirical study using a nonparametric additive regression model. *Sci. Total Environ.* **2021**, *785*, 147109. [CrossRef]

40. Ho, T.K. The random subspace method for constructing decision forests. *IEEE Trans. Pattern Anal. Mach. Intell.* **1998**, *20*, 832–844. [CrossRef]

41. Havlíček, V.; Córcoles, A.D.; Temme, K.; Harrow, A.W.; Kandala, A.; Chow, J.M.; Gambetta, J.M. Supervised learning with quantum-enhanced feature spaces. *Nat. Cell Biol.* **2019**, *567*, 209–212. [CrossRef]

42. Kuncheva, L.I.; Rodriguez, J.J.; Plumpton, C.O.; Linden, D.E.J.; Johnston, S.J. Random Subspace Ensembles for fMRI Classification. *IEEE Trans. Med. Imaging* **2010**, *29*, 531–542. [CrossRef]

43. Pham, B.T.; Bui, D.T.; Prakash, I.; Dholakia, M. Hybrid integration of Multilayer Perceptron Neural Networks and machine learning ensembles for landslide susceptibility assessment at Himalayan area (India) using GIS. *Catena* **2017**, *149*, 52–63. [CrossRef]

44. Ting, K.M.; Witten, I.H. *Stacking Bagged and Dagged Models*; University of Waikato: Hamilton, New Zealand, 1997.

45. Yariyan, P.; Janizadeh, S.; Van Phong, T.; Nguyen, H.D.; Costache, R.; Van Le, H.; Pham, B.T.; Pradhan, B.; Tiefenbacher, J.P. Improvement of Best First Decision Trees Using Bagging and Dagging Ensembles for Flood Probability Mapping. *Water Resour. Manag.* **2020**, *34*, 3037–3053. [CrossRef]

46. Zuo, C.; Chen, Q.; Tian, L.; Waller, L.; Asundi, A. Transport of intensity phase retrieval and computational imaging for partially coherent fields: The phase space perspective. *Opt. Lasers Eng.* **2015**, *71*, 20–32. [CrossRef]

47. Tran, Q.C.; Minh, D.D.; Jaafari, A.; Al-Ansari, N.; Minh, D.D.; Van, D.T.; Nguyen, D.A.; Tran, T.H.; Ho, L.S.; Nguyen, D.H.; et al. Novel Ensemble Landslide Predictive Models Based on the Hyperpipes Algorithm: A Case Study in the Nam Dam Commune, Vietnam. *Appl. Sci.* **2020**, *10*, 3710. [CrossRef]

48. Malek, A.G.; Mansoori, M.; Omranpour, H. Random forest and rotation forest ensemble methods for classification of epileptic EEG signals based on improved 1D-LBP feature extraction. *Int. J. Imaging Syst. Technol.* **2021**, *31*, 189–203. [CrossRef]
49. Jiang, Q.; Shao, F.; Lin, W.; Gu, K.; Jiang, G.; Sun, H. Optimizing Multistage Discriminative Dictionaries for Blind Image Quality Assessment. *IEEE Trans. Multimed.* **2018**, *20*, 2035–2048. [CrossRef]
50. Pham, B.T.; Jaafari, A.; Avand, M.; Al-Ansari, N.; Du, T.D.; Yen, H.P.H.; Van Phong, T.; Nguyen, D.H.; Van Le, H.; Mafi-Gholami, D.; et al. Performance Evaluation of Machine Learning Methods for Forest Fire Modeling and Prediction. *Symmetry* **2020**, *12*, 1022. [CrossRef]
51. Zhang, K.; Zhang, J.; Ma, X.; Yao, C.; Zhang, L.; Yang, Y.; Wang, J.; Yao, J.; Zhao, H. History Matching of Naturally Fractured Reservoirs Using a Deep Sparse Autoencoder. *SPE J.* **2021**, 1–22. [CrossRef]
52. Zhao, C.; Li, J. Equilibrium Selection under the Bayes-Based Strategy Updating Rules. *Symmetry* **2020**, *12*, 739. [CrossRef]
53. Adnan, R.M.; Zounemat-Kermani, M.; Kuriqi, A.; Kisi, O. Machine Learning Method in Prediction Streamflow Considering Periodicity Component. In *Understanding Built Environment*; Springer: Berlin/Heidelberg, Germany, 2020; pp. 383–403.
54. Kisi, O.; Shiri, J.; Karimi, S.; Adnan, R.M. Three different adaptive neuro fuzzy computing techniques for forecasting long-period daily streamflows. In *Big Data in Engineering Applications*; Springer: Singapore, 2018; pp. 303–321.
55. Alizamir, M.; Kisi, O.; Muhammad Adnan, R.; Kuriqi, A. Modelling reference evapotranspiration by combining neuro-fuzzy and evolutionary strategies. *Acta Geophys.* **2020**, *68*, 1113–1126. [CrossRef]
56. Zhao, J.; Liu, J.; Jiang, J.; Gao, F. Efficient Deployment with Geometric Analysis for mmWave UAV Communications. *IEEE Wirel. Commun. Lett.* **2020**, *9*, 1. [CrossRef]
57. Kişi, Ö. River flow forecasting and estimation using different artificial neural network techniques. *Hydrol. Res.* **2008**, *39*, 27–40. [CrossRef]
58. Adnan, R.M.; Yuan, X.; Kisi, O.; Yuan, Y.; Tayyab, M.; Lei, X. Application of soft computing models in streamflow forecasting. In *Proceedings of the Institution of Civil Engineers—Water Management*; Thomas Telford Ltd.: London, UK, 2019; Volume 172, pp. 123–134. [CrossRef]
59. Pham, B.T.; Jaafari, A.; Van Phong, T.; Yen, H.P.H.; Tuyen, T.T.; Van Luong, V.; Nguyen, H.D.; Van Le, H.; Foong, L.K. Improved flood susceptibility mapping using a best first decision tree integrated with ensemble learning techniques. *Geosci. Front.* **2021**, *12*, 101105. [CrossRef]

 *sustainability*

Article

Estimation of Daily Stage–Discharge Relationship by Using Data-Driven Techniques of a Perennial River, India

Manish Kumar [1], Anuradha Kumari [1], Daniel Prakash Kushwaha [1], Pravendra Kumar [1], Anurag Malik [2,*], Rawshan Ali [3] and Alban Kuriqi [4,*]

1   Department of Soil and Water Conservation Engineering, College of Technology, G.B. Pant University of Agriculture & Technology, Pantnagar 263145, India; manishcae2k11@gmail.com (M.K.); anuradhakhushio7@gmail.com (A.K.); danielprakash45499@gmail.com (D.P.K.); pravendrak_05@yahoo.co.in (P.K.)
2   Punjab Agricultural University, Regional Research Station, Bathinda 151001, India
3   Department of Petroleum, Koya Technical Institute, Erbil Polytechnic University, Erbil 44001, Iraq; rawshan.ali@epu.edu.iq
4   CERIS, Instituto Superior Técnico, University of Lisbon, 1649-004 Lisbon, Portugal
*   Correspondence: amalik19@pau.edu (A.M.); alban.kuriqi@tecnico.ulisboa.pt (A.K.)

Received: 9 August 2020; Accepted: 17 September 2020; Published: 23 September 2020

Abstract: Modeling the stage-discharge relationship in river flow is crucial in controlling floods, planning sustainable development, managing water resources and economic development, and sustaining the ecosystem. In the present study, two data-driven techniques, namely wavelet-based artificial neural networks (WANN) and a support vector machine with linear and radial basis kernel functions (SVM-LF and SVM-RF), were employed for daily discharge (Q) estimation. The hydrological data of daily stage (H) and discharge (Q) from June to October for 10 years (2004–2013) at the Govindpur station, situated in the Burhabalang river basin, Orissa, were considered for analysis. For model construction, an optimum number of inputs (lags) was extracted using the partial autocorrelation function (PACF) at a 5% level of significance. The outcomes of the WANN, SVM-LF, and SVM-RF models were appraised over the observed value of Q based on performance indicators, viz., root mean square error (RMSE), Nash–Sutcliffe efficiency (NSE), Pearson's correlation coefficient (PCC), and Willmott index (WI), and through visual inspection (time variation, scatter plot, and Taylor diagram). Results of the evaluation showed that the SVM-RF model (RMSE = 104.426 m^3/s, NSE = 0.925, PCC = 0.964, WI = 0.979) outperformed the WANN and SVM-LF models with the combination of three inputs, i.e., current stage, one-day antecedent stage, and discharge, during the testing period. In addition, the SVM-RF model was found to be more reliable and robust than the other models and having important implications for water resources management at the study site.

Keywords: non-linear modeling; PACF; WANN; SVM-LF; SVM-RF; Govindpur

---

1. Introduction

River discharge and water level observation is an essential issue in hydrological and hydraulic modeling; in addition, it represents a piece of vital source information for water resources planning and management. For instance, accurate stage-discharge estimation is crucial for estimating design flows for different hydraulic infrastructures, such as bridges, culverts, and canals [1]. In very dynamic or compound rivers, direct measurements of flow discharge are very often difficult or not feasible [2]. Moreover, in some cases, neither discharge nor water level may be available or have the same data series record. Therefore, in such circumstances, flow rating curves (FRCs) are the standard and most common procedure to estimate missing information regarding a specific variable. For more than a century, FRCs have been based on calibrated historical records of the stage-discharge rating (i.e., discharge and water level) [3]. FRCs can be constructed by fitting the stage-discharge observation with different polynomial regression functions. Notably, FRCs are most often used for medium and large rivers where making direct measurements may be costly in time and resources [4].

In contrast, for small rivers, in addition to FRCs, both flow discharge and water level can also be measured directly by utilizing current meters or other advanced technologies [5]. Nevertheless, the performance of the FRCs may be influenced by the geometry of the river stage and measurement variability in general, which limits the estimation of high values [6]. Furthermore, polynomial equations used to describe the stage-discharge relationship fail to predict extreme values accurately. In general, most stage-discharge measurements are observed manually during the day, whereas flood peaks often occur at night and are of short duration, which adds uncertainty to the discharge data [7]. It should be emphasized that FRCs perform better when assuming a steady-state hydraulic regime and neglecting hysteresis, which occurs in the discharge–water level relationship during high flow events, notably floods [6,8]. In the cases when flood wave propagation progresses down the river channel, it influences the backwater conditions; therefore, the discharge would be higher for the same water level during the rising level than the falling stage. In such conditions, a single value obtained for the FRC may produce biased discharge [5,8,9]. Many empirical formulas have been developed to smooth the stage-discharge relationship and account for the hysteresis issue, particularly for high flow estimation [1,4].

Nevertheless, the empirical approaches require many measurements along the river reach and are usually site-specific; application for another river or different flow regime type requires additional adjustment and calibration [5,6]. Thus, despite new technologies and methods in streamflow observation, uncertainty persists in the historical data records, which may be influenced by the different factors, such as flow regime [10] and river dynamics near the gauging stations, among others [8,11]. However, different machine learning and data-driven techniques have been shown to provide an accurate prediction of the stage-discharge estimation over different time scales [12–16]. Artificial neural network (ANN) models are the pioneers applied in the field of hydrology and hydraulics in general, and specifically for establishing a stage-discharge relationship [17,18]. Deka and Chandramouli [19] applied and compared conventional methods with three machine learning-based models, finding that the fuzzy neural network provided the best results in terms of performance accuracy. Similar results concerning the performance and prediction accuracy of stage-discharge by using a fuzzy neural network were reported by Lohani et al. [20]. Alizadeh et al. [21] estimated the stage-discharge relationship by utilizing the ensemble empirical mode decomposition algorithm (EEMD), wavelet transform (WT), and mutual information (MI) techniques. They found EEMD and MI performed better than the EEMD and WT models.

Furthermore, Lohani et al. [20] found that the fuzzy logic-based model was able to predict the hysteresis effect more accurately than the ANN and conventional formula. Roushangar et al. [22] applied gene expression programming (GEP) and adaptive neuro-fuzzy inference systems (ANFIS) to predict the discharge coefficient of converging ogee spillways and found that the GEP model performed better than the ANFIS model. Norouzi et al. [23] found that the multilayer perceptron (MLP) provided very accurate results for the estimation of the discharge coefficient of trapezoidal labyrinth weirs. In general, machine learning-based models have been widely applied in water quality modeling [24–26], rainfall prediction [27–30], evapotranspiration [31–33], pan evaporation [34–38], droughts [39–41], and sediment transport, among others [42–46].

However, we noticed that the machine learning-based models generally show robust results, some remain as not widely applied for stage-discharge relationship estimation. Therefore, considering the previous application of efficient machine learning techniques in different hydrologic- and hydraulic-related issues, we were inspired to explore the applicability of related methods to model this complex relationship. In the present study, we investigate the application of some new data-driven models to examine the stage-discharge relationship of some real datasets by using WANN, SVM-LF, and SVM-RF. To the best of our knowledge, these models have not previously been used for stage-discharge relationship estimation; moreover, they have been rarely applied in other hydrologic- or hydraulic-related issues. Therefore, this study attempts to bring to researchers in the water resources community a set of new data-driven models for potential applications in solving different complex problems in the field of hydraulics and hydrology.

The objectives of this study are (i) to indicate the reliability and precision of the applied data-driven models, (ii) to investigate their performance on stage-discharge datasets relationship estimation, and finally (iii) to compare model fits employing some known comparison criteria. The numerical results demonstrate the efficiency of all the proposed models on the seven real datasets considered. The paper is organized as follows: Section 2 presents a brief description of the study site, data acquisition, and the methodological approach, including descriptions of the data-driven models; Section 3 discusses the main results and findings; finally, concluding remarks and recommendations are presented in Section 4.

2. Materials and Methods

2.1. Study Area and Data Collection

The study area NH-5 road bridge Govindpur is commonly known as Govindpur, located in the Balasore district of Orissa State (India) with latitude 21°32′52″ N and longitude 86°55′ 14″ E. The study site is the mainstream of the Burhabalang river which is an east-flowing river and also a part of the Subarnarekha river basin located in Orissa State. The contributing area of the drainage basin is 4495 km^2. Figure 1 illustrates the location map of the study area. The basin is strongly dominated by the south-west monsoon that starts in June and descends in mid-October. The average annual rainfall in the basin is about 1800 mm. The maximum temperature in the plains of the basin varies between 42 and 49 °C during May and goes down 8 to 14 °C during December–January. Geologically, the basin belongs mostly to Archean terrains. The rocks in the basin include Gneisses, Schist, Quartzite, and Amphibolite. Igneous rocks are also seen in the riverbed at some places.

The hydrological data including the daily stage (m) and discharge (m^3/s) of 10 years (1st June 2004–31st October 2013) were obtained from the India-Water Resources Information System (WRIS) portal. The time series plot of the total available datasets of stage and discharge versus time is shown in Figure 2. The whole data were divided into two parts: (i) training dataset consisting of 70% (1st June 2004 to 31st October 2010) of the total data which were used for the development of the model, and (ii) remaining 30% (1st June 2011 to 31st October 2013) of the total data which were used for testing to check the prediction capability of the applied models (Figure 2). Figure 3 shows the relationship between stage and discharge through the rating curve at the study site. In contrast, Figure 4 illustrates the flowchart of the adopted methodology for discharge estimation at the Govindpur site.

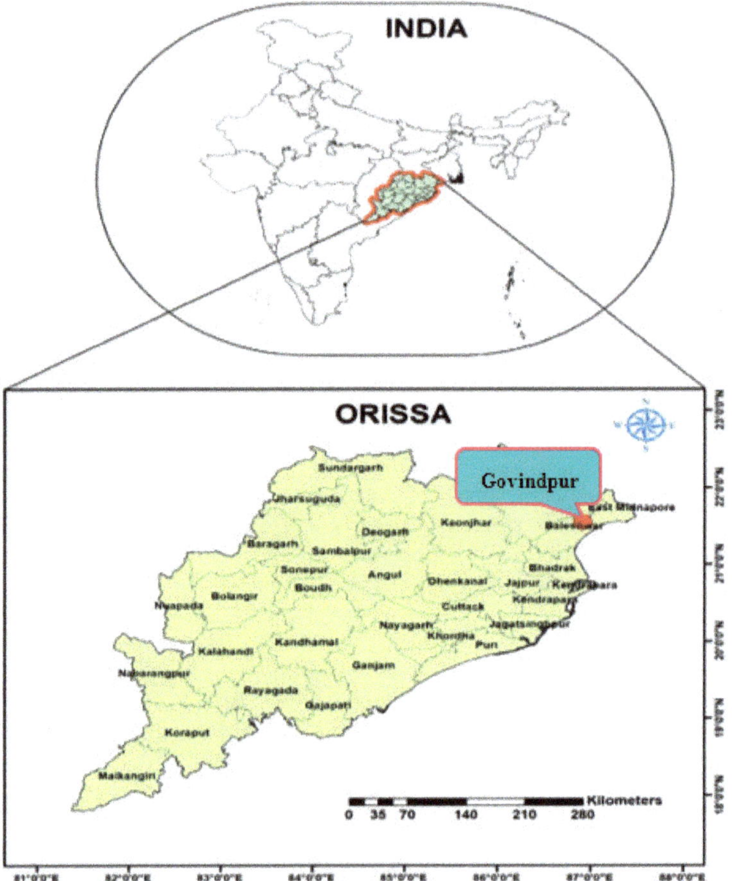

Figure 1. Location Map of the Study Area.

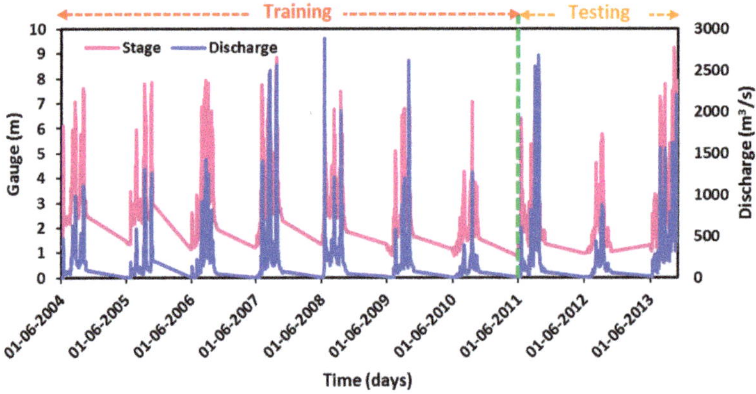

Figure 2. Time Series Plot of Stage and Discharge Datasets at the Study Site.

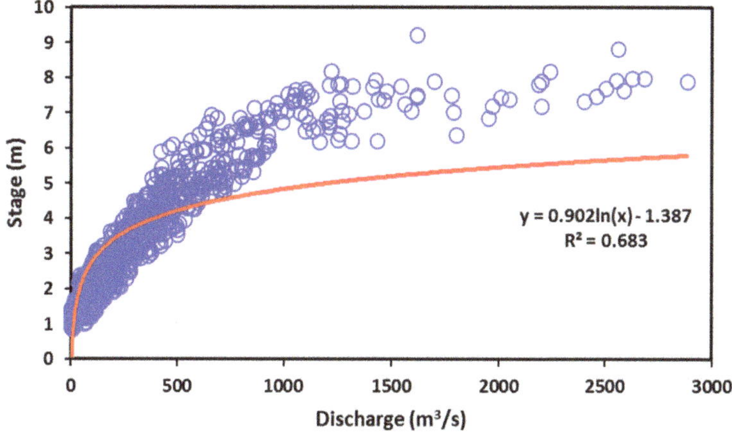

Figure 3. Rating Curve of the Stage-Discharge Relationship at the Study Site.

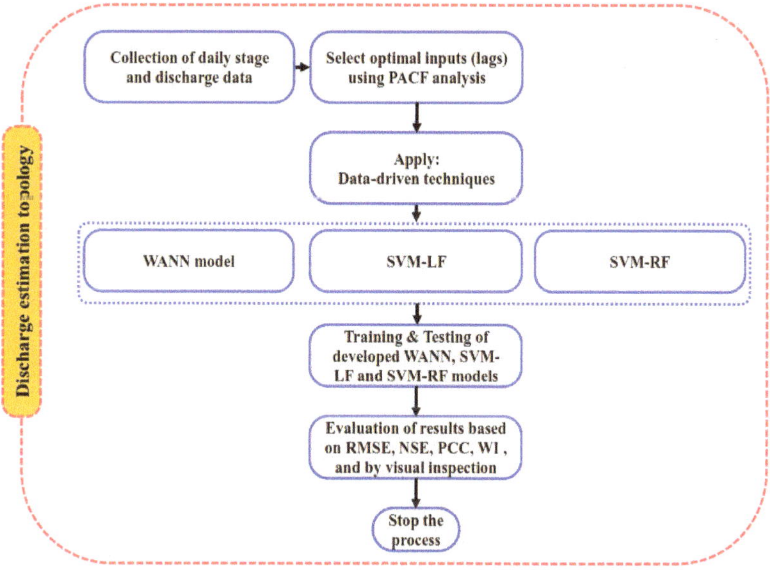

Figure 4. Flowchart of Discharge Estimation Methodology at the Study Site.

2.2. Wavelet Transforms

Wavelet analysis (WA) is a promising time-frequency technique for signal processing with more advantages than Fourier analysis [14]. WA is an enhanced version of Fourier transformation used to detect time features in data [47,48]. Generally, discrete wavelet transformation (DWT) has been used for data decomposition which is advantageous over continuous wavelet transformation (CWT), so that CWT computes wavelet coefficients at every possible scale, which is time-consuming and also produces comprehensive data. DWT is better for analyzing. It reduces the scaling and shifting factors of the fundamental wavelet function to discrete values, maintaining analytical exactness. DWT was notably used in recent years as a computing tool to extract information on non-stationary signals [47,49,50].

The original discrete time-series $C_0(t)$ can be resolved by the Haar à trous decomposition algorithm [51] using Equations (1) and (2):

$$C_r(t) = \sum_{l=0}^{+\infty} h(l) C_{r-1}(t + 2^r)(r = 1, 2, 3, \ldots, n) \tag{1}$$

$$W_r(t) = C_{r-1}(t) - C_r(t)(r = 1, 2, 3, \ldots, n) \tag{2}$$

where $h(l)$ is the discrete low-pass filter, and $C_r(t)$ and $W_r(t)$ ($r = 1, 2, 3, \ldots, n$) are scale and wavelet coefficients at the resolution level. For detailed information regarding wavelet transformation, readers can refer to [52–56].

In the present study, the DWT method was employed for daily discharge estimation. The wavelet transform decomposes the original input time series data of stage and discharges into different frequencies. Three levels of the Haar à trous decomposition algorithm were used in this study. The new decomposed frequencies values act as input for the ANN. The hybridization of the decomposed wavelet value with ANN becomes a wavelet artificial neural network (WANN). The detailed information about ANN can be found in [57]. The Levenberg–Marquardt algorithm was utilized for the training of the model, and the hyperbolic tangent sigmoid transfer function was used to calculate a layer's output from its net input.

2.3. Support Vector Machine (SVM)

Vapnik [58] developed the idea of a support vector machine (SVM). The SVM technology informs an excess glider from the input field that disintegrates a particular training dataset and permits distance on both sides of the hyperplane from the nearest instances. The data showing the maximum margin are referred to as support vectors during the regression analysis. These are the dataset points where approximate errors are equal to or greater than the available tube size of the SVM. There would be a non-linear separation between the training data. Then, it is necessary to construct a non-linear separable boundary. The mapping of the original space to a higher dimension is needed to create a non-linear boundary, and this is called the feature space. A kernel function defines the mapping of the feature space from a given input space. For optimization of the model, a penalty factor (c) has been introduced for misclassification. The total penalty in mapping is obtained by adding the penalties on each misclassification. Several useful applications of the SVM technique have been found in water resources engineering [59–64].

When the SVM algorithm is applied to classification problems, it is called support vector classification (SVC), and when applied to regression problems, it is called support vector regression (SVR) [65,66]. The use of kernel function makes this technique attractive, an excellent generalization, and applicable in the approximation of both linear and non-linear datasets. The lack of an optimal solution is due to the convex nature of the target function and its limitations. The SVM work based on the principle of structural risk minimization was carried out to mitigate the generalization rather than the training error. Consider a training dataset, T, represented using Equation (3):

$$T = \{(x_1, y_1), (x_2, y_2), \ldots, (x_m, y_m)\} \tag{3}$$

where $x \in X \subset R^n$ are the training inputs and $y \in Y \subset R^n$ are the training outputs. Assume a non-linear function $f(x)$ is given by Equation (4):

$$f(x) = w^T \phi(x_i) + b \tag{4}$$

where w is the weight vector, b is the bias, and ϕ is a linearly mapped space with a high-dimensional function, x. Therefore, Equation (4) is transformed into a constrained complex optimization problem using Equations (5) and (6) as:

$$\text{minimize} : \frac{1}{2} w^T w + c \sum_{i=1}^{m} \left(\xi_i + \xi_i^* \right) \tag{5}$$

$$\text{subject to} : \begin{cases} y_i - \left(w^T \Phi(x_i) - b \right) \leq \varepsilon + \xi_i \\ \left(w^T \Phi(x_i) + b \right) - y_i \geq \varepsilon + \xi_i^* \\ \xi_i, \xi_i^* \geq 0, \ i = 1, 2, \dots, m \end{cases} \tag{6}$$

where ξ_i and ξ_i^* are the loose (or slack) parameters, c (>0) is the penalty variable, and ε is the tube size that represents the maximum acceptable deviation. The Lagrangian multipliers are used to solve complex optimization problems [67,68]. The final expansion of SVM is defined using Equation (7) as [69]:

$$f(x) = \sum_{i=1}^{m} \left(a_i^+ - a_i^- \right) K\left(x_i, x_j \right) + b \tag{7}$$

where α_i^+ and α_i^- are the Lagrangian multipliers, and $K\left(x_i, x_j \right)$ is the kernel function. The kernel function of the SVM technique allows solving non-linear approximations into a linear function. The kernel functions used in this study were [69–71]:

- Linear kernel function: the simplest type of kernel function and written by using Equation (8) [72]:

$$K\left(x_i, x_j \right) = \left(x_i, x_j \right) \tag{8}$$

- Radial basis function (RBF): a mapping of RBF that is similar to Gaussian bell-shaped, and expressed by using Equation (9) [72]:

$$K\left(x_i, x_j \right) = \exp\left(-\gamma \| x_i - x_j \|^2 \right) \tag{9}$$

  where γ is the width of the Gaussian RBF kernel parameter. The RBF is widely used among all the kernel functions in the SVM technique. The optimization of SVM in the training phase largely depends on c, γ, and ε parameters. This is because of outstanding features that can effectively tackle the linear and non-linear input-output mapping.

2.4. Model Development and Performance Indicators

The current day streamflow not only depends on the current day conditions but also on the previous days [73]. In this context, lagged input variables are very epochal in time series modeling. However, it is challenging to determine the optimal number of lagged input variables. PACF analysis gives a promising idea to select the optimal number of lags/inputs variables and regression of the time series against its past lagged value, served to remove any dependence on intermediate elements within lags [41,74–76]. In the present study, time-series data of discharge and stage have been lagged based on PACF analysis, so that the actual pattern of PACF among the data could be understood (Figure 5). It was observed that the first three days of lags from the present give more influence on discharge and stage at the 5% significance level. Based on this, lag 1, 2, and 3 from H and Q were selected, and the following three scenarios have been developed in Equations (10)–(12):

$$\text{Scenario} - 1 : Q_t = f(H_t, \ H_{t-1}, \ Q_{t-1}) \tag{10}$$

$$\text{Scenario} - 2 : Q_t = f(H_t, \ H_{t-1}, \ H_{t-2}, \ Q_{t-1}, Q_{t-2}) \tag{11}$$

$$\text{Scenario} - 3 : Q_t = f(H_t, \ H_{t-1}, \ H_{t-2}, \ H_{t-3}, \ Q_{t-1}, Q_{t-2}, Q_{t-3}) \tag{12}$$

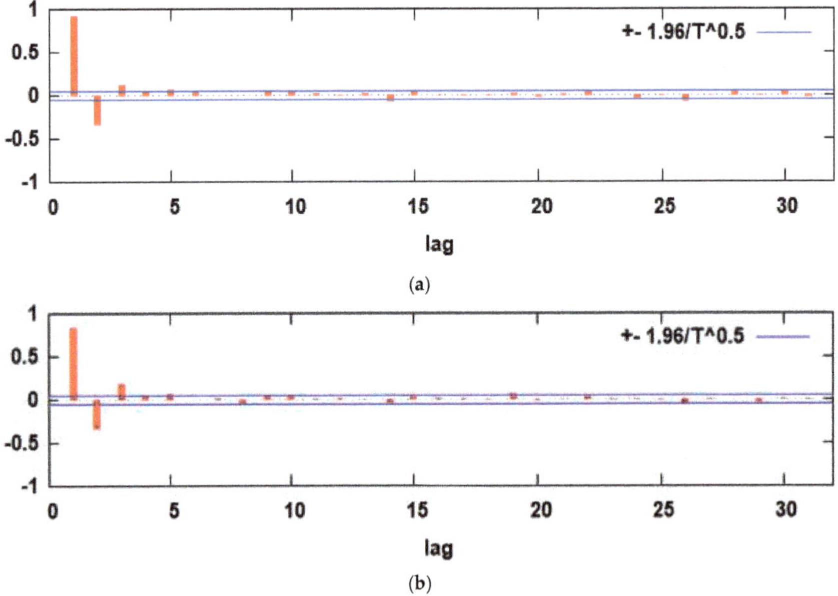

Figure 5. Partial Autocorrelation Function Values of (**a**) Stage, and (**b**) Discharge at the Study Site.

Scenario 1 has a minimum number of inputs, viz., current-day stage, previous 1-day stage, and discharge (Equation (10)). Scenario 2 comprises the average number of inputs, viz., current-day stage, previous 1- and 2-days stage, and discharge (Equation (11)). Meanwhile, scenario 3 includes the maximum number of inputs, namely., current-day stage, previous 1-, 2-, and 3-days stage, and discharge (Equation (12)). All the models have been formulated to predict current day discharge (Q_t) at the study site.

The performance of the scenarios mentioned above was evaluated statistically using root mean square error (RMSE), Nash–Sutcliffe efficiency (NSE), Pearson's correlation coefficient (PCC), and the Willmott index (WI), and through graphical interpretation (time series plot, scatter plot, and Taylor diagram). The advantages and disadvantages of RMSE, NSE, PCC, and WI with definitions are discussed subsequently:

The RMSE measures the difference between observed and estimated values (Equation (13)). The RMSE reports in the same units as the model output and illustrates the size of a typical error. For continuous long-term simulation, RMSE performs well. The RMSE inclines to give more weight to high values than low values because errors in high values are generally more in absolute values than the errors in low values. The RMSE ranges from zero to infinite ($0 < \text{RMSE} < \infty$), so the lower the RMSE, the better the model performance [77,78].

NSE was initially proposed by Nash–Sutcliffe [79] and widely used to evaluate the hydrologic models [78,80,81]. It is the ratio of the mean square error to the variance of observed data during the period under examination, subtracted from unity (Equation (14)). The major limitation of NSE is that the differences between observed and estimated values are calculated as squared values. In other words, it cannot help to identify model bias, differences in magnitudes of peak flows, and the shape of recession curves. Similarly, it cannot be used for single-event simulation [78,80,81]. NSE ranges from minus infinity to one ($-\infty < \text{NSE} < 1$), so the closer to 1, the better the fit. An NSE lower than zero (NSE < 0) shows that the observed mean is as good a predictor as the model, while negative values specify that the observed mean is a better predictor than the model [78,80,81].

The PCC also is known as the correlation coefficient or coefficient of correlation used to measure the degree of collinearity between the observed and estimated variables in hydrological studies [78,81]. The PCC is oversensitive to extreme values and insensitive to additive and proportional variances among model predictions and observed data [81,82]. The PCC varies from minus one to plus one ($-1 < \mathrm{PCC} < 1$), so close to one means a perfect fit (Equation (15)).

The WI, also known as the index of agreement, was developed by Willmott [83] to overcome the insensitivity of NSE and the coefficient of determination (R^2) to the differences in observed and estimated means and variances [81,82]. It represents the ratio of the mean square error and the potential error [83]. The WI varies between zero and one ($0 < \mathrm{WI} \leq 1$), so near to 1 means a perfect agreement/fit, while approaching 0 means complete disagreements between the observed and estimated data (Equation (16)). The main disadvantages of WI are over-sensitivity to extremes values due to the squared differences. The high values of WI were reported even for poor model fits [81,82].

Finally, the RMSE [31,69,77,78], NSE [79], PCC [38,78,81,84], and WI [83] are written as

$$RMSE = \sqrt{\frac{1}{N}\sum_{i=1}^{N}\left(Q_{obs,i} - Q_{est,i}\right)^2} \ (0 < \mathrm{RMSE} < \infty) \tag{13}$$

$$NSE = 1 - \left[\frac{\sum_{i=1}^{N}\left(Q_{obs,i} - Q_{est,i}\right)^2}{\sum_{i=1}^{N}\left(Q_{obs,i} - \overline{Q_{obs}}\right)^2}\right] \tag{14}$$

$$PCC = \frac{\sum_{i=1}^{N}\left(Q_{obs,i} - \overline{Q_{obs}}\right)\left(Q_{est,i} - \overline{Q_{est}}\right)}{\sqrt{\sum_{i=1}^{N}\left(Q_{obs,i} - \overline{Q_{obs}}\right)^2 \sum_{i=1}^{N}\left(Q_{est,i} - \overline{Q_{est}}\right)^2}} \ (-1 < \mathrm{PCC} < 1) \tag{15}$$

$$WI = 1 - \left[\frac{\sum_{i=1}^{N}\left(Q_{est,i} - Q_{obs,i}\right)^2}{\sum_{i=1}^{N}\left(\left|Q_{est,i} - \overline{Q_{obs}}\right| + \left|Q_{obs,i} - \overline{Q_{obs}}\right|\right)^2}\right] \tag{16}$$

where N is the data points, Q_{obs} and Q_{est} are the observed and estimated discharge values for ith observations, and $\overline{Q_{obs}}$ and $\overline{Q_{est}}$ are the means of the observed and estimated discharge values.

3. Results and Discussion

3.1. Statistical Analysis

The statistical analysis of stage (H) and discharge (Q) datasets for training, testing, and the entire period is given in Table 1, which includes various statistical parameters like mean, median, minimum and maximum value, standard deviation (Std. Dev.), coefficient of variation (CV), and skewness. These statistical parameters show the variability of data over time. When dividing the dataset into training and testing subsets, it is necessary to cross-validate the data to have the same statistical population. Due to the high skewness coefficient, there has been a considerable negative effect on model performance. Therefore, skewness coefficients are low for both calibration (1.3012) and validation (1.3441) sets for the given station. This is appropriate for discharge estimation at the study site. The standard deviation for the datasets shows that the values that are farther from zero mean that the variability in the data is higher. Hence, the variation of data from the mean value is higher.

Table 1. Statistics of Stage and Discharge Variables During Training, Testing, and Entire Periods at the Study Station.

Statistical Parameter	Training		Testing		Entire	
	H (m)	Q (m^3/s)	H (m)	Q (m^3/s)	H (m)	Q (m^3/s)
Mean	2.9461	243.50	2.7548	291.60	2.8887	257.93
Median	2.5200	136.80	2.2000	157.12	2.4900	142.61
Minimum	0.8600	1.3690	0.8600	3.5730	0.8600	1.3690
Maximum	8.8400	2885.9	9.2400	2685.6	9.2400	2885.9
Std. Dev.	1.5805	349.15	1.7028	381.48	1.6200	359.71
CV	0.5364	1.4339	0.6181	1.3082	0.5608	1.3946
Skewness	1.3012	3.6133	1.3441	2.8999	1.3013	3.3629

3.2. Evaluation of Results from Various Trails

In the selection process of the best model, several trails have been performed on a single output. The trails of WANN were performed based on the different number of neurons in hidden layers. In contrast, trails of SVM-LF and SVM-RF were performed by taking several values of SVM-g, SVM-c, and SVM-e parameters from scenarios 1 to 3. The best four trails have been listed in Tables 2–4 based on testing results. The results of trail-2, trail-1, and trail-4 of WANN-1, SVM-LF-1, and SVM-RF-1 (Table 2); trail-3, trail-2, and trail-4 of WANN-2, SVM-LF-2, and SVM-RF-2 (Table 3); and trail-2 of WANN-3, SVM-LF-3, and SVM-RF-3 (Table 4) were found to be more promising than the other trails. Out of these trails, a total of nine have been imposed based on techniques and input selections and further evaluated to find the optimal one for daily discharge estimation at the study site (Table 5).

Table 2. Performance Indicators of WANN-1, SVM-LF-1, and SVM-RF-1 Models During Testing at the Study Station.

Model	Performance Indicators			
	RMSE	NSE	PCC	WI
WANN-1				
Trail-1	148.662	0.848	0.924	0.959
Trail-2	127.349	0.888	0.944	0.968
Trail-3	133.695	0.877	0.938	0.968
Trail-4	157.487	0.829	0.927	0.960
SVM-LF-1				
Trail-1	130.404	0.883	0.941	0.967
Trail-2	217.531	0.674	0.952	0.930
Trail-3	135.250	0.874	0.954	0.968
Trail-4	180.688	0.775	0.954	0.948
SVM-RF-1				
Trail-1	108.920	0.918	0.961	0.977
Trail-2	106.227	0.922	0.963	0.978
Trail-3	106.227	0.922	0.963	0.978
Trail-4	104.426	0.925	0.964	0.979

Table 3. Performance Indicators of WANN-2, SVM-LF-2, and SVM-RF-2 Models During Testing at the Study Station.

Model	Performance Indicators			
	RMSE	**NSE**	**PCC**	**WI**
WANN-2				
Trail-1	139.597	0.866	0.931	0.962
Trail-2	139.839	0.866	0.933	0.961
Trail-3	139.559	0.866	0.931	0.963
Trail-4	151.836	0.842	0.935	0.963
SVM-LF-2				
Trail-1	206.840	0.706	0.953	0.938
Trail-2	130.556	0.883	0.942	0.967
Trail-3	135.972	0.873	0.956	0.968
Trail-4	174.246	0.791	0.954	0.952
SVM-RF-2				
Trail-1	111.356	0.915	0.962	0.975
Trail-2	109.005	0.918	0.962	0.977
Trail-3	108.376	0.919	0.963	0.977
Trail-4	106.594	0.922	0.964	0.978

Table 4. Performance Indicators of WANN-3, SVM-LF-3, and SVM-RF-3 Models During Testing at the Study Station.

Model	Performance Indicators			
	RMSE	**NSE**	**PCC**	**WI**
WANN-3				
Trail-1	148.561	0.848	0.925	0.961
Trail-2	130.441	0.883	0.945	0.971
Trail-3	244.984	0.588	0.824	0.901
Trail-4	134.526	0.876	0.939	0.968
SVM-LF-3				
Trail-1	128.384	0.887	0.945	0.968
Trail-2	124.954	0.893	0.950	0.970
Trail-3	139.634	0.866	0.954	0.966
Trail-4	173.277	0.794	0.951	0.953
SVM-RF-3				
Trail-1	130.589	0.883	0.951	0.964
Trail-2	122.262	0.897	0.956	0.969
Trail-3	147.599	0.850	0.939	0.952
Trail-4	124.596	0.893	0.954	0.968

Table 5. Comparison of Best Outputs of WANN, SVM-LF, and SVM-RF Models at the Study Station.

Model	Structure/Parameter	Performance Indicators			
		RMSE	**NSE**	**PCC**	**WI**
WANN-1	12-5-1	127.349	0.888	0.944	0.968
SVM-LF-1	$\gamma = 0.330, \varepsilon = 0.100, c = 10$	130.404	0.883	0.941	0.967
SVM-RF-1	$\gamma = 0.160, \varepsilon = 0.010, c = 10$	104.426	0.925	0.964	0.979
WANN-2	20-9-1	139.559	0.866	0.931	0.963
SVM-LF-2	$\gamma = 0.1428, \varepsilon = 0.010, c = 10$	130.556	0.883	0.942	0.967
SVM-RF-2	$\gamma = 0.120, \varepsilon = 0.010, c = 10$	106.594	0.922	0.964	0.978
WANN-3	28-5-1	130.441	0.883	0.945	0.971
SVM-LF-3	$\gamma = 0.143, \varepsilon = 0.010, c = 10$	124.954	0.893	0.950	0.970
SVM-RF-3	$\gamma = 0.160, \varepsilon = 0.100, c = 10$	122.262	0.897	0.956	0.969

3.3. Quantitative and Qualitative Evaluation of Results

The RMSE, NSE, PCC, and WI values of all of the nine screened models over scenario 1 (S-1), scenario 2 (S-2), and scenario 3 (S-3) are given in Table 5. The model performance was classified as very good (PCC > 0.95, NSE > 0.80), good (0.85 ≤ PCC ≤ 0.95, 0.70 ≤ NSE ≤ 0.80), satisfactory (0.70 ≤ PCC ≤ 0.85, 0.50 ≤ NSE ≤ 0.70), and unsatisfactory (PCC ≤ 0.70, NSE ≤ 0.50), as stated by Moriasi et al. [78], Kouchi et al. [85], and Paul and Negahban-Azar, [86]. After considering all the techniques' best trails from three scenarios (S-1 to S-3), it was noted that the SVM-RF model performed better than the WANN and SVM-LF models based on quantitative performance evaluation indicators. It was also observed that the performance of the SVM-RF model was reduced as the input variables were increased. The values of RMSE (m^3/s), NSE, PCC, and WI were obtained as 104.426, 0.925, 0.964, and 0.979, respectively, for SVM-RF-1, 106.594, 0.922, 0.964, and 0.978 for SVM-RF-2, and 122.262, 0.897, 0.956, and 0.969 for SVM-RF-3. The order of model performance based on NSE from very good to unsatisfactory was attained as SVM-RF-1 (0.925) > SVM-RF-2 (0.922) > SVM-RF-3 (0.897) > SVM-LF-3 (0.893) > WANN-1 (0.888) > SVM-LF-1 (0.883) = SVM-LF-2 (0.883) = WANN-3 (0.883) > WANN-2 (0.866). The order of model performance on the basis of the RMSE from best to inferior was obtained as SVM-RF-1 (104.426) > SVM-RF-2 (106.594) > SVM-RF-3 (122.262) > SVM-LF-3 (124.954) > WANN-1 (127.349) > SVM-LF-1 (130.404) > WANN-3 (130.441) > SVM-LF-2 (130.556) > WANN-2 (139.559). The order of model performance based on the WI from best to inferior was found as SVM-RF-1 (0.979) > SVM-RF-2 (0.978) > WANN-3 (0.971) > SVM-LF-3 (0.970) > SVM-RF-3 (0.969) > WANN-1 (0.968) > SVM-LF-1 (0.967) = SVM-LF-2 (0.967) > WANN-2 (0.963). The comparison of results in Table 5 confirmed the superiority of the SVM-RF model with M-1 (inputs H_t, H_{t-1}, Q_{t-1}) having the lowest value of RMSE = 104.426 m^3/s, and the highest values of NSE = 0.925, PCC = 0.964, and WI = 0.979, closely followed by the SVM-RF-2 model.

The results of the optimal nine models in three different scenarios were plotted between observed and estimated discharge values in the form of time variation and scatter plots through Figures 6–8. It was noted that from these figures the high discharge values are under-estimated (>180 m^3/s), whereas low discharge (<180 m^3/s) values are over-estimated by WANN, SVM-LF, and SVM-RF models during the testing period. The quantity of explained variation out of the total variation (R^2: coefficient of determination) was obtained as excellent for SVM-RF-1 and SVM-RF-2 models. Based on R^2 values, the 'order of the model performance from very satisfactory to unsatisfactory [87] was found as SVM-RF-1 (0.930) = SVM-RF-2 (0.930) > SVM-RF-3 (0.914) > SVM-LF-3 (0.903) > WANN-3 (0.894) > WANN-1 (0.890) > SVM-LF-2 (0.887) > SVM-LF-1 (0.886) > WANN-2 (0.867).

Figure 9a–c demonstrates the Taylor diagrams of WANN, SVM-LF, and SVM-RF corresponding to S-1, S-2, and S-3 during the testing period at the study site. The concept of the Taylor diagram was given by Taylor [88] to represent the spatial distribution of estimated values (i.e., test field) concerning the observed (reference field) by compiling the RMSE, standard deviation, and correlation coefficient in the polar system. It can be seen from these figures that the SVM-RF model is close to the observed (reference) field. Moreover, the SVM-RF-1 model has the lowest RMSE, less standard deviation, and a higher correlation in comparison to other models, and is nominated as an optimal model for daily discharge estimation with H_t, $H_{t-1,}$ Q_{t-1} inputs at the study site.

Further, to support the finding of this study, the results were compared with the recent literature [89–94]. Adnan et al. [95] applied the group method of data handling-neural network (GMDH-NN), dynamic evolving neural-fuzzy inference system (DENFIS), and multivariate adaptive regression splines (MARS) for monthly streamflow prediction at the Kalam and Chakdara stations of the Swat river basin, Pakistan. They found better performance of the DENFIS at the Kalam site (RMSE = 18.9 m^3/s, MAE = 13.1 m^3/s, NSE = 0.94), and MARS at the Chakdara site (RMSE = 47.5 m^3/s, MAE = 31.6 m^3/s, NSE = 0.91). Ali and Shahbaz [96] evaluated the performance of an ANN for daily streamflow prediction in the Jhelum river basin, Pakistan. The results of the analysis revealed the better suitability of ANN in daily streamflow prediction with RMSE = 127.70 m^3/s, PCC = 0.98, and NSE = 0.96 during the testing period. Mohammadi et al. [97] predicted the monthly streamflow of the Vu Gia Thu Bon river (Vietnam) using a standalone ANFIS and hybrid ANFIS coupled with the shuffled frog leaping algorithm (ANFIS-SFLA). The results of the perusal displayed the superior performance of the ANFIS-SFLA model with RMSE = 141.39 m^3/s, NSE = 0.88, and PCC = 0.88 over the ANFIS model (RMSE = 167.81 m^3/s, NSE = 0.83, PCC = 0.83). Mohammadi et al. [98] applied classical MLP and their hybrid integrated with particle swarm (MLP-PSO), PSO-multi-verse optimizer (MLP-PSO-MVO), and bi-linear (MLP-BL) to predict the daily streamflow at four stations, i.e., Brantford and Galt located in Grand River, Canada, and Macon and Elkton positioned in Ocmulgee and Umpqua rivers, United States. The results of the comparison revealed that the MLP-BL models (RMSE = 6.426/ 6.067/ 24.441/ 34.535 m^3/s, MAE = 3.530/ 3.190/ 11.825/ 14.878 m^3/s, and R^2 = 0.994/ 0.990/ 0.990/ 0.986) outperformed the other models at the Brantford, Galt, Macon, and Elkton stations, respectively. Tripura et al. [99] forecasted hourly streamflow of Barak riven basin, Assam (India) by employing the standalone co-active neuro-fuzzy inference system (CANFIS) and a hybrid of CANFIS optimized with the genetic algorithm (CANFIS-GA) and firefly algorithm (CANFIS-FA). They found that the CANFIS-FA model provides better results than the other models. The results of these studies support the application of artificial intelligence (AI) techniques in monthly and daily streamflow/discharge prediction. Likewise, the results of the current research are in fair agreement with the utility of the SVM-RF technique for daily discharge prediction at Govindpur station.

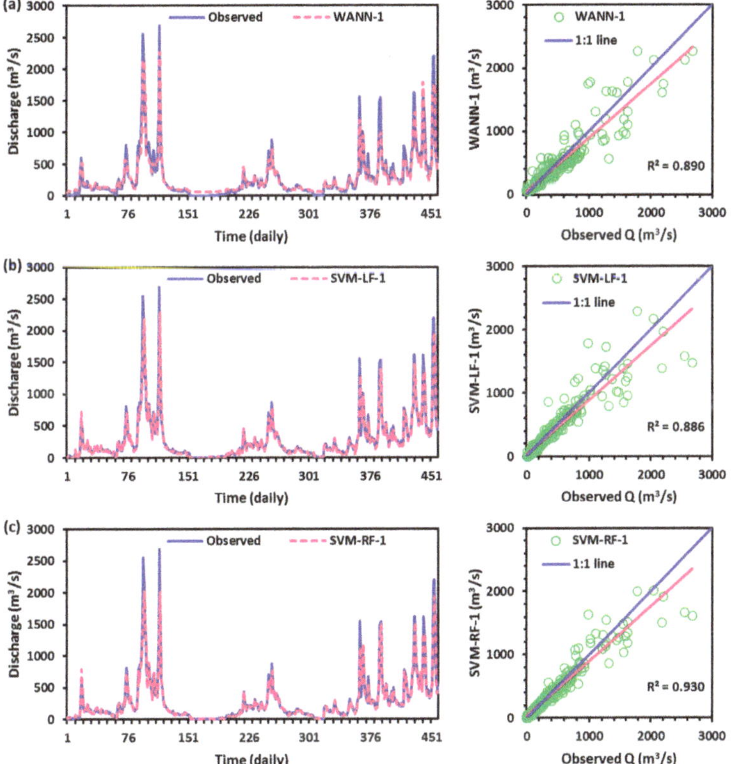

Figure 6. Observed Versus Estimated Discharge of Best (**a**) WANN-1, (**b**) SVM-LF-1, and (**c**) SVM-RF-1 Models During the Testing Period at the Study Site.

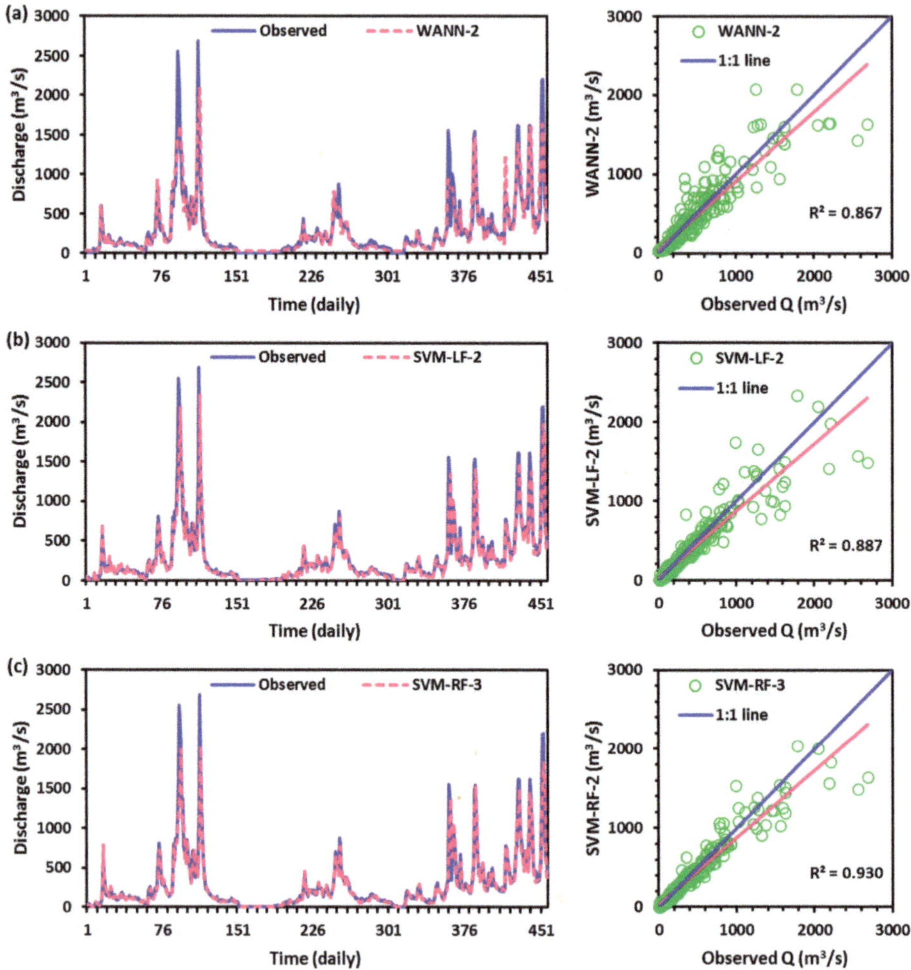

Figure 7. Observed versus estimated discharge of best (**a**) WANN-2, (**b**) SVM-LF-2, and (**c**) SVM-RF-2 models during the testing period at the study site.

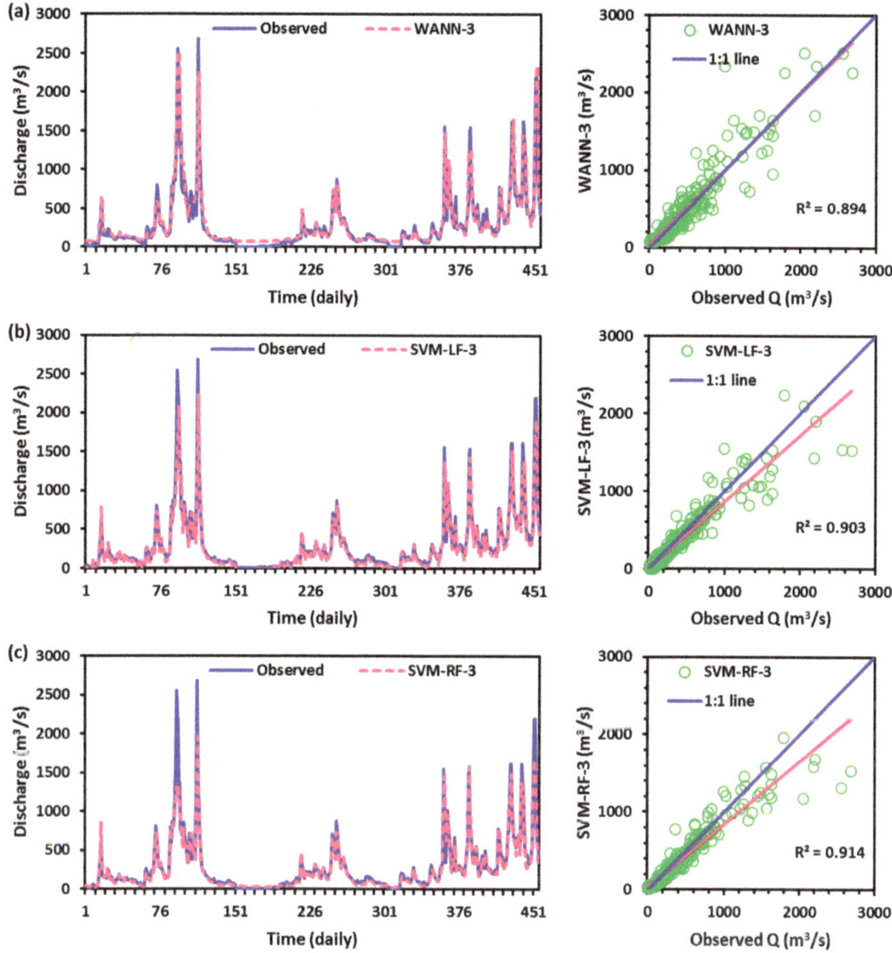

Figure 8. Observed Versus Estimated Discharge of Best (**a**) WANN-3, (**b**) SVM-LF-3, and (**c**) SVM-RF-3 Models During the Testing Period at the Study Site.

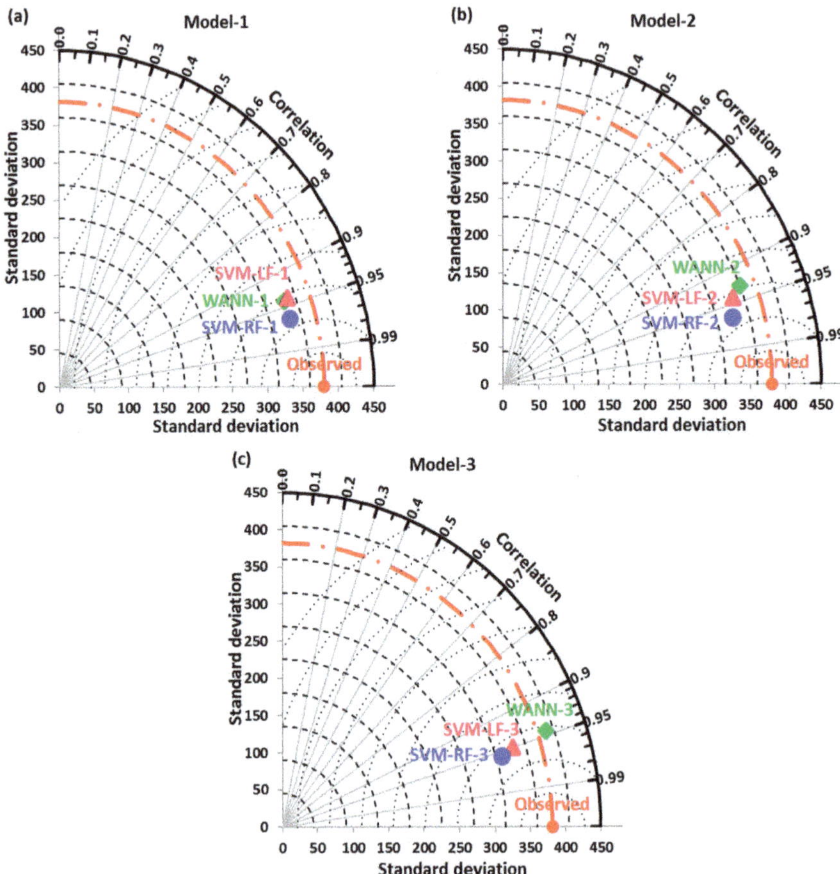

Figure 9. Taylor diagram of WANN, SVM-LF, and SVM-RF corresponding to (**a**) scenario 1, (**b**) scenario 2, and (**c**) scenario 3 during the testing period at the study site.

4. Conclusions

Prediction of discharge on daily, weekly, and monthly timescales is vital for short- and long-term water resources management, particularly in extreme events like floods and drought. Thus, the present study was projected to predict the daily stage-discharge relationship at Govindpur station located at the Burhabalang river basin, Orissa (India), by employing wavelet-based artificial neural networks (WANN) and a support vector machine (SVM) optimized with linear and radial basis kernel functions. The PACF analysis gives an appropriate idea to select the optimum numbers on input variables in time series-based modeling. Data with more variability have been chosen for training, and remaining data have been utilized to test the model performance. Based on performance indicators and by visual inspection, the results revealed that the SVM-RF model with H_t, H_{t-1}, Q_{t-1} inputs perform superior to the WANN and SVM-LF models for daily discharge estimation during monsoon season at the study site. Also, it was noted that as the input variable increases, the computation process becomes more difficult, time-consuming, and sometimes produces inferiority in the results. The best performance of the SVM-RF technique can help researchers to use highly variable discharge data for such modeling in the future. Researchers are also suggested to take as many trails as possible to avoid any bias and related problems of over- and under-estimation for highly variable data.

Author Contributions: Conceptualization, M.K., D.P.K. and A.M.; methodology, M.K. and A.K. (Anuradha Kumari); software, M.K.; validation, M.K., A.K. (Anuradha Kumari), P.K., D.P.K. and A.M.; formal analysis, M.K. and A.M.; investigation, M.K., A.K. (Anuradha Kumari), D.P.K., P.K., A.M., R.A. and A.K. (Alban Kuriqi); writing—original draft preparation, M.K., A.K. (Anuradha Kumari), D.P.K., P.K., A.M., R.A. and A.K. (Alban Kuriqi); writing—review and editing, M.K., A.K. (Anuradha Kumari), D.P.K., P.K., A.M., R.A. and A.K. (Alban Kuriqi); visualization, P.K., A.M. and A.K. (Alban Kuriqi); supervision, P.K., A.M. and A.K. (Alban Kuriqi); project administration, A.K. (Alban Kuriqi); funding acquisition, A.K. (Alban Kuriqi). All authors have read and agreed to the published version of the manuscript.

Funding: This research received no external funding.

Acknowledgments: The authors appreciate the comments of anonymous reviewers which helped to improve this paper further. Alban Kuriqi was supported by a Ph.D. scholarship granted by Fundação para a Ciência e a Tecnologia, I.P. (FCT), Portugal, under the Ph.D. Program FLUVIO–River Restoration and Management, grant number PD/BD/114558/2016.

Conflicts of Interest: The authors declare no conflict of interest.

References

1. Gericke, O.J.; Smithers, J.C. Review of methods used to estimate catchment response time for the purpose of peak discharge estimation. *Hydrol. Sci. J.* **2014**, *59*, 1935–1971. [CrossRef]

2. Mohanty, P.K.; Mohanty, L.P.; Khatua, K.K. Discharge estimation in wide meandering compound channels. *ISH J. Hydraul. Eng.* **2019**, *25*, 1–15. [CrossRef]

3. Schmidt, A.R.; Garcia, M.H. Theoretical Examination of Historical Shifts and Adjustments to Stage-Discharge Rating Curves. In Proceedings of the World Water & Environmental Resources Congress 2003, American Society of Civil Engineers, Reston, VA, USA, 23–26 June 2003; pp. 1–10.

4. Schmidt, A.R.; Yen, B.C. Theoretical Development of Stage-Discharge Ratings for Subcritical Open-Channel Flows. *J. Hydraul. Eng.* **2008**, *134*, 1245–1256. [CrossRef]

5. Ardıçlıoğlu, M.; Kuriqi, A. Calibration of channel roughness in intermittent rivers using HEC-RAS model: Case of Sarimsakli creek, Turkey. *SN Appl. Sci.* **2019**, *1*, 1080. [CrossRef]

6. Manfreda, S.; Pizarro, A.; Moramarco, T.; Cimorelli, L.; Pianese, D.; Barbetta, S. Potential advantages of flow-area rating curves compared to classic stage-discharge-relations. *J. Hydrol.* **2020**, *585*, 124752. [CrossRef]

7. Westerberg, I.; Guerrero, J.-L.; Seibert, J.; Beven, K.J.; Halldin, S. Stage-discharge uncertainty derived with a non-stationary rating curve in the Choluteca River, Honduras. *Hydrol. Process.* **2011**, *25*, 603–613. [CrossRef]

8. Petersen-Øverleir, A. Modelling stage—Discharge relationships affected by hysteresis using the Jones formula and nonlinear regression. *Hydrol. Sci. J.* **2006**, *51*, 365–388. [CrossRef]

9. Rojas, M.; Quintero, F.; Young, N. Analysis of Stage–Discharge Relationship Stability Based on Historical Ratings. *Hydrology* **2020**, *7*, 31. [CrossRef]

10. Kuriqi, A.; Ardiçlioğlu, M. Investigation of hydraulic regime at middle part of the Loire River in context of floods and low flow events. *Pollack Period.* **2018**, *13*, 145–156. [CrossRef]

11. Kuriqi, A.; Koçileri, G.; Ardiçlioğlu, M. Potential of Meyer-Peter and Müller approach for estimation of bed-load sediment transport under different hydraulic regimes. *Model. Earth Syst. Environ.* **2020**, *6*, 129–137. [CrossRef]

12. Ghorbani, M.A.; Deo, R.C.; Kim, S.; Hasanpour Kashani, M.; Karimi, V.; Izadkhah, M. Development and evaluation of the cascade correlation neural network and the random forest models for river stage and river flow prediction in Australia. *Soft Comput.* **2020**, *24*, 12079–12090. [CrossRef]

13. Bhattacharya, B.; Solomatine, D.P. Neural networks and M5 model trees in modelling water level–discharge relationship. *Neurocomputing* **2005**, *63*, 381–396. [CrossRef]

14. Adamowski, J.; Fung Chan, H.; Prasher, S.O.; Ozga-Zielinski, B.; Sliusarieva, A. Comparison of multiple linear and nonlinear regression, autoregressive integrated moving average, artificial neural network, and wavelet artificial neural network methods for urban water demand forecasting in Montreal, Canada. *Water Resour. Res.* **2012**, *48*, W01528. [CrossRef]

15. Aggarwal, S.K.; Goel, A.; Singh, V.P. Stage and Discharge Forecasting by SVM and ANN Techniques. *Water Resour. Manag.* **2012**, *26*, 3705–3724. [CrossRef]

16. Kalteh, A.M. Monthly river flow forecasting using artificial neural network and support vector regression models coupled with wavelet transform. *Comput. Geosci.* **2013**, *54*, 1–8. [CrossRef]

17. Supharatid, S. Application of a neural network model in establishing a stage-discharge relationship for a tidal river. *Hydrol. Process.* **2003**, *17*, 3085–3099. [CrossRef]

18. Londhe, S.; Panse-Aglave, G. Modelling Stage–Discharge Relationship using Data-Driven Techniques. *ISH J. Hydraul. Eng.* **2015**, *21*, 207–215. [CrossRef]

19. Deka, P.; Chandramouli, V. A fuzzy neural network model for deriving the river stage—Discharge relationship. *Hydrol. Sci. J.* **2003**, *48*, 197–209. [CrossRef]

20. Lohani, A.K.; Goel, N.K.; Bhatia, K.K.S. Takagi–Sugeno fuzzy inference system for modeling stage–discharge relationship. *J. Hydrol.* **2006**, *331*, 146–160. [CrossRef]

21. Alizadeh, F.; Faregh Gharamaleki, A.; Jalilzadeh, R. A two-stage multiple-point conceptual model to predict river stage-discharge process using machine learning approaches. *J. Water Clim. Chang.* **2020**, *11*, 1–18. [CrossRef]

22. Roushangar, K.; Foroudi Khowr, A.; Saneie, M. Experimental study and artificial intelligence-based modeling of discharge coefficient of converging ogee spillways. *ISH J. Hydraul. Eng.* **2019**, *25*, 1–8. [CrossRef]

23. Norouzi, R.; Daneshfaraz, R.; Ghaderi, A. Investigation of discharge coefficient of trapezoidal labyrinth weirs using artificial neural networks and support vector machines. *Appl. Water Sci.* **2019**, *9*, 148. [CrossRef]

24. Najah Ahmed, A.; Binti Othman, F.; Abdulmohsin Afan, H.; Khaleel Ibrahim, R.; Ming Fai, C.; Shabbir Hossain, M.; Ehteram, M.; Elshafie, A. Machine learning methods for better water quality prediction. *J. Hydrol.* **2019**, *578*, 124084. [CrossRef]

25. Muharemi, F.; Logofătu, D.; Leon, F. Machine learning approaches for anomaly detection of water quality on a real-world data set. *J. Inf. Telecommun.* **2019**, *3*, 294–307. [CrossRef]

26. Di, Z.; Chang, M.; Guo, P. Water Quality Evaluation of the Yangtze River in China Using Machine Learning Techniques and Data Monitoring on Different Time Scales. *Water* **2019**, *11*, 339. [CrossRef]

27. Moon, S.-H.; Kim, Y.-H.; Lee, Y.H.; Moon, B.-R. Application of machine learning to an early warning system for very short-term heavy rainfall. *J. Hydrol.* **2019**, *568*, 1042–1054. [CrossRef]

28. Bojang, P.O.; Yang, T.-C.; Pham, Q.B.; Yu, P.-S. Linking Singular Spectrum Analysis and Machine Learning for Monthly Rainfall Forecasting. *Appl. Sci.* **2020**, *10*, 3224. [CrossRef]

29. Pham, Q.B.; Abba, S.I.; Usman, A.G.; Linh, N.T.T.; Gupta, V.; Malik, A.; Costache, R.; Vo, N.D.; Tri, D.Q. Potential of Hybrid Data-Intelligence Algorithms for Multi-Station Modelling of Rainfall. *Water Resour. Manag.* **2019**, *33*, 5067–5087. [CrossRef]

30. Pour, S.H.; Wahab, A.K.A.; Shahid, S. Physical-empirical models for prediction of seasonal rainfall extremes of Peninsular Malaysia. *Atmos. Res.* **2020**, *233*, 104720. [CrossRef]

31. Malik, A.; Kumar, A.; Ghorbani, M.A.; Kashani, M.H.; Kisi, O.; Kim, S. The viability of co-active fuzzy inference system model for monthly reference evapotranspiration estimation: Case study of Uttarakhand State. *Hydrol. Res.* **2019**, *50*, 1623–1644. [CrossRef]

32. Alizamir, M.; Kisi, O.; Muhammad Adnan, R.; Kuriqi, A. Modelling reference evapotranspiration by combining neuro-fuzzy and evolutionary strategies. *Acta Geophys.* **2020**, *68*, 1113–1126. [CrossRef]

33. Yamaç, S.S.; Todorovic, M. Estimation of daily potato crop evapotranspiration using three different machine learning algorithms and four scenarios of available meteorological data. *Agric. Water Manag.* **2020**, *228*, 105875. [CrossRef]

34. Malik, A.; Kumar, A. Pan Evaporation Simulation Based on Daily Meteorological Data Using Soft Computing Techniques and Multiple Linear Regression. *Water Resour. Manag.* **2015**, *29*, 1859–1872. [CrossRef]

35. Malik, A.; Kumar, A.; Kisi, O. Monthly pan-evaporation estimation in Indian central Himalayas using different heuristic approaches and climate based models. *Comput. Electron. Agric.* **2017**, *143*, 302–313. [CrossRef]

36. Ashrafzadeh, A.; Malik, A.; Jothiprakash, V.; Ghorbani, M.A.; Biazar, S.M. Estimation of daily pan evaporation using neural networks and meta-heuristic approaches. *ISH J. Hydraul. Eng.* **2018**, *24*, 1–9. [CrossRef]

37. Malik, A.; Kumar, A.; Kisi, O. Daily Pan Evaporation Estimation Using Heuristic Methods with Gamma Test. *J. Irrig. Drain. Eng.* **2018**, *144*, 04018023. [CrossRef]

38. Malik, A.; Rai, P.; Heddam, S.; Kisi, O.; Sharafati, A.; Salih, S.Q.; Al-Ansari, N.; Yaseen, Z.M. Pan Evaporation Estimation in Uttarakhand and Uttar Pradesh States, India: Validity of an Integrative Data Intelligence Model. *Atmosphere* **2020**, *11*, 553. [CrossRef]

39. Rahmati, O.; Falah, F.; Dayal, K.S.; Deo, R.C.; Mohammadi, F.; Biggs, T.; Moghaddam, D.D.; Naghibi, S.A.; Bui, D.T. Machine learning approaches for spatial modeling of agricultural droughts in the south-east region of Queensland Australia. *Sci. Total Environ.* **2020**, *699*, 134230. [CrossRef]

40. Das, P.; Naganna, S.R.; Deka, P.C.; Pushparaj, J. Hybrid wavelet packet machine learning approaches for drought modeling. *Environ. Earth Sci.* **2020**, *79*, 221. [CrossRef]

41. Malik, A.; Kumar, A.; Singh, R.P. Application of Heuristic Approaches for Prediction of Hydrological Drought Using Multi-scalar Streamflow Drought Index. *Water Resour. Manag.* **2019**, *33*, 3985–4006. [CrossRef]

42. Malik, A.; Kumar, A.; Piri, J. Daily suspended sediment concentration simulation using hydrological data of Pranhita River Basin, India. *Comput. Electron. Agric.* **2017**, *138*, 20–28. [CrossRef]

43. Malik, A.; Kumar, A.; Kisi, O.; Shiri, J. Evaluating the performance of four different heuristic approaches with Gamma test for daily suspended sediment concentration modeling. *Environ. Sci. Pollut. Res.* **2019**, *26*, 22670–22687. [CrossRef] [PubMed]

44. Zounemat-Kermani, M.; Mahdavi-Meymand, A.; Alizamir, M.; Adarsh, S.; Yaseen, Z.M. On the complexities of sediment load modeling using integrative machine learning: Application of the great river of Loíza in Puerto Rico. *J. Hydrol.* **2020**, *585*, 124759. [CrossRef]

45. Kisi, O.; Dailr, A.H.; Cimen, M.; Shiri, J. Suspended sediment modeling using genetic programming and soft computing techniques. *J. Hydrol.* **2012**, *450–451*, 48–58. [CrossRef]

46. Kumar, D.; Pandey, A.; Sharma, N.; Flügel, W.-A. Daily suspended sediment simulation using machine learning approach. *CATENA* **2016**, *138*, 77–90. [CrossRef]

47. Daubechies, I. The wavelet transform, time-frequency localization and signal analysis. *IEEE Trans. Inf. Theory* **1990**, *36*, 961–1005. [CrossRef]

48. Rioul, O.; Vetterli, M. Wavelets and signal processing. *IEEE Signal Process. Mag.* **1991**, *8*, 14–38. [CrossRef]

49. Kim, C.-K.; Kwak, I.-S.; Cha, E.-Y.; Chon, T.-S. Implementation of wavelets and artificial neural networks to detection of toxic response behavior of chironomids (Chironomidae: Diptera) for water quality monitoring. *Ecol. Model.* **2006**, *195*, 61–71. [CrossRef]

50. Dash, P.K.; Majumder, I.; Nayak, N.; Bisoi, R. Point and Interval Solar Power Forecasting Using Hybrid Empirical Wavelet Transform and Robust Wavelet Kernel Ridge Regression. *Nat. Resour. Res.* **2020**, *29*, 2813–2841. [CrossRef]

51. Wang, W.; Ding, J. Wavelet Network Model and Its Application to the Prediction of Hydrology. *Nat. Sci.* **2003**, *1*, 67–71.

52. Bhardwaj, S.; Chandrasekhar, E.; Padiyar, P.; Gadre, V.M. A comparative study of wavelet-based ANN and classical techniques for geophysical time-series forecasting. *Comput. Geosci.* **2020**, *138*, 104461. [CrossRef]

53. Graf, R.; Zhu, S.; Sivakumar, B. Forecasting river water temperature time series using a wavelet–neural network hybrid modelling approach. *J. Hydrol.* **2019**, *578*, 124115. [CrossRef]

54. Ghazvinei, P.T.; Shamshirband, S.; Motamedi, S.; Hassanpour Darvishi, H.; Salwana, E. Performance investigation of the dam intake physical hydraulic model using Support Vector Machine with a discrete wavelet transform algorithm. *Comput. Electron. Agric.* **2017**, *140*, 48–57. [CrossRef]

55. Zhou, F.; Liu, B.; Duan, K. Coupling wavelet transform and artificial neural network for forecasting estuarine salinity. *J. Hydrol.* **2020**, *588*, 125127. [CrossRef]

56. Zhang, J.; Zhang, X.; Niu, J.; Hu, B.X.; Soltanian, M.R.; Qiu, H.; Yang, L. Prediction of groundwater level in seashore reclaimed land using wavelet and artificial neural network-based hybrid model. *J. Hydrol.* **2019**, *577*, 123948. [CrossRef]

57. Haykin, S. *Neural Networks—A Comprehensive Foundation*, 2nd ed.; Prentice-Hall: Up Saddle River, NJ, USA, 1999; pp. 26–32.

58. Vapnik, V.N. *The Nature of Statistical Learning Theory*; Springer: New York, NY, USA, 1995; p. 314.

59. Asefa, T.; Kemblowski, M.; Urroz, G.; McKee, M. Support vector machines (SVMs) for monitoring network design. *Ground Water* **2005**, *43*, 413–422. [CrossRef]

60. Raghavendra, N.S.; Deka, P.C. Support vector machine applications in the field of hydrology: A review. *Appl. Soft Comput.* **2014**, *19*, 372–386. [CrossRef]

61. Hipni, A.; El-shafie, A.; Najah, A.; Karim, O.A.; Hussain, A.; Mukhlisin, M. Daily Forecasting of Dam Water Levels: Comparing a Support Vector Machine (SVM) Model With Adaptive Neuro Fuzzy Inference System (ANFIS). *Water Resour. Manag.* **2013**, *27*, 3803–3823. [CrossRef]

62. Nguyen, L. Tutorial on support vector machine. *Appl. Comput. Math.* **2017**, *6*, 1–15.

63. Misra, D.; Oommen, T.; Agarwal, A.; Mishra, S.K.; Thompson, A.M. Application and analysis of support vector machine based simulation for runoff and sediment yield. *Biosyst. Eng.* **2009**, *103*, 527–535. [CrossRef]
64. Gholami, R.; Fakhari, N. Support Vector Machine: Principles, Parameters, and Applications. In *Handbook of Neural Computation*; Elsevier: Amsterdam, The Netherlands, 2017; pp. 515–535.
65. Mohammadi, B.; Mehdizadeh, S. Modeling daily reference evapotranspiration via a novel approach based on support vector regression coupled with whale optimization algorithm. *Agric. Water Manag.* **2020**, *237*, 106145. [CrossRef]
66. Banadkooki, F.B.; Ehteram, M.; Panahi, F.; Sammen, S.S.; Othman, F.B.; EL-Shafie, A. Estimation of total dissolved solids (TDS) using new hybrid machine learning models. *J. Hydrol.* **2020**, *587*, 124989. [CrossRef]
67. Su, H.; Li, X.; Yang, B.; Wen, Z. Wavelet support vector machine-based prediction model of dam deformation. *Mech. Syst. Signal Process.* **2018**, *110*, 412–427. [CrossRef]
68. Panahi, M.; Sadhasivam, N.; Pourghasemi, H.R.; Rezaie, F.; Lee, S. Spatial prediction of groundwater potential mapping based on convolutional neural network (CNN) and support vector regression (SVR). *J. Hydrol.* **2020**, *588*, 125033. [CrossRef]
69. Tikhamarine, Y.; Malik, A.; Souag-Gamane, D.; Kisi, O. Artificial intelligence models versus empirical equations for modeling monthly reference evapotranspiration. *Environ. Sci. Pollut. Res.* **2020**, *27*, 30001–30019. [CrossRef]
70. Zhang, X.; Wang, J.; Zhang, K. Short-term electric load forecasting based on singular spectrum analysis and support vector machine optimized by Cuckoo search algorithm. *Electr. Power Syst. Res.* **2017**, *146*, 270–285. [CrossRef]
71. Ansari, H.R.; Gholami, A. An improved support vector regression model for estimation of saturation pressure of crude oils. *Fluid Phase Equilib.* **2015**, *402*, 124–132. [CrossRef]
72. Han, D.; Chan, L.; Zhu, N. Flood forecasting using support vector machines. *J. Hydroinform.* **2007**, *9*, 267–276. [CrossRef]
73. Cobaner, M.; Unal, B.; Kisi, O. Suspended sediment concentration estimation by an adaptive neuro-fuzzy and neural network approaches using hydro-meteorological data. *J. Hydrol.* **2009**, *367*, 52–61. [CrossRef]
74. Deo, R.C.; Tiwari, M.K.; Adamowski, J.F.; Quilty, J.M. Forecasting effective drought index using a wavelet extreme learning machine (W-ELM) model. *Stoch. Environ. Res. Risk Assess.* **2017**, *31*, 1211–1240. [CrossRef]
75. Malik, A.; Kumar, A.; Salih, S.Q.; Kim, S.; Kim, N.W.; Yaseen, Z.M.; Singh, V.P. Drought index prediction using advanced fuzzy logic model: Regional case study over Kumaon in India. *PLoS ONE* **2020**, *15*, e0233280. [CrossRef] [PubMed]
76. Malik, A.; Kumar, A. Meteorological drought prediction using heuristic approaches based on effective drought index: A case study in Uttarakhand. *Arab. J. Geosci.* **2020**, *13*, 276. [CrossRef]
77. Gan, T.Y.; Dlamini, E.M.; Biftu, G.F. Effects of model complexity and structure, data quality, and objective functions on hydrologic modeling. *J. Hydrol.* **1997**, *192*, 81–103. [CrossRef]
78. Moriasi, D.N.; Wilson, B.N.; Douglas-Mankin, K.R.; Arnold, J.G.; Gowda, P.H. Hydrologic and Water Quality Models: Use, Calibration, and Validation. *Trans. ASABE* **2012**, *55*, 1241–1247. [CrossRef]
79. Nash, J.E.; Sutcliffe, J.V. River flow forecasting through conceptual models part I—A discussion of principles. *J. Hydrol.* **1970**, *10*, 282–290. [CrossRef]
80. Willmott, C.; Matsuura, K. Advantages of the mean absolute error (MAE) over the root mean square error (RMSE) in assessing average model performance. *Clim. Res.* **2005**, *30*, 79–82. [CrossRef]
81. Krause, P.; Boyle, D.P.; Bäse, F. Comparison of different efficiency criteria for hydrological model assessment. *Adv. Geosci.* **2005**, *5*, 89–97. [CrossRef]
82. Legates, D.R.; McCabe, G.J. Evaluating the use of "goodness-of-fit" Measures in hydrologic and hydroclimatic model validation. *Water Resour. Res.* **1999**, *35*, 233–241. [CrossRef]
83. Willmott, C.J. On the validation of models. *Phys. Geogr.* **1981**, *2*, 184–194. [CrossRef]
84. Malik, A.; Kumar, A.; Kim, S.; Kashani, M.H.; Karimi, V.; Sharafati, A.; Ghorbani, M.A.; Al-Ansari, N.; Salih, S.Q.; Yaseen, Z.M.; et al. Modeling monthly pan evaporation process over the Indian central Himalayas: Application of multiple learning artificial intelligence model. *Eng. Appl. Comput. Fluid Mech.* **2020**, *14*, 323–338. [CrossRef]
85. Kouchi, D.H.; Esmaili, K.; Faridhosseini, A.; Sanaeinejad, S.H.; Khalili, D.; Abbaspour, K.C. Sensitivity of Calibrated Parameters and Water Resource Estimates on Different Objective Functions and Optimization Algorithms. *Water* **2017**, *9*, 384. [CrossRef]

86. Paul, M.; Negahban-Azar, M. Sensitivity and uncertainty analysis for streamflow prediction using multiple optimization algorithms and objective functions: San Joaquin Watershed, California. *Model. Earth Syst. Environ.* **2018**, *4*, 1509–1525. [CrossRef]

87. Shamseldin, A.Y. Application of a neural network technique to rainfall-runoff modelling. *J. Hydrol.* **1997**, *199*, 272–294. [CrossRef]

88. Taylor, K.E. Summarizing multiple aspects of model performance in a single diagram. *J. Geophys. Res. Atmos.* **2001**, *106*, 7183–7192. [CrossRef]

89. Singh, A.; Malik, A.; Kumar, A.; Kisi, O. Rainfall-runoff modeling in hilly watershed using heuristic approaches with gamma test. *Arab. J. Geosci.* **2018**, *11*, 261. [CrossRef]

90. Tikhamarine, Y.; Souag-Gamane, D.; Kisi, O. A new intelligent method for monthly streamflow prediction: Hybrid wavelet support vector regression based on grey wolf optimizer (WSVR–GWO). *Arab. J. Geosci.* **2019**, *12*, 540. [CrossRef]

91. Tikhamarine, Y.; Souag-Gamane, D.; Najah Ahmed, A.; Kisi, O.; El-Shafie, A. Improving artificial intelligence models accuracy for monthly streamflow forecasting using grey Wolf optimization (GWO) algorithm. *J. Hydrol.* **2020**, *582*, 124435. [CrossRef]

92. Tikhamarine, Y.; Souag-Gamane, D.; Ahmed, A.N.; Sammen, S.S.; Kisi, O.; Huang, Y.F.; El-Shafie, A. Rainfall-runoff modelling using improved machine learning methods: Harris hawks optimizer vs. particle swarm optimization. *J. Hydrol.* **2020**, *589*, 125133. [CrossRef]

93. Hussain, D.; Khan, A.A. Machine learning techniques for monthly river flow forecasting of Hunza River, Pakistan. *Earth Sci. Inform.* **2020**, *13*, 939–949. [CrossRef]

94. Khatibi, R.; Ghorbani, M.A.; Naghshara, S.; Aydin, H.; Karimi, V. A framework for 'Inclusive Multiple Modelling' with critical views on modelling practices–Applications to modelling water levels of Caspian Sea and Lakes Urmia and Van. *J. Hydrol.* **2020**, *587*, 124923. [CrossRef]

95. Adnan, R.M.; Liang, Z.; Parmar, K.S.; Soni, K.; Kisi, O. Modeling monthly streamflow in mountainous basin by MARS, GMDH-NN and DENFIS using hydroclimatic data. *Neural Comput. Appl.* **2020**, *32*, 1–19. [CrossRef]

96. Ali, S.; Shahbaz, M. Streamflow forecasting by modeling the rainfall–streamflow relationship using artificial neural networks. *Model. Earth Syst. Environ.* **2020**, *6*, 1645–1656. [CrossRef]

97. Mohammadi, B.; Linh, N.T.T.; Pham, Q.B.; Ahmed, A.N.; Vojteková, J.; Guan, Y.; Abba, S.; El-Shafie, A. Adaptive neuro-fuzzy inference system coupled with shuffled frog leaping algorithm for predicting river streamflow time series. *Hydrol. Sci. J.* **2020**, *65*, 1738–1751. [CrossRef]

98. Mohammadi, B.; Ahmadi, F.; Mehdizadeh, S.; Guan, Y.; Pham, Q.B.; Linh, N.T.T.; Tri, D.Q. Developing Novel Robust Models to Improve the Accuracy of Daily Streamflow Modeling. *Water Resour. Manag.* **2020**, *34*, 3387–3409. [CrossRef]

99. Tripura, J.; Roy, P.; Barbhuiya, A.K. Simultaneous streamflow forecasting based on hybridized neuro-fuzzy method for a river system. *Neural Comput. Appl.* **2020**, *32*, 1–13. [CrossRef]

 *sustainability*

Article

Estimation of Fuzzy Parameters in the Linear Muskingum Model with the Aid of Particle Swarm Optimization

Mike Spiliotis [1,*], Alvaro Sordo-Ward [2] and Luis Garrote [2]

[1] Department of Civil Engineering, School of Engineering, Democritus University of Thrace, 671 00 Xanthi, Greece

[2] Department of Civil Engineering: Hydraulics, Energy and Environment, Universidad Politécnica de Madrid, 28040 Madrid, Spain; alvaro.sordo.ward@upm.es (A.S.-W.); l.garrote@upm.es (L.G.)

* Correspondence: m.spiliotis@gmail.com or mspiliot@civil.duth.gr

Abstract: The Muskingum method is one of the widely used methods for lumped flood routing in natural rivers. Calibration of its parameters remains an active challenge for the researchers. The task has been mostly addressed by using crisp numbers, but fuzzy seems a reasonable alternative to account for parameter uncertainty. In this work, a fuzzy Muskingum model is proposed where the assessment of the outflow as a fuzzy quantity is based on the crisp linear Muskingum method but with fuzzy parameters as inputs. This calculation can be achieved based on the extension principle of the fuzzy sets and logic. The critical point is the calibration of the proposed fuzzy extension of the Muskingum method. Due to complexity of the model, the particle swarm optimization (PSO) method is used to enable the use of a simulation process for each possible solution that composes the swarm. A weighted sum of several performance criteria is used as the fitness function of the PSO. The function accounts for the inclusive constraints (the property that the data must be included within the produced fuzzy band) and for the magnitude of the fuzzy band, since large uncertainty may render the model non-functional. Four case studies from the references are used to benchmark the proposed method, including smooth, double, and non-smooth data and a complex, real case study that shows the advantages of the approach. The use of fuzzy parameters is closer to the uncertain nature of the problem. The new methodology increases the reliability of the prediction. Furthermore, the produced fuzzy band can include, to a significant degree, the observed data and the output of the existent crisp methodologies even if they include more complex assumptions.

Keywords: flood routing; Muskingum method; extension principle; calibration; fuzzy sets and systems; particle swarm optimization

Citation: Spiliotis, M.; Sordo-Ward, A.; Garrote, L. Estimation of Fuzzy Parameters in the Linear Muskingum Model with the Aid of Particle Swarm Optimization. *Sustainability* **2021**, *13*, 7152. https://doi.org/10.3390/su13137152

Academic Editor: Ozgur Kisi

Received: 5 May 2021
Accepted: 18 June 2021
Published: 25 June 2021

Publisher's Note: MDPI stays neutral with regard to jurisdictional claims in published maps and institutional affiliations.

1. Introduction

Flood risk management is a key component in sustainable water resources management. Floods impact both individuals and communities, producing harmful consequences of social, economic, and environmental implications. Negative impacts of flooding include loss of human life, destruction of crops, damage to essential infrastructure, and disruption of the value chain. Many flood risk management strategies are based on operational early waring schemes that predict the arrival of the flood wave and enable civil protection actions to safeguard life and property. The prediction of the outflow hydrograph for a river's reach given a specific inflow hydrograph is, therefore, an important issue in water resources management. Hydrological forecasting requires a sound understanding of the physical mechanisms that control the propagation of flood waves along rivers. Mathematical models for flood routing are based on the unsteady flow Saint-Venant equations. Numerical solutions of the flow equations are complex, and thus, simplified versions are preferred in operational flood forecasting schemes. Various methods have been developed for this purpose, which may be categorized into three distinct groups [1,2]: (1) the distributed (or hydraulic) methods (e.g., dynamic wave, diffusion wave, and kinematic wave models), which are based on the mass conservation and momentum transport equations; (2) the lumped (or hydrologic) methods (in which the linear and nonlinear Muskingum models [3] are widely used); and (3) the semi-distributed or hybrid methods (e.g., Muskingum–Cunge family models [4–6]). This work is focused on enhancing models of the third category using fuzzy sets and logic.

Lumped models estimate the flood hydrograph at a downstream section of the river from the flood hydrograph at the upstream section without considering explicitly the river characteristics between the upstream and the downstream sections. These models use the one-dimensional continuity equation along with a storage equation in the calibration and verification steps.

The Muskingum method for natural streamflow routing, first proposed by McCarthy (1938) [3] in the Muskingum River basin in Ohio, is a widely applied hydrologic method [7]. It is a lumped method which cannot provide the outflow at the intermediate sections. However, these models are very popular since, in most cases, the scarcity of field data prevents the use of the Saint-Venant equations to route floods in rivers, as occurs in the second category of river-reach routing [8].

The widely-used linear Muskingum method depends on two parameters: K and x [9]. Originally, the graphical method was developed to determine the parameters of the widely-used Muskingum linear equation. As is widely known, the graphical solution is based on the graphical representation of the storage versus the weighted discharge as a function of x. The preferred value of x is the one that produces the narrowest loop (e.g., Wilson, 1974 [10]). Next, some techniques based either on linear programming or on linear regression were developed [11]. However, the use of the regression-based techniques can lead to unreasonable values for K, x (e.g., [12]), due to an overtraining behavior.

The use of the nonlinear storage equation within the Muskingum method increases the number of model parameters, but it is closer to reality. However, the nonlinear storage equation increases the difficulty of the calibration process. Various mathematical-hydrological methodologies are developed to calibrate the river-routing models by using a nonlinear approach regarding the relation of the reach storage. For instance, according to the segmented least-squares method (S-LSM) (Gill 1978) [13], the whole available data are subdivided into several groups based on the storage values. It is then assumed that K and x remain constant in each subgroup, and hence, for each subgroup, the least square analysis is used. Therefore, the optimization procedure is based on the minimization between the estimated storage and the existent storage.

Next, a directly (global) nonlinear formulation of the Muskingum method (regarding the reach storage) was proposed by several authors with the use of a nonlinear form for the storage. For instance, [14] used several mathematical techniques to minimize the sum of the squares of deviations between observed channel storage and computed channel storage.

Similarly, the nonlinear least squares method (NL-LSM) [15], the Lagrange multiplier method (LMM) [7], and the Broyden–Fletcher–Goldfarb–Shanno (BFGS) technique [16] have been exploited for assessing the parameter values of the models. Das (2004) [7] and Geem (2006) [15] proposed an optimization model whose objective function emphasizes the determination of a set of parameter values that minimizes the error between model-predicted and observed outflows. Das (2004) [7] considered the validation of the storage-discharge relation by using the Lagrange multipliers.

However, these techniques have the drawbacks of a complex derivative requirement and/or good initial vector consideration [17,18]. For example, the BFGS technique (Geem 2006) [16], although it reached the best solution ever found, relies heavily on the consideration of the initial vector [17]. Hence, several researchers have proposed various phenomenon-mimicking algorithms where an initial population of possible solutions is used instead of an initial vector. For instance, the genetic algorithm (GA) [19] (Mohan 1997), harmony search (HS) (Kim et al., 2001) [20], particle swarm optimization (PSO) (Chu and Chang 2009) [21], differential evolution (DE) (Xu et al., 2011) [22], parameter-setting free harmony search (PSFHS) algorithm [23], and modified honey bee mating optimization (MHBMO) algorithm (Niazkar and Afzali 2015) [24] were developed to achieve the calibration of the nonlinear Muskingum method (Niazkar and Afzali 2016) [25].

In order to improve the optimum solution, several authors proposed hybrid optimization algorithms that combine the heuristic and the derivatives-based algorithms from a mathematical point of view. Two steps are mainly considered. In the first step, one of search-based, phenomenon-mimicking algorithms, which requires no initial guess, is exploited to assess the Muskingum parameters. Afterwards, the obtained values for nonlinear Muskingum parameters are utilized as the first initial vector for the second step in which a deterministic, derivatives-based method continues the routing simulation optimization (Niazkar and Afzali 2016) [25]. For instance, in Karahan et al., 2015 [18], the HS algorithm (global search) searched the optimum solution with multiple solution vectors, and then the BFGS algorithm (local search) adjusted the results of the HS algorithm by getting its results as the new initial solution [18]. A critical point is that most of the suitability measures are based on the comparison between the observed and the predicted outflow, whilst the other older works, such as [13,14], are focused on the storage equation (as can be seen from the objective function used).

In addition, several authors proposed other innovations in the nonlinear Muskingum relation. Karahan et al., 2015 [18] proposed the use of the cuckoo search algorithm (which also uses an initial population) during the calibration of the model parameters by also including the lateral flow. Easa (2015) [26] proposed to increase performance using continuous and discontinuous parameters expressed as a function of a dimensionless inflow variable. In addition, Farzin et al., 2018 [27] proposed the multi-reach Muskingum method to enhance the accuracy of the Muskingum method. The river under study was divided into several smaller reaches, and hence, for each reach, routing was applied separately. However, if there are no data in intermediate sections, it is not obvious that the intermediate outcomes can be used to estimate the outflow at these sections, and therefore, the method could be characterized as a black box model rather than a conceptual model.

Regarding the application of the fuzzy sets and logic methods in these types of problems, most of the proposed models are based on the adaptive network-based fuzzy inference systems (ANFIS) implementation. ANFIS is a neuro-adaptive learning technique which provides a hybrid method for the fuzzy modeling procedure to learn information based on a data set. Therefore, it can be seen as a neuro-fuzzy approach. It is very popular because of the corresponding toolbox made available in MATLAB. In many hydraulic applications, the ANFIS method is widely used, providing very good results; however, it seems that sometimes it was solely utilized as a black-box model without any specific utilization of the produced fuzzy rules. Chu, 2009 [28] proposed the combined application of fuzzy inference system (FIS) and Muskingum model in flood routing. The implementation is based on the ANFIS toolbox of MATLAB. Three points of this methodology must be noted. The first one is that the Muskingum method is used indirectly in the structure of the model (neither the evaluation of K or x takes place). The second point is that the output (outflow) for each time step is a crisp number and not a fuzzy number. Thirdly, based on the ANFIS method, a large amount of data is required.

Another idea is to use the fuzzy linear regression to calibrate the linear Muskingum method instead of the crisp linear regression as O'Donnel (1985) [11] proposed. Spiliotis and Garrote (2017) [29] proposed the calibration of the Muskingum parameters based on fuzzy linear regression. Therefore, instead of the crisp coefficient of the O'Donnel (1985) method [11], the coefficients are proposed to be fuzzy symmetric triangular numbers. This implies that, in contrast with the widely used ANFIS, the output (outflow) of the proposed model is a fuzzy number. Furthermore, based on the Tanaka (1987) [30] fuzzy regression implementation, all the observed data (regarding the outflow) are forced to be included in the produced fuzzy band aiming at its minimum width. Instead of the identification of the parameters K, x, the problem is oriented to the determination of the corresponding regression coefficient. Another point of view is that, based on the problem itself, the constant term of the usual regression model must be removed [29]. The relevant feature of articles mentioned in this paragraph is that they contain methodologies that produce a fuzzy outflow as a final output for the channel-routing problem.

However, the fuzzy linear regression of Tanaka (1987) [30] is heavily influenced by the existence of outliers, which can extremely enlarge the spread of the model. Hence, Spiliotis and Garrote, 2017 [29] proposed the use of a fuzzy linear regression model that is modified by incorporating ideas from the field of goal programming. Spiliotis et al., 2018 [12] expanded the methodology by including the lateral flow on the basis of the O'Donnel article [11]. As it was noted by the authors, an interesting point is that sometimes the produced fuzzy coefficients cannot correspond to the physical meaning of the storage equation, and hence, the model leads to a rather metric or black-box method.

In this work, a hybrid, fuzzy-based methodology is proposed. The proposed method consists of two main ideas. Firstly, the linear Muskingum model is adopted. The parameters K, x and α (where lateral flow occurs) are selected to be fuzzy symmetrical triangular numbers to avoid any irrational training as may occur in the case of the fuzzy linear regression. Hence, the methods produce a fuzzy estimation of the outflow based on the (crisp) Muskingum function to predict the outflow. Mathematically, the application of a crisp function using fuzzy parameters can be treated based on the extension principle of the fuzzy sets and logic. Secondly, the calibration is addressed by using a hybrid optimization procedure. In this article, the particle swarm optimization method (PSO) is used. Hence, there is no need to predict an initial solution. For each possible solution based on the (considered) values of the fuzzy parameters K, x and α, the outflow is calculated. The outflow will be a fuzzy number, whilst its determination is based on the extension principle. Unfortunately, the implementation of the extension principle leads to some sub-optimization problems but without difficult constraints. The solvability of these sub-optimization problems is discussed in the methodology section.

Each possible solution is evaluated based on the degree of inclusion of the real data within the produced fuzzy band. This criterion should be included in the formulation of the fitness function of the heuristic optimization method. However, other criteria will be added which will be discussed in the methodology section. For instance, a fuzzy solution with a very large band will contain all the data, but this information is non-functional. Therefore, the fitness function must contain not only the inclusion degree but the magnitude of the fuzzy band, etc. In PSO, the values of the fitness function of the swarm and the use of random numbers determine the movement of the swarm, that is, the positions of the new possible solutions. The procedure is finished by considering a maximum number of iterations.

Three classical case studies from the references are widely used to benchmark the variations of the Muskingum method. These three case studies include: (1) smooth outflow (Wilson 1974 [10]), (2) double-peak outflow (Viessman and Lewis 2003) [9], and (3) non-smooth outflow with lateral flow (O'Donnel 1985) data sets [11]. The implementation of the proposed methodology is validated using the three benchmark examples. The validation exercise is focused on the following questions: (1) Do the achieved fuzzy values of K, x have physical meaning? (2) Is the maximum outflow included satisfactorily within the produced fuzzy band? (3) Does the produced fuzzy band include the data with a rational width? (4) Does the produced fuzzy band include the aforementioned results of the crisp (deterministic, without uncertainty) methods?

A fourth case study taken from real-life data in a basin in Spain was used to validate the approach.

2. Materials and Methods

2.1. Principles of Fuzz Set and Logic

From a mathematical point of view, a fuzzy set can be described as a mapping from a general set X to the closed interval [0, 1] [31]. The membership function (MF) is a key concept which describes the fuzzy sets. Hence, the membership function declares to what degree an element belongs to the specific set. A membership degree of 0 denotes that the element does not belong to the set, and a membership degree of 1 denotes that the element belongs fully to the set. Subsequently, an element with a membership degree between 0 and 1 will partially belong to the examined set. A classical (crisp or precise or deterministic) set can be considered as a special case of a fuzzy set with a MF that only takes values 0 and 1.

A fuzzy number is a special case of the fuzzy set satisfying additionally the properties of convexity and normality. It is defined in the axis of real numbers, and its MF is a piecewise continuous function (Figure 1).

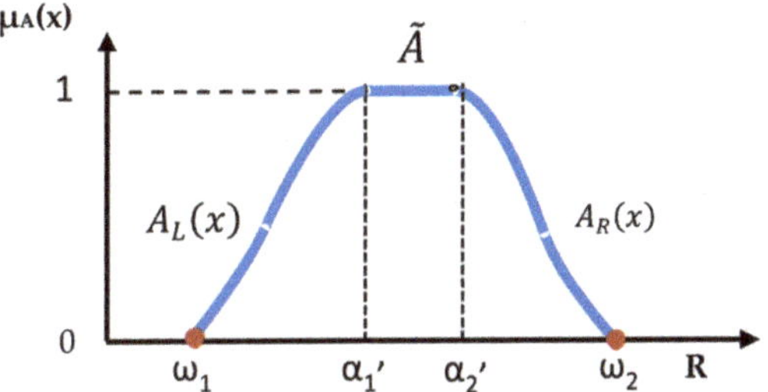

Figure 1. Representation of the membership function for a fuzzy number.

In general, the definition of fuzzy numbers can be found in Klir and Yuan, 1995. It is proven (Klir and Yuan, 1995) [31] that the MF of a fuzzy number (Figure 1) follows the mathematical expression (Equation (1)):

$$\mu_A(x) = \begin{cases} 0 \ for \ x < \omega_1 \\ A_L(x) \ for \ \omega_1 \leq x \leq \alpha_1' \\ 1 \ for \ \alpha_1' \leq x \leq \alpha_2' \\ A_R(x) \ for \ \alpha_2' \leq x \leq \omega_2 \\ 0 \ for \ x > \omega_2 \end{cases} \tag{1}$$

where $A_L : [\omega_1, a_1'] \rightarrow [0,1]$ and $A_R : [a_2', \omega_2] \rightarrow [0,1]$ are the left and right parts of the membership function regarding the fuzzy number $\tilde{A}$. In addition, A_L is increasing and continuous from the right, and A_R is decreasing and continuous from the left. The interval $[\alpha_1', \alpha_2']$ can be an interval or a point, but it cannot be an empty set [31].

A simple fuzzy number for representing the parameters K and x is the fuzzy symmetrical triangular number (Figure 2), which is a kind of fuzzy number [12]. The symbols $\bar{\alpha}$, w denote the central value (a single point where $\mu = 1$) and the width of the fuzzy number.

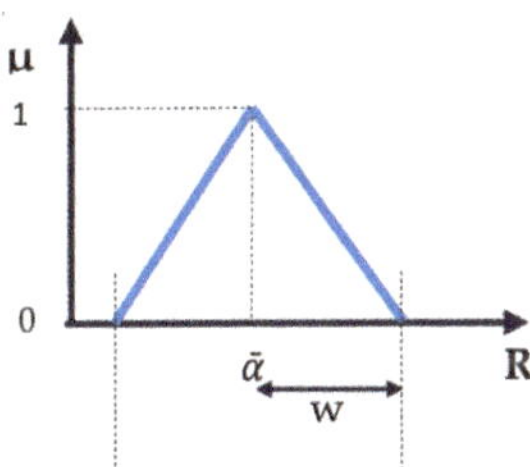

Figure 2. Fuzzy triangular symmetrical number.

The *h-cut* set of the fuzzy number A (with $0 < h \leq 1$) is the key idea to move from the fuzzy to the precise (crisp) sets, and it is defined as follows [31,32] (Equation (2)):

$$\tilde{A}_h = \{x | \mu_{\tilde{A}}(x) \geq h, x \in \Re\}. \tag{2}$$

The *h-cut* set is a crisp set determined from the fuzzy set according to a selected value of the membership function, α, and, reciprocally, a fuzzy set can be derived from a significant number of α-cut sets. In the case of a fuzzy triangular number, the α-cut is (Equation (3)):

$$\widetilde{A}_h = \left[A_h^L, A_h^R \right]$$

or equivalently:

$$\widetilde{A}_h = \left[\overline{A} - (1 - h) \cdot (\overline{A} - A^-), \overline{A} + (1 - h) \cdot (A^+ - \overline{A}) \right]. \tag{3}$$

The crisp set including all the elements with non-zero membership function is the 0-strongcut, which is defined as follows [33,34] (Equation (4)):

$$\widetilde{A}_{0^+} = \{ x | \mu_A(x) > 0, x \in \Re \}. \tag{4}$$

The following symbols are used for the zero-cut (Equation (5)):

$$\widetilde{A}_{0^+} = (A^-, A^+). \tag{5}$$

More analytically, according to Equation (5), the 0-cut is an open interval and does not contain the boundaries. For this reason (and to have a closed interval containing the boundaries), Hanss (2005) [35] proposed the phrase "worst-case interval W", which is the union of the 0-strongcut and the boundaries [35,36].

We can now extend the operation of the usual crisp functions in cases where the inputs are fuzzy sets, based on the extension principle that is briefly presented below.

Let X be a Cartesian product of universe $X = X_1 \times X_2 \times \ldots \times X_n$ and $\widetilde{A}_1, \widetilde{A}_2, \ldots, \widetilde{A}_n$ be defined in the universe sets $X_1, X_2, \ldots, X_n$, respectively. Let f be a (crisp) mapping from X to a universe Y, $y = f(x_1, x_2, \ldots, x_n)$. The mapping f for these particular input sets can now be defined as $\widetilde{B} = \{ (y, \mu_{\widetilde{B}}(y)) | y = f(x_1, x_2, \ldots, x_n), (x_1, x_2, \ldots, x_n) \in X \}$ in which the membership function of the image $\widetilde{B}$ can be defined (Zimmermann 1991) [32] by (Equation (6)):

$$\mu_{\widetilde{B}}(y) = \sup_{(x_1, x_2, \ldots, x_n) \in f^{-1}(y)} \min(\mu_{A_n}(x_1), \ldots, \mu_{A_n}(x_n)) \tag{6}$$

where f^{-1} is the inverse image of f.

The above principle is known as the extension principle. Its implementation provides the means to use a crisp function even if the inputs are fuzzy numbers. With most input variables, it is preferable to use a number of *h-cuts* in fuzzy analysis instead of using a MF based directly on the above definition ([37]). If f is a continuous function in the extension principle, the use of *h-cuts* can be also extended by determining the *h-cuts* of the function f, as follows ([33,34]) (Equation (7)):

$$\begin{cases} f^L\left(\widetilde{A}_1, \widetilde{A}_2, \widetilde{A}_3\right)_h = \min\left\{ f(x_1, x_2, x_3, h) \middle| x_1 \in \widetilde{A}_{1h}, x_2 \in \widetilde{A}_{2h}, x_3 \in \widetilde{A}_{3h} \right\}, \\[2mm] f^R\left(\widetilde{A}_1, \widetilde{A}_2, \widetilde{A}_3\right)_h = \max\left\{ f(x_1, x_2, x_3, h) \middle| x_1 \in \widetilde{A}_{1h}, x_2 \in \widetilde{A}_{2h}, x_3 \in \widetilde{A}_{3h} \right\}. \end{cases} \tag{7}$$

It is customary to use fuzzy numbers as inputs (here, the parameters K and x) to the crisp function (here, the Muskingum equation, which calculates the outflow, Q), and hence, the boundaries of the decision space will be closed (because of Equation (1)). Then, from the theorem of global existence for maxima and minima of functions with many variables, it is known that if the domain of a real function is closed and bounded, and the real function is continuous, then the function will have its absolute minimum and maximum values at some points in the domain [38]. Based on this theorem, in cases where fuzzy triangular numbers appear as inputs [37], the *h-cut* for any real continuous function with real variables in this domain can be determined. The determination of each *h-cut* is concluded to a double sub-optimization problem. Hence, the fuzzy output can be described by determining several representative *h-cuts*. Summarizing, in the case of fuzzy parameters, the simulation problem (that is, the determination of the outflow based on the Muskingum method) leads to the problem of determining several *h-cuts*, which is implemented through a double optimization procedure. In case of lateral flow, then a fuzzy estimation of the parameter α with the use of the *h-cuts* can be added.

2.2. Formulation of the Muskingum Method

The lumped hydrological model of Muskingum is based on the mass balance equation (Equation (8)) and the storage equation (Equation (9)).

$$I - Q = \frac{dS}{dt} \tag{8}$$

where I is the inflow rate to the reach, Q is the outflow rate from the reach, and S the storage in the reach. In the case where the relationship between storage and flow through a reach is linear, the Muskingum storage relationship can be written as (Equation (9)):

$$S = K[xI + (1-x)Q] \tag{9}$$

where I is the inflow rate to the reach, Q is the outflow rate from the reach, K is the storage time constant for the reach, and x is a weighting factor that varies between 0 and 0.5 [9].

The mass balance equation can be expressed in discrete form as follows (Equation (10)):

$$\frac{(I_j + I_{j-1})}{2} \cdot \Delta t - \frac{(Q_j + Q_{j-1})}{2} \cdot \Delta t = (S_j - S_{j-1}) \tag{10}$$

By combining Equations (8) and (9), the Muskingum routing equation is obtained (Equation (11)):

$$Q_j = C_0 I_j + C_1 I_{j-1} + C_2 Q_{j-1}$$

$$\text{where } C_0 = \frac{-Kx + 0.5\Delta t}{K - Kx + 0.5\Delta t} \text{ ; } C_1 = \frac{(Kx + 0.5\Delta t)}{K - Kx + 0.5\Delta t} \text{ ; } C_2 = \frac{(K - Kx - 0.5\Delta t)}{K - Kx + 0.5\Delta t} \tag{11}$$

$$\text{with } C_0 + C_1 + C_2 = 1$$

The linear Muskingum model can be modified to include the lateral flow. O'Donnell, 1985 [11] suggested a simple approach: "Possibly, the simplest model is one which assumes that the rate at which lateral inflow enters the reach is directly proportional to the rate of inflow I into the reach, with a proportionality factor α."

Following this approach, the mass balance equation and the empirical storage equations can be written as (Equations (12) and (13)):

$$I(1 + \alpha) - Q = \frac{dS}{dt} \tag{12}$$

$$S = K[x(1 + \alpha)I + (1-x)Q] \tag{13}$$

After some algebraic operations, the following equation can be used (Equation (14)) [11]:

$$d_1 I_j + d_2 I_{j+1} + d_3 Q_j = Q_{j+1}, d_1 = (1+\alpha)C_1, d_2 = (1+\alpha)C_0, \, d_3 = C_2 \tag{14}$$

Or equivalently [12] (Equation (15)):

$$Q_j = (1+a)\frac{\Delta T + 2Kx}{\Delta T + 2K(1-x)} I_{j-1} + (1+a)\frac{\Delta T - 2Kx}{\Delta T + 2K(1-x)} I_j$$
$$+ \frac{-\Delta T + 2K(1-x)}{\Delta T + 2K(1-x)} Q_{j-1} \tag{15}$$

The focus of this work is the crisp function that determines the current outflow when the parameters K and x are fuzzy numbers. Whereas in the case of no lateral flow, the Equation (11) describes this function, in the case of lateral flow, the Equation (15) is applied.

2.3. Implementation of the Muskingum Method with Fuzzy Parameters

In this study, the parameters K and x (and α in case of lateral flow) are selected by assuming them as fuzzy triangular symmetrical numbers (Figure 2). The simple linear Muskingum equation is applied with fuzzy parameters. As aforementioned, the application of a crisp function with fuzzy parameters is achieved by applying the extension principle. In practice, several *h-cuts* can be determined in order to describe the values of the examined function (in this application, the function f describes the outflow). In case of no lateral flow (based on Equation (11)), the boundaries of each *h-cut* of the outflow can be determined as follows (Equation (16)):

$$
\left\{
\begin{array}{l}
Q_j{}^L\left(\widetilde{K},\widetilde{x}\right)_h = \min\left\{
\begin{array}{l}
\frac{\Delta T+2x_1x_2}{\Delta T+2x_1(1-x_2)}I_{j-1} + \frac{\Delta T-2x_1x_2}{\Delta T+2x_1(1-x_2)}I_j + \\
\frac{-\Delta T+2x_1(1-x_2)}{\Delta T+2x_1(1-x_2)}Q_{j-1}\Big| x_1 \in \widetilde{K}_h\,, x_2 \in \widetilde{x}_h
\end{array}
\right\}, \\[2em]
Q_j{}^R\left(\widetilde{K},\widetilde{x}\right)_h = \max\left\{
\begin{array}{l}
\frac{\Delta T+2x_1x_2}{\Delta T+2x_1(1-x_2)}I_{j-1} + \frac{\Delta T-2x_1x_2}{\Delta T+2x_1(1-x_2)}I_j + \\
\frac{-\Delta T+2x_1(1-x_2)}{\Delta T+2x_1(1-x_2)}Q_{j-1}\Big| x_1 \in \widetilde{K}_h\,, x_2 \in \widetilde{x}_h
\end{array}
\right\},
\end{array}
\right. \tag{16}
$$

where I is the inflow rate to the reach, Q is the outflow rate from the reach, K is the storage time constant for the reach, and x is a weighting factor. The index j is referred to the time step, and the indexes L and R are referred to the left and the right hand of the produced *h-cut* regarding the examined fuzzy number.

In cases of lateral flow, based on Equation (15) it holds (Equation (17)):

$$
\left\{
\begin{array}{l}
Q_j{}^L\left(\widetilde{K},\widetilde{x},\widetilde{a}\right)_h = \min\left\{
\begin{array}{l}
(1+x_3)\frac{\Delta T+2x_1x_2}{\Delta T+2x_1(1-x_2)}I_{j-1} + (1+x_3)\frac{\Delta T-2x_1x_2}{\Delta T+2x_1(1-x_2)}I_j \\
+ \frac{-\Delta T+2x_1(1-x_2)}{\Delta T+2x_1(1-x_2)}Q_{j-1}\Big| x_1 \in \widetilde{K}_h\,, x_2 \in \widetilde{x}_h, x_3 \in \widetilde{a}_h
\end{array}
\right\}, \\[2em]
Q_j{}^R\left(\widetilde{K},\widetilde{x},\widetilde{a}\right)_h = \max\left\{
\begin{array}{l}
(1+x_3)\frac{\Delta T+2x_1x_2}{\Delta T+2x_1(1-x_2)}I_{j-1} + (1+x_3)\frac{\Delta T-2x_1x_2}{\Delta T+2x_1(1-x_2)}I_j \\
+ \frac{-\Delta T+2x_1(1-x_2)}{\Delta T+2x_1(1-x_2)}Q_{j-1}\Big| x_1 \in \widetilde{K}_h\,, x_2 \in \widetilde{x}_h, x_3 \in \widetilde{a}_h
\end{array}
\right\},
\end{array}
\right. \tag{17}
$$

For the case of non-lateral flow, to calibrate the Muskingum method, the parameters K and x need to be estimated. The central values ($h = 1$) and the left and the right hands of the zero-cut (or the worst-case interval W) regarding the outflow are exploited. The zero-cut can be used to express the concept of inclusion, that is, to check the property that the fuzzy band contains the observed data at least to a high degree. This is set as a key idea in order to calibrate the proposed fuzzy Muskingum model. The determination of the boundaries of the worst-case interval W is finally achieved by following an optimization procedure, which, in the examined case, will have a solution since the inputs are fuzzy numbers and the used crisp function Q is continuous. The determination of the worst-case interval W composes the sub-optimization problem that arises during the calibration of the model.

The problem of the calibration process is to assess the central values and the semi-width of the parameters K, x, and α so that a selected criterion is minimized (including the inclusion constraints). Because of the model complexity, the calibration is achieved by applying a heuristic optimization model. In this article, the particle swarm optimization (PSO) method is selected. This method can be coupled with a simulation model; in this study, the Muskingum equation enhanced with the use of the extension principle for fuzzy parameters is applied.

2.4. Particle Swarm Optimization Method

Particle swarm optimization (PSO) is a stochastic global optimization method based on the simulation of the swarm. As in genetic algorithms (GA), PSO exploits a population of possible candidate solutions to probe the search area. The PSO algorithm can be characterized as one of the population-based algorithms (Parsopoulos and Vrahatis, 2002) [39]. Although PSO is an effective and widely used method, it is simpler than other heuristic optimization algorithms (e.g., GA) because the crossover and mutation operations in the original version of the GA [40,41] are not used.

PSO can deal with nonlinear optimization problems in non-convex domains [42]. Each candidate solution is called a particle, and the set of potential possible solutions in each iteration creates the swarm [41,43]. A swarm has a dimension N′, in which N′ is the number of potential solutions. Each potential solution is comprised of D variables, in which D is the dimension of the problem [41,44]. In this article, in case of non-lateral flow, D = 4 $\left(\overline{K}, w_K, \overline{x}, w_x\right)$, where $\overline{K}$ is the central value of $\widetilde{K}$, w_K the semi-width of $\widetilde{K}$, $\overline{x}$ is the central value of $\widetilde{x}$, w_x is the semi-width of $\widetilde{x}$.

Analytically, the population dynamics in PSO simulates the behavior of a bird flock, where social sharing of information takes place and individuals benefit from the discoveries and previous experience of all other companions during their search for food. Thus, two variants of the PSO algorithm were developed considering either a local neighborhood or a global neighborhood. In the former, the partial optimum of the particle is usually applied [40]. In the latter, each particle moves towards its best previous position and towards the best particle in the whole swarm [39,41,45].

Several modifications to the original version of the PSO method of Kennedy and Eberhart (1995) [45] have been proposed, such as the adaptation of inertia term and the consideration of the maximum velocity (e.g., [46]).

The basic PSO algorithm is detailed below (e.g., [44,46,47]:

1: Initialize a population array of particles with random positions and velocities on D dimensions in the search area.
2: Loop
3: For each particle, evaluate the desired optimization fitness function in D variables.
4: Compare particle fitness evaluation with its best previously visited position (p_i). If the current value is better than p_i, then set p_i equal to the current value.
5: Identify the particle with the best fitness function value of the swarm p_g.
6: Change the velocity and position of the particle (x_i) according to the Equation (18):

$$\begin{cases} v_i(t+1) = \omega v_i + c_1\rho_1 \cdot (p_i - x_i(t)) + c_2\rho_2 \cdot (p_g - x_i(t)) \\ x_i(t+1) = x_i(t) + v_i(t+1) \end{cases} \tag{18}$$

7: If a criterion is met (usually a sufficiently good fitness or a maximum number of iterations), exit loop.
8: End loop

where $\rho()$ is a vector of random numbers uniformly distributed in the open interval 0, 1 that is generated at each iteration, and for each particle, p_i is the best previously visited position of the ith particle (partial optimum), and p_g is the global best previously visited position of all particles (global optimum). Furthermore, the term $c_1\rho_1 \cdot \left(\overrightarrow{p}_i - \overrightarrow{x}_i\right)$ that associates the particle's own experience with its current position is weighted by the constant c_1 and is called individuality (cognitive acceleration). The term $c_2\rho_2 \cdot \left(\overrightarrow{p}_g - \overrightarrow{x}_i\right)$ is associated with the social interaction between the particles of the swarm and weighted by the constant c_2, and is called sociality (social acceleration). As v the velocity is meant. The velocity, by supposing that $\Delta t = 1$, modulates the new position $x_i(t+1)$.

For the basic PSO, the coefficients c_1 and c_2 were allowed to take values in the interval 1.5 to 2.5 [41].

Clerc and Kennedy's analysis [48] based on the initial PSO method proposed the following modified equation for the new positions (Equation (19)):

$$\begin{cases} v_i(t+1) = \chi\left[v_i(t) + c_1\rho_1 \cdot (p_i - x_i(t)) + c_2\rho_2 \cdot (p_g - x_i(t))\right] \\ x_i(t+1) = x_i(t) + v_i(t+1) \end{cases} \tag{19}$$

where χ is a parameter called constriction coefficient or constriction factor (Equation (20)):

$$\chi = \frac{2}{\left| 2 - \varphi - \sqrt{\varphi^2 - 4\varphi} \right|}, \varphi = c_1 + c_2, \varphi > 4 \tag{20}$$

On the other hand, based on the literature, the following parameters are proposed ([48]) (Equation (21)):

$$\chi = 0.729, c_1 = c_2 = 2.05 \tag{21}$$

In this study, the proposals of Clerc & Kennedy, 2002 are adopted in the implementation of the PSO.

It is worth noting that, in some cases, the PSO method converges rapidly to a local optimum. This problem can be overcome by the strategy of Salehizadeh et al. (2009) [49] by dividing the population or by using other hybrid methods, e.g., by combining the PSO together with simulated annealing behavior. For instance, [41] applied a modified PSO to achieve the identification of optimal hedging rules for complex real systems of operating reservoirs with many parameters, seeking to mitigate the drought impacts.

3. Proposed Calibration and Performance Measures

A procedure is followed to implement the model calibration (PSO) method together with the extension principle. First, a swarm of possible solutions is randomly created within the decision space. For non-lateral flow, each possible optimum solution (member of the swarm) contains the values of the parameters $\left(\overline{K}, \overline{x}, w_K, w_x \right)$ that describe the membership functions of $\widetilde{K}, \widetilde{x}$. Then, a fuzzy simulation method is applied based on the extension principle. For each possible solution, the frontiers of the 0-cut and the central values are calculated. Finally, we apply the PSO methodology, and the objective is calculated. The objective function is identical with the fitness function, since by using the simulation stage, the use of constraints could be avoided.

The key question is about the formulation of the objective function (fitness function), which will be used during the calibration process. The objective function will differ from those selected by the crisp calibrations, since the output of the model will be a fuzzy band that expresses the outflow hydrograph. In this article, an objective function that contains a weighted sum of four performance measures is proposed (Equations (22)–(24)):

$$f = w_1 \left[\sum_{j=1}^{M} a_{R_j} \left(Q_j^{observed} - Q_j^+ \right)^2 + \sum_{j=1}^{M} a_{L_j} \left(Q_j^- - Q_j^{observed} \right)^2 \right] + \frac{1}{M} \left(\sum_{j=1}^{M} \left(Q_j^{observed} - \overline{Q}_j \right)^2 \right) +$$

$$\underbrace{\hspace{4cm}}_{including\ all\ data} \qquad \underbrace{\hspace{3cm}}_{central\ values\ near\ to\ data}$$

$$\frac{1}{M} \left(\sum_{j=1}^{M} \left(Q_j^+ - Q_j^- \right)^2 \right) + \left(a_{max} \left(Q^{max} - Q_p^+ \right)^2 \right) \tag{22}$$

$$\underbrace{\hspace{3cm}}_{fuzzy\ width} \qquad \underbrace{\hspace{3.5cm}}_{including\ the\ maximum\ value}$$

where

$$a_{R_j} = \begin{cases} 0\ if\ Q_j^+ \geq Q_j^{observed} \\ 1\ if\ Q_j^+ \leq Q_j^{observed} \end{cases}, \quad a_{L_j} = \begin{cases} 0\ if\ Q_j^- \leq Q_j^{observed} \\ 1\ if\ Q_j^- \geq Q_j^{observed} \end{cases} \tag{23}$$

The first term expresses the divergence of the produced fuzzy band to include all data. In other words, the squared penalty term, E_1, is activated only whether the observed data are not included within the produced fuzzy band. This procedure was initially proposed by Ishibuchi et al. (1993) [50] as a cost function to be minimized in the learning process regarding a neural network with interval weights, whilst in this work, it is used as a term of the fitness function (Equation (24)):

$$E_1 = \left[\sum_{j=1}^{M} a_{R_j} \left(Q_j^{observed} - Q_j^+ \right)^2 + \sum_{j=1}^{M} a_{L_j} \left(Q_j^- - Q_j^{observed} \right)^2 \right] \left(\frac{m^3}{s} \right)^2 \tag{24}$$

The above measure represents the idea of inclusion of the fuzzy linear regression of Tanaka [30]. According to this model, all the data must be included within the produced fuzzy band [51,52].

The second term, E_2, expresses the distance between the central values ($\alpha = 1$) and the observed data (Equation (25)):

$$E_2 = \left(\sum_{j=1}^{M} \left(Q_j^{observed} - \overline{Q}_j \right)^2 \right) \left(\frac{m^3}{s} \right)^2 \tag{25}$$

The third term, E_3, expresses the magnitude of the produced fuzzy width. A large width leads to an unfunctional fuzzy band, which cannot be exploited in real applications (Equation (26)):

$$E_3 = \left(\sum_{j=1}^{M} \left(Q_j^+ - Q_j^- \right)^2 \right) \left(\frac{m^3}{s} \right)^2 \tag{26}$$

The use of this distance comes from the fuzzy regression of Tanaka where the problem of the fuzzy linear regression concludes to a constrained optimization problem aiming to minimize the total width of the produced fuzzy band (e.g., [51,52])

Finally, the last term, E_4, expresses the distance between the maximum value of the outflow and the right hand of the estimated outflow (Equations (27) and (28)):

$$E_4 = \left(a_{max} \left(Q^{max} - Q_p^+ \right)^2 \right) \left(\frac{m^3}{s} \right)^2 \tag{27}$$
$$\text{including the maximum value}$$

$$\text{where } a_{max} = \begin{cases} 1 & if\ q^{max} \geq q_p^+ \\ 0 & otherwise \end{cases} ,\ p : time\ when\ q_{max} = max(Q_1, \ldots, Q_j, \ldots, Q_M) \tag{28}$$

This distance is activated only if the observed data are above the produced fuzzy band as it is indicated by the coefficient α_{max} where p is the time, which is referred to as the time where the maximum value occurs. This criterion is useful in river routing. For flood protection, it is critical that the produced fuzzy band (from the proposed model) includes the observed maximum value even in real time. Therefore, since this criterion expresses the aforementioned property, it is obviously useful. This measure is proposed by [44] where a fuzzified version of the fuzzy unit hydrograph was developed.

The terms $Q_j^{observed}$, Q_j^-, and Q_j^+ express the observed value at time j, the left, and the right-hand boundary of the zero-cut, respectively.

These behavioral parameters are user defined and control the optimization process. The adopted values were taken mainly from the literature because they are shown to lead to good performance. Their effectiveness is later analyzed and compared to other alternatives in the validation section.

The weight w_1 is selected to give more emphasis on the first term of the fitness function. Indeed, the main advantage of the fuzzy model, similarly with the fuzzy regression model (e.g., [30]), is the inclusion of the observed data into the produced fuzzy band. However, in the case of a precise satisfaction of the inclusion constraints, then an outlier can significantly enlarge the size of the produced fuzzy band. Hence, the larger the w_1, the closer to the absolute satisfaction of the inclusion constraints the model becomes. If a non-useful fuzzy band is produced, then the value of w_1 is reduced. In fact, the first measure strengthens the satisfaction of the inclusion constraints. E_1 conflicts with E_3; however, an unfunctional fuzzy band must be rejected. E_2 is similar but not identical with the crisp measures SSQ, since apart from the central value, the output of the fuzzy model is the entire fuzzy number.

4. Results and Discussion

In order to check the proposed methodology, four case studies from the references are used to benchmark the proposed method. The first three case studies include: (1) the routing of a smooth hydrograph (Wilson 1974 [10]), (2) a double peak hydrograph (Viessman and Lewis 2003 [9]), and (3) non-smooth hydrograph with lateral flow (O'Donnel 1985) data sets [11]. The last case study is located in the Ebro River Basin with more data available. Therefore, the validation of the solution achieved is studied in the last case study.

4.1. Smooth Hydrograph

The first data set is the data set presented by [10]. For these kinds of heuristic optimization problems, a maximum of 100 iterations were selected. Furthermore, the swarm consists of 50 members, whilst the expression of Clerc and Kennedy's is adopted for training (Equations (19)–(21)). The results were not practically changed even if the SWARM was activated again. The fitness or objective function consists of the four performance measures (Equation (22)). These selections were adopted for all the examined examples.

Since no lateral flow occurs, the Equation (16) is used to determine the boundary of the worst-case interval W, which is used to check the width of the produced fuzzy band and the degree of inclusion. Because of the sub-optimization problems (according to extension principle, Equation (16)) a significant computational time is required for the calibration process. Initially, a significant value of the w_1 is considered ($w_1 = M^2$). By following the aforementioned calibration procedure, the following results are achieved:

$$\widetilde{K} = (1.2482, 0.6533)\text{days and } \widetilde{x} = (0.2972, 0.0580) \text{ and } E_1 = 3.31$$

where M is the number of data, whilst in the bracket, the first term means the central value and the second term the semi-width. The brackets symbolize the fuzzy symmetrical triangular numbers.

Wilson, 1974 [10] proposed the following crisp values for the examined parameters: $K = 1.5$ days and $x = 0.25$, which are included from the produced corresponding quantities.

Therefore, the produced solution has physical sense and incorporates all the data to a high degree. However, the above suggestion holds only if the width of the produced fuzzy band is functional. For flood protection, the most important factor is the uncertainty within the neighborhood of the peak flow. Figure 3 shows that a range between the left and the right boundaries of the outflow is lower than 10 m^3/s, which seems a reasonable range. However, the greater values have a high dispersion in time according to the proposed fuzzy solution.

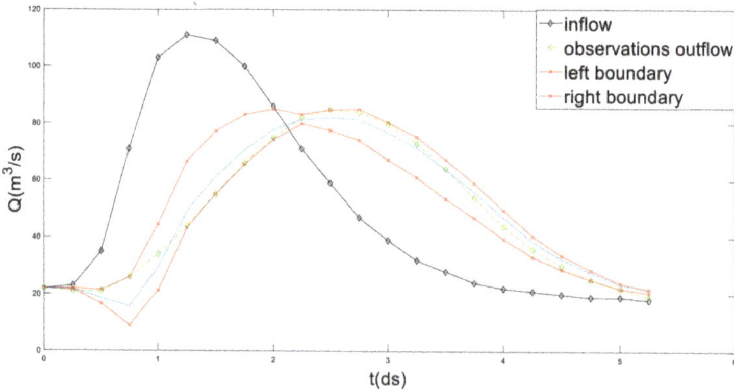

Figure 3. Simulation of the Wilson (1974) [10] example based on the proposed fuzzy method for $w_1 = M^2$ in the case of the Wilson (1974) data.

The crisp formulation uses the measure $SSQ = \sum_{j=1}^{M} \left(Q_j^{observed} - Q_j \right)^2$, which cannot be compared directly with the performance evaluation measures, since according to the proposed methodology, the output of the fuzzy Muskingum simulation will be a fuzzy number in contrast with the crisp formulation where the output is a crisp number. In case a significantly very small weight w_1 is selected, an almost crisp solution is obtained. In this case, the value of E_1, E_2, and SSQ are practically identical. However, the comparison between the SSQ and E_1 indicates that E_1 takes the smallest value (Table 1 in Niazkar and Afzali, 2016) of the SSQ.

Table 1. The logical test that is true if the fuzzy solution contains the other crisp methods. P (green box) means that the model "passed" the test, and F (red box) means that the model "failed".

Models																						

Models	0	6	12	18	24	30	36	42	48	54	60	66	72	78	84	90	96	102	108	114	120	126
Wilson-trial	P	P	P	P	P	P	P	P	P	P	P	P	P	P	P	P	P	P	P	P	P	P
Regression	P	P	P	P	P	P	P	P	P	P	P	P	P	P	P	P	P	F	F	F	F	F
NL-LSM	P	P	F	F	P	P	P	P	P	P	P	P	P	P	F	P	P	P	P	F	F	F
S-LSM	P	P	F	F	P	P	P	P	P	P	P	P	P	P	P	P	P	P	F	F	F	F
LMM	P	P	F	F	P	P	P	P	P	P	P	P	P	P	P	P	P	P	F	F	F	F
HJ+DFP	P	P	P	P	P	P	P	P	P	P	P	P	P	P	P	P	P	P	P	P	P	P
GA	P	P	P	P	P	P	P	P	P	P	P	P	P	P	P	P	P	P	P	P	P	P
BFGS	P	P	P	P	P	P	P	P	P	P	F	P	P	P	P	P	P	P	P	P	P	P
BFGS-HS	P	P	P	P	P	P	P	P	P	P	F	P	P	P	P	P	P	P	P	P	P	P
NLMM-L	P	P	P	P	P	P	P	P	P	P	P	P	P	P	P	P	P	P	P	F	F	F
NLI (SSQ)	P	P	P	P	P	P	P	P	P	P	P	P	P	P	P	P	P	P	P	F	F	F
NLII (SSQ)	P	P	P	P	P	P	P	P	P	P	P	P	P	P	P	P	P	P	P	P	P	P
NLIII (SSQ)	P	P	P	P	P	P	P	P	P	P	P	P	P	P	P	P	P	P	P	P	P	F
NLI (MARE)	P	P	P	P	P	P	P	P	P	F	P	P	P	P	P	P	P	P	P	P	P	P
NLII (MARE)	P	P	P	P	P	P	P	P	P	F	P	P	P	P	P	P	P	P	P	P	P	P
NLIII (MARE)	P	P	P	P	P	P	P	P	P	F	P	P	P	P	P	P	P	P	P	P	P	F

To summarize, it is proposed that, in the case E_1 has a small value (preferably smaller than the SSQ of the crisp simulation) and a logical width, then the fuzzy calibration can be accepted. The time-step of Wilson 1974 [10], 6 h, can be characterized as sufficient, since in 6 h, some emergency measures can be implemented.

Analytically, for $w_1 = M^2$, the proposed measures, for which their weighted sum composes the objective function, have the following values:

$$E_1 = \left[\sum_{j=1}^{M} a_{R_j} \left(Q_j^{observed} - Q_j^+ \right)^2 + \sum_{j=1}^{M} a_{L_j} \left(Q_j^- - Q_j^{observed} \right)^2 \right] = 3.31$$

including all data

$$E_2 = \left(\sum_{j=1}^{M} \left(Q_j^{observed} - \overline{Q}_j \right)^2 \right) = 344.9$$

central values near to data

$$E_3 = \left(\sum_{j=1}^{M} \left(Q_j^+ - Q_j^- \right)^2 \right) = 3,410.8$$

fuzzy width

$$E_4 = \left(a_{max} \left(Q^{max} - Q_p^+ \right)^2 \right) = 0.02$$

including the maximum value

Two additional calibrations are presented in Figure 4 for (a) $w_1 = 1$ and (b) $w_1 = 1/M$. In cases where smaller weights are selected, the produced fuzzy band becomes thinner and smoother (Figure 4a,b), but it cannot contain a significant number of data (especially Figure 4b, where a smaller weight is selected). However, the maximum discharge is practically included in all cases. In all figures, the blue line expresses the central value, that is, the value which corresponds to unit membership function.

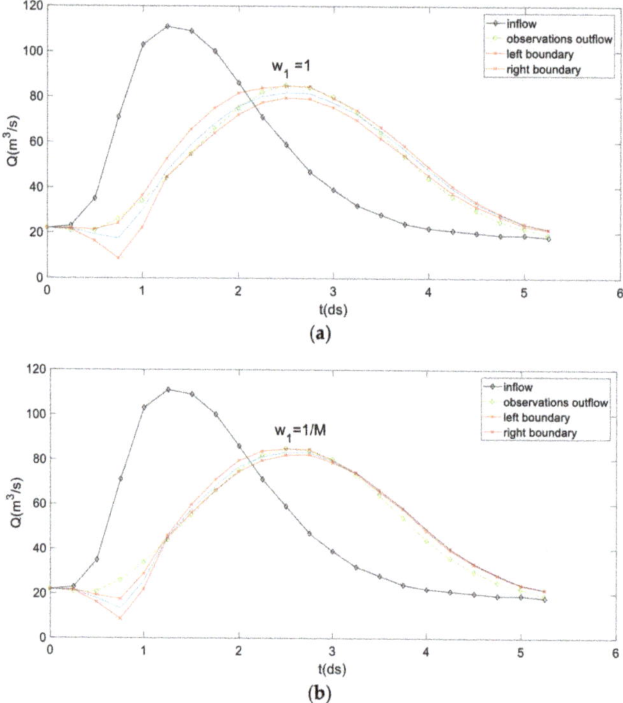

Figure 4. Simulation of the Wilson (1974) [10] example based on the proposed fuzzy method for (**a**) $w_1 = 1$ and (**b**) $w_1 = 1/M$ in the case of the Wilson (1974) data.

For illustrative purposes, Figure 5 shows the evolution of the member of the swarm. Since there are four decision variables, only the parameters central values of K and x are represented. The blue circles indicate the swarm of the possible solutions and the diamond the global optimum solution for each generation. After the 50th generation, the swarm converts rapidly near the global optimum, and the optimum solution remains stable after the 89th iteration.

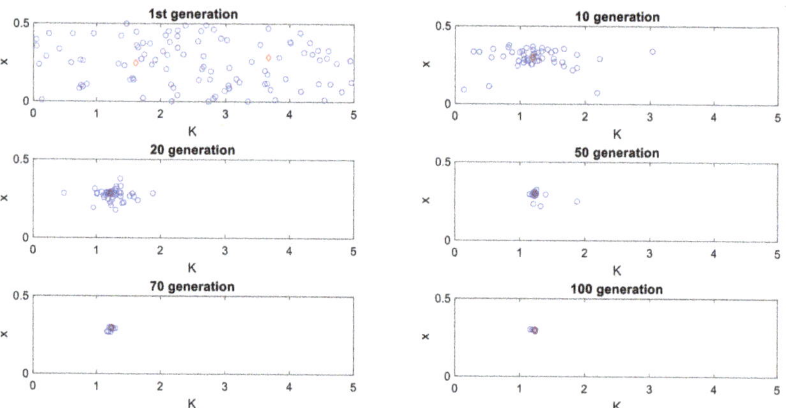

Figure 5. Simulation of the Wilson (1974) [10] example based on the proposed fuzzy method for $w_1 = M^2$ in case of the Wilson (1974) data. The blue circles indicate the swarm of the possible solution and the diamond the global optimum solution for each generation.

After the end of the calibration procedure (and hence, the PSO algorithm), for each time step, the fuzzy number which corresponds to the outflow can be separately calculated by considering a very large number of *h-cuts*. For each *h-cut*, the extension principle is used to determine the boundaries of each *h-cut*. Figure 6 represents the produced outflow when the maximum outflow occurs in the observed data. As it can be seen, the output remains a fuzzy number (since it satisfies Equation (1)), but the linearity and the symmetry (which appears in case of the parameters K and x) were lost because the used crisp functions (Equation (16)) do not remain linear.

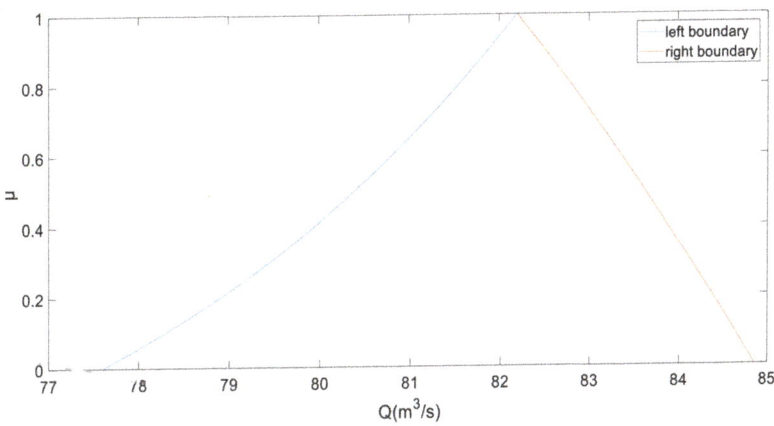

Figure 6. Produced outflow when the maximum outflow occurs in the real data in the case of the Wilson (1974) data for $w_1 = M^2$.

By comparing the proposed fuzzy solution with many (crisp) methodologies, it is concluded that the fuzzy solution contains the majority of the described solutions (see the literature session). Indicatively, the Wilson-trial approach [10], the regression model, the NL-LSM (Yoon and Padmanabhan 1993) [15], the S-LSM (Gill 1978) [13], the LMM (Das 2004) [7], the HJ+DFP (Tung 1985) [14], the GA, the BFGS (Geem 2006) [16], the BFGS-HS (Karahan et al., 2013), the NLMM-L (Karahan et al., 2013), the NLI (SSQ) (Karahan et al., 2013), the NLII (SSQ) (Karahan et al., 2013), the NLIII (SSQ) (Karahan et al., 2013), the NLI (MARE) (Karahan et al., 2013), the NLII (MARE) (Karahan et al., 2013), the NLIII (MARE) (Karahan et al., 2013) [53], the CS (Karahan et al., 2015) [18], and the CM (Easa 2015) [26] models are figured together with the fuzzy solution. The fuzzy solution is depicted by the left-hand and the right-hand bound of the zero-cut and the central values (Figure 7).

The Table 1 contains the comparison between the fuzzy solution with the aforementioned crisp models. The logical test is passed when the previous crisp solution for the outflow is included within the produced fuzzy band. A tolerance of 1 m³/s is permitted. From Table 1, it is shown that the majority of the crisp solutions are included within the produced fuzzy band apart from the last (decreasing) part of the hydrograph. The yellow lines indicate the time steps near the time where the maximum occurs.

In addition, the produced fuzzy band is separately compared with the graphical solution of Wilson, 1974 [10] (Figure 8). The graphical solution is based on the graphical representation of the storage versus the weighted discharge and with x as the preferred value, which produces the narrowest loop. An important point is that by comparing the central values of the fuzzy solution and the graphical solution, these two solutions are close, and moreover, the initial region of the outflows has a similar decreasing behavior.

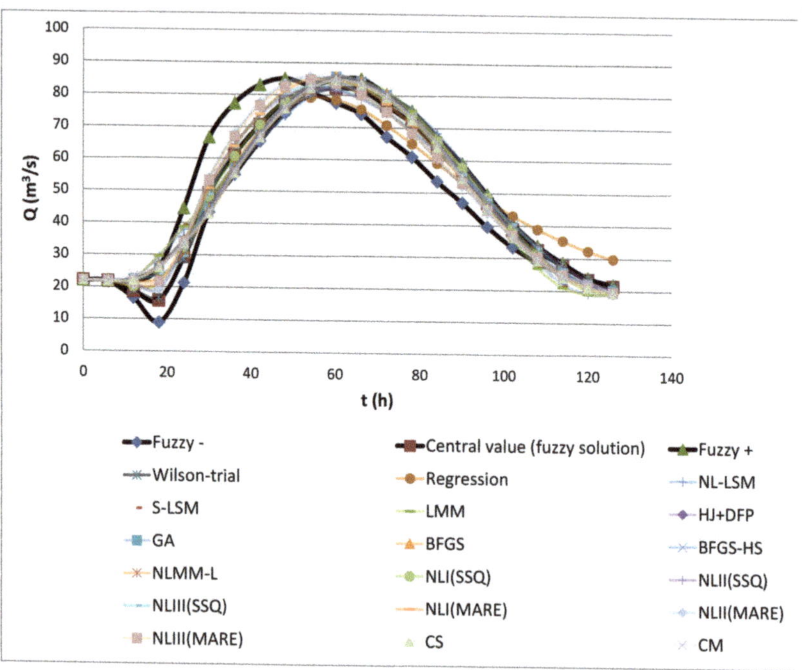

Figure 7. Other crisp simulations, the observed data, and the produced fuzzy band in case of the Wilson (1974) [10] data.

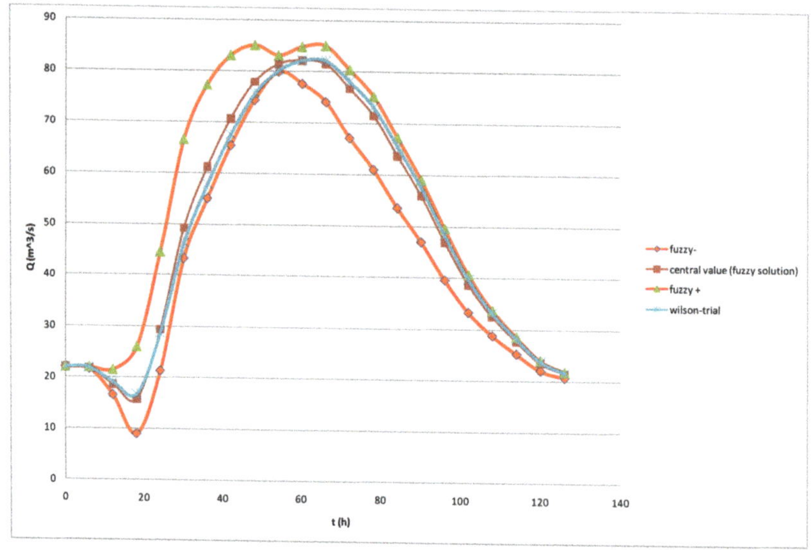

Figure 8. The graphical solution and the produced fuzzy band in the case of the Wilson (1974) data.

4.2. Two-Peak Hydrograph

The second case study is a multi-peak flood hydrograph [9]. By selecting $w_1 = M^2$, $\widetilde{K} = (3.5744, 1.5672)$, and $\widetilde{x} = (0.3460, 0.1399)$, whilst $E_1 = 0.51$ (Figure 9), where M is the number of data for the examined set.

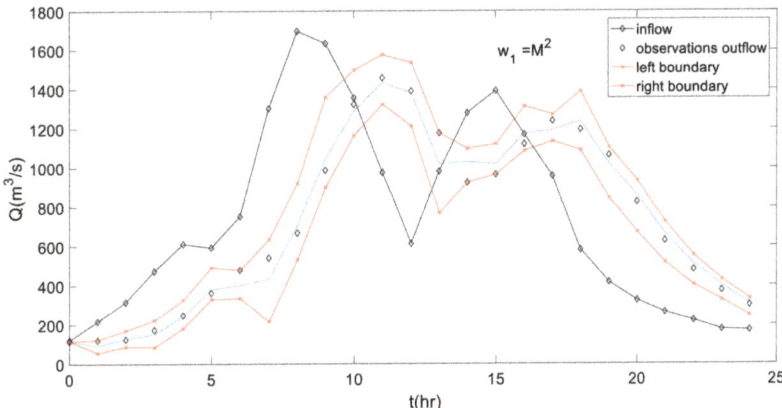

Figure 9. Simulation of the [9] example based on the proposed fuzzy method for $w_1 = M^2$. The blue line means the central values.

By selecting $w_1 = 1$, a functional approach with a smaller fuzzy band but with some points out of the fuzzy band is produced (Figure 10). The value $E_1 = 8974.5$, which is sufficient. $\widetilde{K} = (3.6975, 0.3954)$ hr and $\widetilde{x} = (0.3134, 0.1067)$. An interesting perspective is that the two solutions are very close regarding the central values.

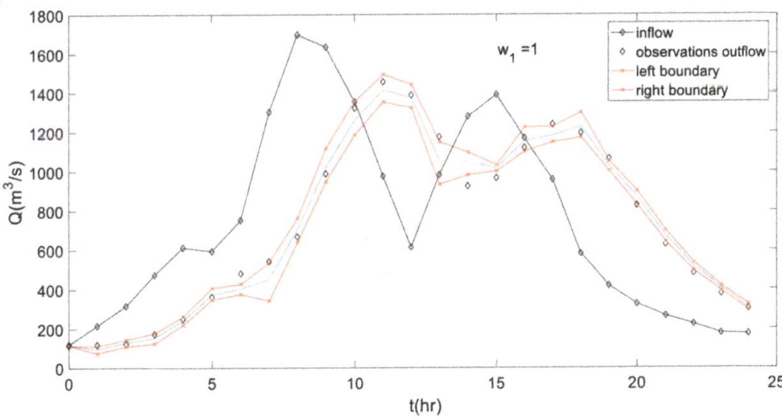

Figure 10. Simulation of the [9] example based on the proposed fuzzy method for $w_1 = 1$. The blue line means the central values.

Furthermore, the produced fuzzy band is compared with BFGS (Karahan 2014) [54] and MHBMO (Niazkar and Afzali 2015) [25] (Figure 11). Although the crisp models are rather complex with many parameters, the fuzzy model based on the Muskingum linear method contains practically all the values for $w_1 = M^2$ (Figure 11a) and the majority of the values when $w_1 = 1$ (Figure 11b), whilst the fuzziness is within a rational value. The method of BFGS seems to be away from the data.

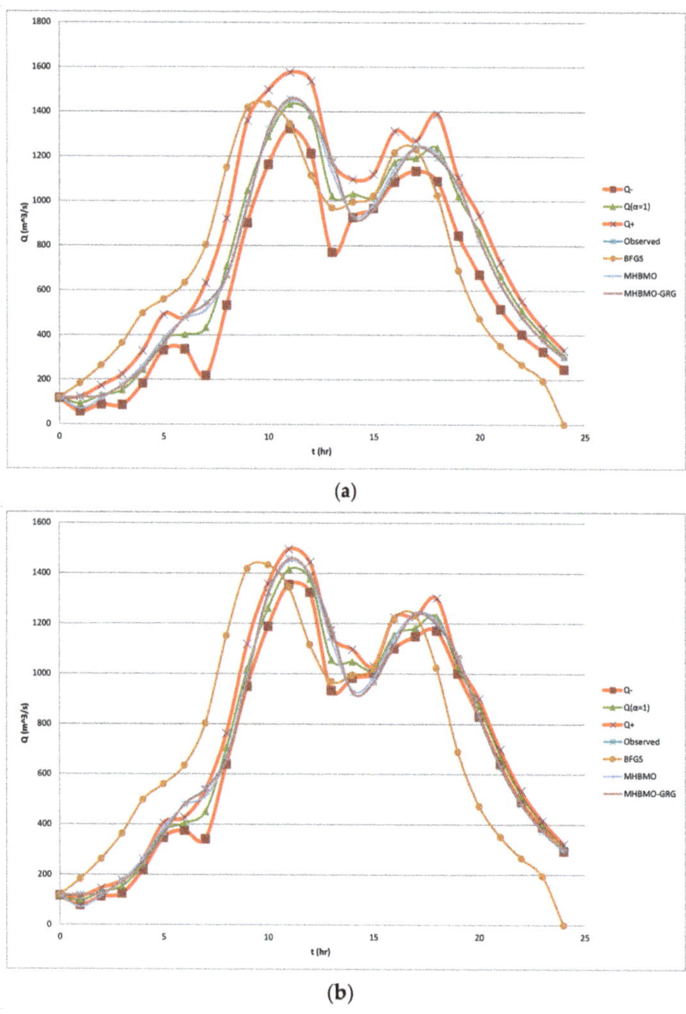

(a)

(b)

Figure 11. Other crisp simulations, the observed data ,and the produced fuzzy band in case of the [9] data for (**a**) $w_1 = M^2$ (**b**) $w_1 = 1$.

4.3. Non-Smooth Hydrograph with Lateral Flow

A non-smooth outflow hydrograph with lateral flow, previously presented by O'Donnel (1985) [11], was analyzed. The data are based on the event on the River Wyre, 20–21 October 1982. The river flows into the Irish Sea at Fleetwood. It is approximately 28 miles (45 km) in length. The river is a county Biological Heritage Site. According to O' Donnel (1985), there was a considerable increase in the flood volume between the inflow and outflow sections (some 25 km apart), and furthermore, it also had a multi-peaked inflow. Initially, a significant value of the w_1 is considered ($w_1 = M^2$). Since lateral flow occurs, Equation (17) is used to determine the worst-case interval. By selecting $w_1 = M^2$, it is concluded that $\widetilde{K} = (4.9405, 2.1945)$, $\widetilde{x} = (0.0593, 0.1035)$, and $\widetilde{a} = (2.5960, 0.1035)$ (Figure 12). However, the values of x approach the value corresponding to the reservoir, whilst small negative values are expected to be irrational.

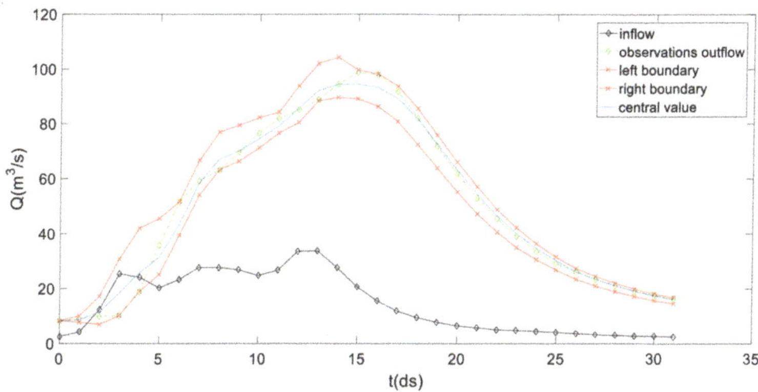

Figure 12. The observed data and the produced fuzzy band in the case of the River Wyre, 20–21 October 1982 data for $w_1 = M^2$.

By selecting $w_1 = 1$, (Figure 13) it is concluded that $\tilde{K} = (6.7211, 2.6006)$, $\tilde{x} = (0.0997, 0.0002)$, and $\tilde{a} = (2.9648, 0.1848)$ and $E = 23.4438$. These values approximate the crisp values that are provided by [11]. By comparing the observed data and the produced fuzzy band, it seems that although a new parameter was added, the performance of the fuzzy solution can be characterized as sufficient since the produced fuzzy band includes to a high degree the observed data of the outflow without an irrational width.

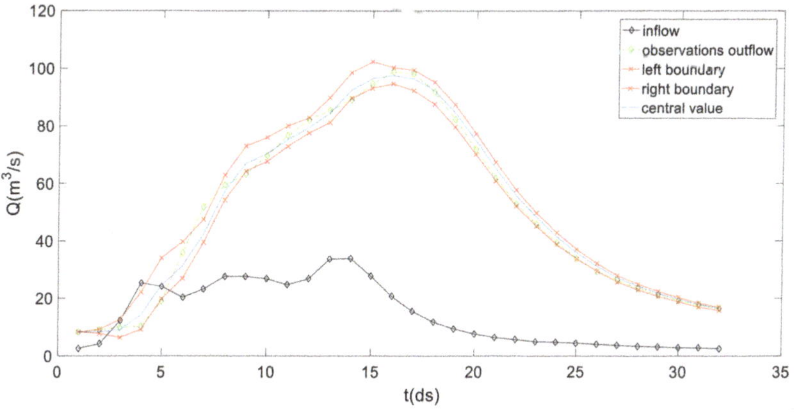

Figure 13. The observed data and the produced fuzzy band in the case of the River Wyre, 20–21 October 1982 data for $w_1 = 1$.

Based on the three examples analyzed, results show advantages of the proposed methodology to be highlighted: (1) the use of a fuzzy estimation with the aim of fuzzy parameters is closer to observed; (2) the new methodology increases the safety of the prediction including the uncertainty; (3) the produced fuzzy band include, to a significant degree, the observed data; and (4) the output of the successful existent crisp methodologies even if they include more complex assumptions.

4.4. Validation with Real-Life Data

The validation case study is located in the Ebro River Basin (Spain, Figure 14). The gauge stations (automatic flow measures) belong to the SAIH network (Automatic Hydrologic Information System) managed by the Ebro Basin Water Authority. The reach under analysis is located in the Aragón River between stations 9271 (elevation 1040 m, basin area

101 km^2) and 9018 (elevation 793 m, basin area 238 km^2). The reach length is 18.09 km, and the mean slope of the reach is 1.365%. The method proposed in this study is validated with four hydrographs recently recorded in the reach. The time step of the original data is equal to 15 min. In order to reduce computational time and for stability reasons, the original data were resampled at $\Delta t = 1$ h time step. The selected hydrographs can be characterized as rather complex since they include important lateral flow, and their shapes are not smooth.

Figure 14. Case study in Aragón River, Ebro Basin, Spain. The studied reach starts at Canfranc gauge station (red dot, Id 9217) and ends at Jaca gauge station (red dot, Id 9018).

The performance of the method is tested by fitting the Muskingum parameters in one hydrograph (Hydrograph 1) and validating the prediction obtained with the fitted parameters for the three remaining hydrographs (Hydrographs 2, 3, and 4). The predictive capability of the method was tested for two values of parameter w_1: 1 and 0.1. The value of w_1 equal to one gives more weight on the inclusion of observations within the fuzzy band, while the value of w_1 equal to 0.1 gives more weight on the reduction of the uncertainty of the forecast. The results obtained for $w_1 = 1$ after 100 iterations are $\widetilde{K} = (5.1605, 3.5130)$h, $\widetilde{x} = (0.1773, 0.0495)$, and $\widetilde{a} = (1.5160, 0.4579)$. The results obtained for $w_1 = 0.1$ are $\widetilde{K} = (3.8328, 1.5637)$h, $\widetilde{x} = (0.2273, 0.0006)$, and $\widetilde{a} = (1.8568, 0.1930)$. The fuzzy bands obtained for the outflow hydrographs are presented in Figure 15. In both cases, the most relevant values of the outflow hydrograph (near the peak discharge) are included within the produced fuzzy band. The wider fuzzy band for $w_1 = 1$ includes a larger fraction of values, but the narrower fuzzy band for $w_1 = 0.1$ reduces the uncertainty of the forecast.

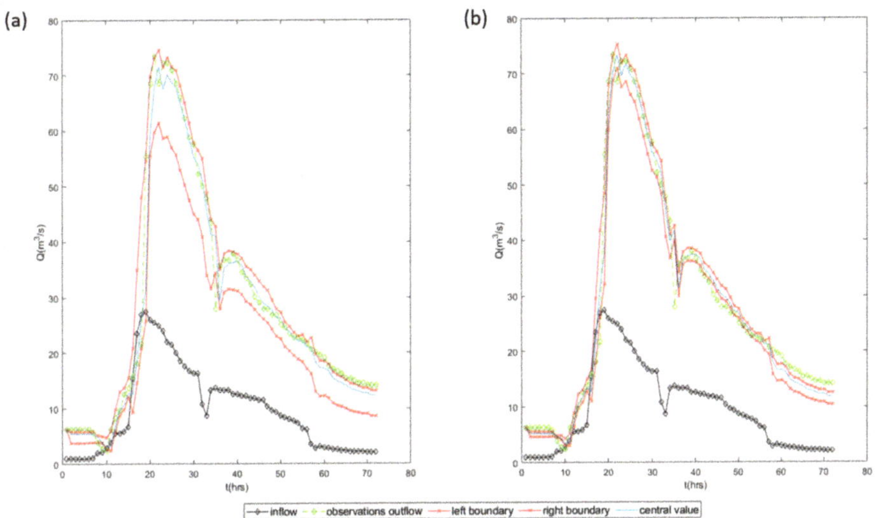

Figure 15. Inflow and outflow hydrographs (data corresponding to Hydrograph 1) and the predicted fuzzy band of the Ebro River based on the proposed method for (**a**) $w_1 = 1$ and (**b**) $w_1 = 0.1$.

Afterwards, the fuzzy parameter values obtained for the two solutions were validated by applying them on three other real events, named Hydrograph 2, 3, and 4. The results are shown in Figure 16. For the fuzzy parameters obtained with hydrograph 1 and $w_1 = 1$, the stronger emphasis on the inclusion during the calibration process leads to results for other events where the observed outflow is almost included within the fuzzy band produced.

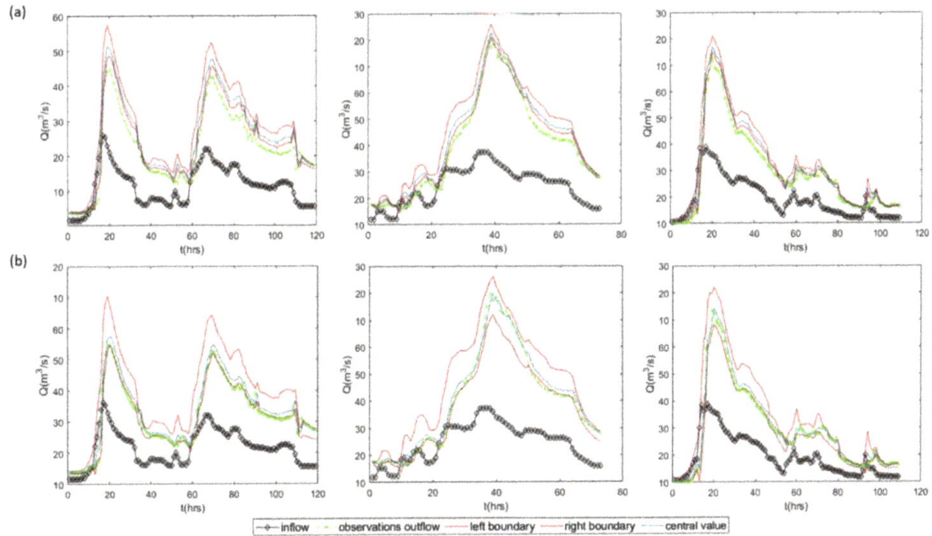

Figure 16. Validation of the achieved solution based on three real available hydrographs for (**a**) $w_1 = 0.1$ and (**b**) $w_1 = 1$. Left column corresponds to Hydrograph 2, central column to Hydrograph 3, and right column to Hydrograph 4.

This aspect is quantified with the following modified measure E_1:

$$\overline{E}_1 = \frac{1}{M}\left[\sum_{j=1}^{M} a_{R_j}\left(Q_j^{observed} - Q_j^+\right)^2 + \sum_{j=1}^{M} a_{L_j}\left(Q_j^- - Q_j^{observed}\right)^2\right],$$

including all data

The values of the modified measure obtained for Hydrographs 2, 3, and 4 are 0.6556, 0.3304, and 0.2526. These values are small, indicating a good coverage of the observations, although most of the observed discharges are below the central values of the fuzzy band.

For the fuzzy parameters obtained with Hydrograph 1 and $w_1 = 0.1$, the observed outflow is not included within the produced fuzzy band in most cases (Figure 16a). However, the shape of the observed outflow is similar to the produced fuzzy bands, suggesting that the deviation may be due to the different contribution of incremental flow in each event. The values of the modified measure are 6.8537, 10.8548, and 4.8134 (Table 2). As aforementioned, the property of inclusion (which is expressed by the modified measure E_1) is in conflict with the goal of low uncertainty, which is expressed by the measure E_3 (Table 2).

Table 2. Modified measure E_1 (property of inclusion) and E_3 (uncertainty) by testing weights $w_1 = 0.1$ and $w_1 = 1$ and applied to Hydrographs 1 (training) and 2, 3, and 4 (validation).

	Training	Validation		
	Hydrograph 1	**Hydrograph 2**	**Hydrograph 3**	**Hydrograph 4**
		$w_1 = 0.1$		
E_1	3.8382	6.8537	10.8548	4.8134
E_3	18.8438	24.9167	41.3546	20.9019
		$w_1 = 1$		
E_1	0.7458	0.6556	1.1253	0.4815
E_3	84.5021	79.5274	144.0766	75.1660

These fuzzy parameters showed a good skill to produce operational forecasts based only on inflow to the reach. The results were encouraging, particularly considering that incremental flow was important in comparison with the inflow to the reach. The parameter w_1 could be used to control whether the emphasis should be placed on inclusion of the observations within the fuzzy band or in narrowing the uncertainty of the predictions. It should be noted that important hydrograph's characteristics were simulated with a high degree of accuracy, for example, the volume of the hydrograph, the peak flow, the lag time, and the duration of the hydrographs, among others.

In this validation exercise, the effectiveness of behavioral parameters was tested against an alternative parameter configuration proposed by Pedersen and Chipperfield [55], who suggested to simplify the approach by eliminating the use of the particle's previous best-known position by setting $c_1 = 0$. The performance of this option was lower than that of the traditional values for behavioral parameters in the examined case. The small improvement in computation time was not found to be relevant because the large computation time is caused by the simulation (based on the extension principle) and not by the topology of the network and the corresponding update of the positions of the swarm. Therefore, the simplification of the topology of the swarm in this problem cannot win computational time and leads to poorer results.

A critical point of which these kind of studies should be aware is the number of data. For instance, the approach of the ANFIS system based on the Wilson, 1974 data [10] is not a safe choice since the ANFIS requires several variables. Furthermore, the use of sophisticated non-linear Muskingum models with either 10 [54] or significantly large [56] calibrated parameters, with only 18 sets of data available, have the risk of overtraining.

A disadvantage of the proposed methodology is the high computational time consumption required and secondly, the selection of the proper weight w_1, which differentiates the solution. However, (1) the production of a functional uncertainty (2) with parameters that make physical sense and (3) the inclusion of the observed data to a high degree, especially near the maximum value, will be the determinant criteria in the selection of the weight. For instance, in the case of the third example, a large quantity of w_1 leads to an unfunctional uncertainty as well as an irrational values for the parameter x.

An extension of the proposed methodology could be the application of a nonlinear form regarding the Muskingum model. However, two limitations must be taken into account. The first one is that a significant number of data is required for calibration, and the second is the difficulty in the incorporation of the fuzziness. Indeed, by using the linear Muskingum theory, an explicit relation exists to determine the outflow with respect to the parameters K and x (and α for lateral flow), while in the case of the nonlinear relation, a process with many steps is required. Hence, in the first case, the extension principle can be directly applied to determine the selected *h-cuts*. The more traditional optimization problems are gradient-based and local search algorithms, and hence the final optimum solutions, may depend on the consideration of the initial point. The evolutionary algorithms, such as genetic algorithm (GA) and swarm intelligence, has an advantage that overcomes this difficulty. Such global optimizers are in most cases simple, flexible, and efficient. However, it lacks in-depth understanding of how such algorithms may converge and how quickly they can converge to the global optimum, and hence, this point is a topic for further investigation [57]. The PSO method is selected because of its simplicity and since it has been applied in the examined problem of river routing with satisfactory results ([21,58,59]). Indeed, [60] suggested that their outcomes confirmed that the PSO algorithm estimated the parameters in a complex nonlinear Muskingum model with high accuracy along with a fast rate of convergence. In addition, in the examined case, which is a calibration problem, we can suppose a possible range of the parameters, and hence, a randomly created initial swarm can be easily constructed. The use of other heuristic algorithms is challenging especially in more complex simulation where a lateral flow exist, and the no linear Muskingum model may be adopted. For instance, the use of hybrid simulated annealing-PSO methods [61]) seem promising alternatives for further work. As it is written in [61], many algorithms introduce ideas that could be easily exported to other methods with varying degrees of compatibility. In [62], these hybrid models were proposed in the case of a water-based algorithm (e.g., [62]). For instance, by a similar way with the PSO, rain-fall optimization algorithm (RFO) has been applied as a new, naturally inspired algorithm based on behavior of raindrops with an effective approach [63].

The utility of the method might be extended in the cases of the rain-runoff models since some conceptual models use the well-known Muskingum method to express the quick flow development (instead of the unit hydrograph theory) and, with other parameters, the slow flow development [1].

5. Concluding Remarks

A general methodology to assess the parameter of the linear Muskingum model for river routing by considering parameters as fuzzy symmetrical triangular numbers is presented in this study. The expanded linear Muskingum storage method presented by O'Donnel (1985) [11] is used also in cases where a lateral flow occurs. Since the calibration model is an optimization of a case model, the PSO is used since it enables us to use a simulation process for each possible solution that composes the swarm. Hence, for each candidate solution, the extension principle of fuzzy sets and logic is activated in order to determine a fuzzy band of the outflow.

A fitness function is established that expresses the divergence of the produced fuzzy band to include all data. It takes into account the distance between the central values and the observed data, the total width of the fuzzy band and the distance between the maximum value of the outflow, and the right hand of the estimated outflow. A critical point is that the aim of divergence of the produced fuzzy band to include all data (property of inclusion) is in conflict with the aim of minimization of the total width of the produced fuzzy band. A weighted sum of the above goals composes the fitness function, whilst a functional width that contains most of the observed data is applied to select the weights of the fitness function.

Four case studies from the references are used to benchmark the proposed method, including smooth and non-smooth hydrographs with lateral flow and a double peak hydrograph. The last case study includes a complicated, real case study, and it is used for validation purposes. The proposed methodology improve results with the use of a fuzzy estimation with the aim of fuzzy parameters closer to nature; the new methodology increases the safety of the prediction; and furthermore, the produced fuzzy band can include, to a significant degree, the observed data and the output of the existent crisp methodologies even if they include a more complex formulation. Regarding the PSO model, results suggest that 100 iterations of a swarm with 50 members is sufficient to approach the final solution, while the candidates solutions converge to the total optimum.

Author Contributions: Conceptualization, M.S., L.G., and A.S.-W.; methodology, M.S.; data processing, M.S. and A.S.-W., software, M.S. and A.S-W.; validation, A.S.-W. and L.G.; writing—original draft preparation, M.S.; writing—review and editing, A.S.-W. and L.G.; visualization, A.S.-W.; supervision, L.G. All authors have read and agreed to the published version of the manuscript.

Funding: This research was funded by the Spanish Ministry of Science and Innovation, grant number PID2019-105852RA-I00: "Simulation of climate scenarios and adaptation in water resources systems (SECA-SRH)".

Institutional Review Board Statement: Not applicable.

Data Availability Statement: Three classical case studies from the references are used to bench-mark the variations of the Muskingum method. These three case studies include: (1) smooth out-flow (Wilson 1974 [10]), (2) double-peak outflow (Viessman and Lewis 2003) [9], and (3) non-smooth outflow with lateral flow (O'Donnel 1985) data sets [11]. The data can be found in the references. In addition, the validation case study is located in the Ebro River Basin (Spain, Figure 14). The gauge stations (automatic flow measures) belong to the SAIH network (Automatic Hydrologic Information System) managed by the Ebro Basin Water Authority. Data from the SAIH network of the Ebro basin can be requested in the following web page: http://www.saihebro.com/saihebro/index.php?url=/autoservicio/inicio.

Conflicts of Interest: The authors declare no conflict of interest. The funders had no role in the design of the study; in the collection, analyses, or interpretation of data; in the writing of the manuscript, or in the decision to publish the results.

References

1. Baymani-Nezhad, M.; Han, D. Hydrological modeling using Effective Rainfall routed by the Muskingum method (ERM). *J. Hydroinformatics* **2013**, *15*, 1437–1455. [CrossRef]
2. Niazkar, M.; Afzali, S.H. Streamline performance of Excel in stepwise implementation of numerical solutions. *Comput. Appl. Eng. Educ.* **2016**, *24*, 555–566. [CrossRef]
3. McCarthy, G.T. *The Unit Hydrograph and Flood Routing*; Conf. of North Atlantic Division, U.S. Army Corps of Engineers, Engineer Department at New London: New London, CT, USA, 24 June 1938.
4. Cunge, J.A. On the subject of a flood propagation computation method (Muskingum method). *J. Hydraul. Res.* **1969**, *7*, 205–230. [CrossRef]
5. Ponce, V.M.; Yevjevich, V. Muskingum-Cunge Method with Variable Parameters. *J. Hydraul. Div.* **1978**, *104*, 1663–1667. [CrossRef]
6. Ponce, V.; Lohani, A.; Scheyhing, C. Analytical verification of Muskingum-Cunge routing. *J. Hydrol.* **1996**, *174*, 235–241. [CrossRef]
7. Das, A. Parameter Estimation for Muskingum Models. *J. Irrig. Drain. Eng.* **2004**, *130*, 140–147. [CrossRef]
8. Akbari, G.H.; Barati, R. Comprehensive analysis of flooding in unmanaged catchments. *Proc. Inst. Civil Eng. Water Manag.* **2012**, *165*. [CrossRef]
9. Viessman, J.; Lewis, G.L. *Introduction to Hydrology*; Pearson Education, Inc.: Upper Saddle River, NJ, USA, 2011.
10. Wilson, E.M. *Engineering Hydrology*; Macmillan Education LTD: Hampshire, UK, 1974.
11. O'Donnel, T. A direct three-parameter Muskingum procedure incorporating lateral inflow. *Hydrol. Sci. J.* **1985**, *30*, 479–496. [CrossRef]

12. Spiliotis, M.; Sordo-Word, A.; Garrote, L. Estimation of the Muskingum Routing Coefficients Including Lateral inflow by using Fuzzy Linear Regression. In Proceedings of the 5th IAHR EUROPE CONGRESS, "New Challenges in Hydraulic Research and Engineering", Trento, Italy, 12–14 June 2018; Armanini, A., Nucci, E., Eds.; The International Association for Hydro-Environment Engineering and Research: Beijing, China, 2018. [CrossRef]

13. Gill, M.A. Flood routing by the Muskingum method. *J. Hydrol.* **1978**, *36*, 353–363. [CrossRef]

14. Tung, Y.-K. River Flood Routing by Nonlinear Muskingum Method. *J. Hydraul. Eng.* **1985**, *111*, 1447–1460. [CrossRef]

15. Yoon, J.; Padmanabhan, G. Parameter Estimation of Linear and Nonlinear Muskingum Models. *J. Water Resour. Plan. Manag.* **1993**, *119*, 600–610. [CrossRef]

16. Geem, Z.W. Parameter Estimation for the Nonlinear Muskingum Model Using the BFGS Technique. *J. Irrig. Drain. Eng.* **2006**, *132*, 474–478. [CrossRef]

17. Geem, Z.W. Parameter Estimation of the Nonlinear Muskingum Model using Parameter-Setting-Free Harmony Search Algorithm. *J. Hydraul. Res.* **2011**, *16*, 684–688.

18. Karahan, H.; Gürarslan, G.; Geem, Z.W. A new nonlinear Muskingum flood routing model incorporating lateral flow. *Eng. Optim.* **2014**, *47*, 737–749. [CrossRef]

19. Mohan, S. Parameter Estimation of Nonlinear Muskingum Models Using Genetic Algorithm. *J. Hydraul. Eng.* **1997**, *123*, 137–142. [CrossRef]

20. Kim, J.H.; Geem, Z.W.; Kim, E.S. Parameter Estimation of the Nonlinear Muskingum Model Using Harmony Search. *JAWRA J. Am. Water Resour. Assoc.* **2001**, *37*, 1131–1138. [CrossRef]

21. Chu, H.-J.; Chang, L.-C. Applying Particle Swarm Optimization to Parameter Estimation of the Nonlinear Muskingum Model. *J. Hydrol. Eng.* **2009**, *14*, 1024–1027. [CrossRef]

22. Xu, D.-M.; Qiu, L.; Chen, S.-Y. Estimation of Nonlinear Muskingum Model Parameter Using Differential Evolution. *J. Hydrol. Eng.* **2012**, *17*, 348–353. [CrossRef]

23. Geem, Z.W.; Roper, W.E. Various continuous harmony search algorithms for web-based hydrologic parameter optimisation. *Int. J. Math. Model. Numer. Optim.* **2010**, *1*, 213–226. [CrossRef]

24. Niazkar, M.; Afzali, S.H. Assessment of Modified Honey Bee Mating Optimization for Parameter Estimation of Nonlinear Muskingum Models. *J. Hydrol. Eng.* **2015**, *20*, 04014055. [CrossRef]

25. Niazkar, M.; Afzali, S.-H. Application of New Hybrid Optimization Technique for Parameter Estimation of New Improved Version of Muskingum Model. *Water Resour. Manag.* **2016**, *30*, 4713–4730. [CrossRef]

26. Easa, S.M. Evaluation of nonlinear Muskingum model with continuous and discontinuous exponent parameters. *KSCE J. Civ. Eng.* **2015**, *19*, 2281–2290. [CrossRef]

27. Farzin, S.; Singh, V.P.; Karami, H.; Farahani, N.; Ehteram, M.; Kisi, O.; Allawi, M.F.; Mohd, N.S.; El-Shafie, A. Flood Routing in River Reaches Using a Three-Parameter Muskingum Model Coupled with an Improved Bat Algorithm. *Water* **2018**, *10*, 1130. [CrossRef]

28. Chu, H.-J. The Muskingum flood routing model using a neuro-fuzzy approach. *KSCE J. Civ. Eng.* **2009**, *13*, 371–376. [CrossRef]

29. Spiliotis, M.; Garrote, L. Estimation of the Muskingum routing coefficients by using fuzzy regression. *Eur. Water* **2017**, *57*, 133–140.

30. Tanaka, H. Fuzzy data analysis by possibilistic linear models. *Fuzzy Sets Syst.* **1987**, *24*, 363–375. [CrossRef]

31. Klir, G.; Yuan, B. *Fuzzy Sets and Fuzzy Logic Theory and its Applications*; Prentice Hall: New York, NY, USA, 1995.

32. Zimmermann, H.J. *Fuzzy Set Theory and Its Applications*; Springer Seience+Business Media: New York, NY, USA, 2001.

33. Buckley, J.; Eslami, E. *Introduction to Fuzzy Logic and Fuzzy Sets (Advances in Soft Computing)*; Springer: Berlin/Heidelberg, Germany, 2002; Volume 13.

34. Buckley, J.; Eslami, E.; Feuring, T. Solving fuzzy equations. In *Fuzzy Mathematics in Economics and Engineering*; Buckley, J., Eslami, E., Feuring, T., Eds.; Springer: Berlin/Heidelberg, Germany, 2002; pp. 19–46.

35. Hanss, M. *Applied Fuzzy Arithmetic, an Introduction with Engineering Applications*; Springer: Berlin, Germany, 2005.

36. Spiliotis, M.; Angelidis, P.; Papadopoulos, B. A hybrid probabilistic bi-sector fuzzy regression based methodology for normal distributed hydrological variable. *Evol. Syst.* **2020**, *11*, 255–268. [CrossRef]

37. Tsakiris, G.; Spiliotis, M. Embankment dam break: Uncertainty of outflow based on fuzzy representation of breach formation param eters. *J. Intell. Fuzzy Syst.* **2014**, *27*, 2365–2378. [CrossRef]

38. Marsden, J.; Tromba, A. *Vector Calculus*, 5th ed.; W.H. Freeman and Company: New York, NY, USA, 2003.

39. Parsopoulos, K.E.; Vrahatis, M.N. Recent approaches to global optimization problems through particle swarm optimization. *Nat. Comput.* **2002**, *1*, 235–306. [CrossRef]

40. Spiliotis, M. A particle swarm optimization PSO heuristic for water distribution system analysis. *Water Util. J.* **2014**, *8*, 47–56.

41. Spiliotis, M.; Mediero, L.; Garrote, L. Optimization of Hedging Rules for Reservoir Operation During Droughts Based on Particle Swarm Optimization. *Water Resour. Manag.* **2016**, *30*, 5759–5778. [CrossRef]

42. Ostadrahimi, L.; Mariño, M.A.; Afshar, A. Multi-reservoir Operation Rules: Multi-swarm PSO-based Optimization Approach. *Water Resour. Manag.* **2011**, *26*, 407–427. [CrossRef]

43. Papadopoulos, K.; Papagianni, C.; Gkonis, P.; Venieris, I.; Kaklamani, D. Particle Swarm Optimization of Antenna Arrays with Efficiency Constraints. *Prog. Electromagn. Res. M* **2011**, *17*, 237–251. [CrossRef]

44. Spiliotis, M.; Garrote, L. Unit hydrograph identification based on fuzzy regression analysis. *Evol. Syst.* **2021**, 1–22. [CrossRef]

45. Eberhart, R.C.; Kennedy, J. A new optimizer using particle swarm theory. In Proceedings of the Sixth Symposium on Micro Machine and Human Science, Nagoya, Japan, 4–6 October 1995; pp. 39–43. [CrossRef]

46. Poli, P.; Keenedy, J.; Blackwell, T. Particle swarm optimization. *Swarm Intell.* **2007**, *1*, 33–57. [CrossRef]

47. Shi, Y.; Eberhart, R. A modified particle swarm optimizer. In Proceedings of the 1998 IEEE International Conference on Evolutionary Computation Proceedings, Anchorage, AK, USA, 4–9 May 1998; pp. 69–73.

48. Clerc, M.; Kennedy, J. The particle swarm—Explosion, stability, and convergence in a multidimensional complex space. *IEEE Trans. Evol. Comput.* **2002**, *6*, 58–73. [CrossRef]

49. Salehizadeh, S.; Yadmellat, P.; Menhaj, M. Local Optima Avoidable Particle Swarm Optimization. In Proceedings of the IEEE Swarm Intelligence Symposium, Nashville, TN, USA, 30 March–2 April 2009; pp. 16–21.

50. Ishibuchi, H.; Tanaka, H.; Okada, H. An architecture of neural networks with interval weights and its application to fuzzy regression analysis. *Fuzzy Sets Syst.* **1993**, *57*, 27–39. [CrossRef]

51. Spiliotis, M.; Hrissanthou, V. Fuzzy and crisp regression analysis between sediment transport rates and stream discharge in the case of two basins in northeastern Greece. In *Conventional and Fuzzy Regression: Theory and Engineering Applications*; Hrissanthou, V., Spiliotis, M., Eds.; Nova Science Publishers: New York, NY, USA, 2018; pp. 1–46.

52. Tzimopoulos, C.; Papadopoulos, K.; Papadopoulos, B. Fuzzy Regression with Applications in Hydrology. *Int. J. Eng. Innov. Technol.* **2016**, *5*, 22.

53. Karahan, H.; Gürarslan, G.; Geem, Z.W. Parameter Estimation of the Nonlinear Muskingum Flood-Routing Model Using a Hybrid Harmony Search Algorithm. *J. Hydrol. Eng.* **2013**, *18*, 352–360. [CrossRef]

54. Karahan, H. Discussion of "Improved Nonlinear Muskingum Model with Variable Exponent Parameter" by Said M. Easa. *J. Hydrol. Eng.* **2014**, *19*, 07014007. [CrossRef]

55. Pedersen, M.; Chipperfield, A. Simplifying Particle Swarm Optimization. *Appl. Soft Comput.* **2010**, *10*, 618–628. [CrossRef]

56. Easa, S.M. Improved nonlinear Muskingum model with variable exponent parameter. *J. Hydrol. Eng.* **2013**, *18*, 1790–1794. [CrossRef]

57. Yang, X.-S. Nature-inspired optimization algorithms: Challenges and open problems. *J. Comput. Sci.* **2020**, *46*, 101104. [CrossRef]

58. Ouyang, A.; Tang, Z.; Li, K.; Sallam, A.; Sha, E. Estimating Parameters of Muskingum Model Using an Adaptive Hybrid pso Algorithm. *Int. J. Pattern Recognit. Artif. Intell.* **2014**, *28*, 1459003. [CrossRef]

59. Norouzi, H.; Bazargan, J. Flood routing by linear Muskingum method using two basic floods data using particle swarm optimization (PSO) algorithm. *Water Supply* **2020**, *20*, 1897–1908. [CrossRef]

60. Moghaddam, A.; Behmanesh, J.; Farsijani, A. Parameters Estimation for the New Four-Parameter Nonlinear Muskingum Model Using the Particle Swarm Optimization. *Water Resour. Manag.* **2016**, *30*, 2143–2160. [CrossRef]

61. Sudibyo, S.; Murat, M.N.; Aziz, N. Simulated annealing-Particle Swarm Optimization (SA-PSO): Particle distribution study and application in Neural Wiener-based NMPC. In Proceedings of the 2015 10th Asian Control Conference (ASCC), Kota Kinabalu, Malaysia, 31 May–3 June 2015; pp. 1–6.

62. Rubio, F.; Rodríguez, I. Water-Based Metaheuristics: How Water Dynamics Can Help Us to Solve NP-Hard Problems. *Complexity* **2019**, *2019*, 1–13. [CrossRef]

63. Clark, D. (Ed.) Kaboli, HRA Rain-Fall Inspired Optimization Algorithm For Optimal Load Dispatch In Power System. In *Robust and Constrained Optimization: Methods and Applications*; Nova Science Publishers: New York, NY, USA, 2020.

 *sustainability*

Article

Modeling and Uncertainty Analysis of Groundwater Level Using Six Evolutionary Optimization Algorithms Hybridized with ANFIS, SVM, and ANN

Akram Seifi [1,*]**, Mohammad Ehteram** [2]**, Vijay P. Singh** [3] **and Amir Mosavi** [4,5,6,7,*] 

[1]   Department of Water Science & Engineering, Vali-e-Asr University of Rafsanjan, Rafsanjan, Iran

[2]   Department of Water Engineering and Hydraulic Structures, Faculty of Civil Engineering, Semnan University, Semnan 35131-19111, Iran; eh.mohammad@yahoo.com

[3]   Department of Biological and Agricultural Engineering & Zachry Department of Civil Engineering Texas A&M University College Station, Texas, TX 77843-2117, USA; vsingh@tamu.edu

[4]   Thuringian Institute of Sustainability and Climate Protection, 07743 Jena, Germany

[5]   Institute of Automation, Obuda University, 1034 Budapest, Hungary

[6]   Department of Mathematics and Informatics, J. Selye University, 94501 Komarno, Slovakia

[7]   Institute of Structural Mechanics, Bauhaus-Universität Weimar, 99423 Weimar, Germany

*   Correspondence: a.seifi@vru.ac.ir (A.S.); amir.mosavi@uni-weimar.de (A.M.)

Received: 14 April 2020; Accepted: 6 May 2020; Published: 14 May 2020

Abstract: In the present study, six meta-heuristic schemes are hybridized with artificial neural network (ANN), adaptive neuro fuzzy interface system (ANFIS), and support vector machine (SVM), to predict monthly groundwater level (GWL), evaluate uncertainty analysis of predictions and spatial variation analysis. The six schemes, including grasshopper optimization algorithm (GOA), cat swarm optimization (CSO), weed algorithm (WA), genetic algorithm (GA), krill algorithm (KA), and particle swarm optimization (PSO), were used to hybridize for improving the performance of ANN, SVM, and ANFIS models. Groundwater level (GWL) data of Ardebil plain (Iran) for a period of 144 months were selected to evaluate the hybrid models. The pre-processing technique of principal component analysis (PCA) was applied to reduce input combinations from monthly time series up to 12-month prediction intervals. The results showed that the ANFIS-GOA was superior to the other hybrid models for predicting GWL in the first piezometer (RMSE:1.21, MAE:0.878, NSE:0.93, PBIAS:0.15, R^2:0.93), second piezometer (RMSE:1.22, MAE:0.881, NSE:0.92, PBIAS:0.17, R^2:0.94), and third piezometer (RMSE:1.23, MAE:0.911, NSE:0.91, PBIAS:0.19, R^2:0.94) in the testing stage. The performance of hybrid models with optimization algorithms was far better than that of classical ANN, ANFIS, and SVM models without hybridization. The percent of improvements in the ANFIS-GOA versus standalone ANFIS in piezometer 10 were 14.4%, 3%, 17.8%, and 181% for RMSE, MAE, NSE, and PBIAS in training stage and 40.7%, 55%, 25%, and 132% in testing stage, respectively. The improvements for piezometer 6 in train step were 15%, 4%, 13%, and 208% and in test step were 33%, 44.6%, 16.3%, and 173%, respectively, that clearly confirm the superiority of developed hybridization schemes in GWL modelling. Uncertainty analysis showed that ANFIS-GOA and SVM had, respectively, the best and worst performances among other models. In general, GOA enhanced the accuracy of the ANFIS, ANN, and SVM models.

Keywords: groundwater; artificial intelligence; hydrologic model; groundwater level prediction; machine learning; principal component analysis; spatiotemporal variation; uncertainty analysis; hydroinformatics; support vector machine; big data; artificial neural network

1. Introduction

One of the most important sources of water supply for industrial, drinking, and irrigation purposes is groundwater (GW). GW has a significant role in economic development, environmental management, and ecosystem sustainability [1,2]. However, in recent years undue exploitation has caused a tremendous pressure on GW resources, resulting in GW crisis [3]. As a result, the GW level (GWL) in different regions of the world has been decreasing rapidly. Further, widespread pollution of surface water is severely affecting GW. A decrease in GWL can also be caused by climate factors and can lead to a number of eco-environmental problems [4]. For proper water resources management, particularly effective utilization and sustainable management of groundwater resources, accurate and reliable prediction of GWL is essential [5,6]. Thus, it is necessary to predict the Ardebil groundwater level for water resources management. Mathematical models incorporating GW dynamics are applied to predict GWL for optimizing groundwater use, optimal management, and development of conservation plans [5,7]. Since such models are costly, time-consuming, and data-intensive, their use in practice is limited because of data-scarcity [8,9]. In such cases, when geological and hydro-geological data are insufficient, soft computing models become an attractive option [10]. Artificial neural network (ANN), adaptive neuro-fuzzy interface (ANFIS), genetic programming (GP), support vector machine (SVM), and decision tree models are among the important soft computing models that are suited for modeling dynamic and uncertain nonlinear systems [7].

Recently, soft computing models have been widely used worldwide to predict GWL. Jalal Kameli et al. [11] evaluated neuro-fuzzy (NF) and ANN models to estimate GWL using rainfall, air temperature, and GWLs in neighboring wells, and showed that the NF model performed better than the ANN model. Identifying the lag time of time series for observed rainfall by correlation analysis, Trichakis et al. [12] used the ANN model to predict GWL and found the ANN model to be useful to model Karst aquifers that are difficult to simulate using numerical models. Using evaporation, rainfall, and water levels in observation levels as input, Fallah-Mehdipour et al. [13] applied the ANFIS and genetic programming models for predicting GWL and showed that GP decreased the value of mean root square error (RMSE) compared to the RMSE by the ANFIS. Moosavi et al. [14] evaluated the ANN, ANFIS-wavelet, and ANN-wavelet models and showed that predicted GWL was more accurate for 1 and 2 months ahead than for 3 and 4 months ahead. Predicting GWL in the Bastam plain by ANFIS and ANN models in Emamgholizadeh et al. [15] study confirmed that if the water shortage of the aquifer remained equal to the pumping rate of water from wells, the minimum reduction of GWL occurred. Suryanarayana et al. [16] proposed a hybrid model integrating the SVM model with the wavelet transform and indicated that the SVM-wavelet model was more accurate in predicting GWL. Using rainfall, pan evaporation, and river stage as input, Mohanty et al. [17] indicated that the ANN model was better using shorter lead times for GWL predictions than the larger lead times. Yoon et al. [18] demonstrated that the SVM model was superior to the ANN model in predicting GWL. Zho et al. [19] found that the wavelet-SVM model was better than the wavelet-ANN model for modelling GWL. Comparing ANN and autoregressive integrated moving average (ARIMA), Choubin and Malekian [20] showed that the ARIMA model was more accurate than ANN in modelling GWL. Das et al. [21] found ANFIS to be better than ANN for predicting GWL.

Literature review shows that although soft computing models are capable for predicting groundwater level, they have weaknesses and uncertainties [22]. The ANN models have different parameters, such as weight connections, bias, and need training algorithms to fine-tune their parameters. ANFIS and SVM models have nonlinear and linear parameters and use different kinds of training algorithms, such as backpropagation algorithm, descent gradient method, etc. However, the standard training algorithms have two major defects: slow convergence and getting trapped in local optima [22]. Recently, nature-based optimization algorithms have been developed for finding the appropriate values of model parameters to improve ANN, ANFIS, and SVM models. Jalalkamali and Jalalkamali [23] applied a hybrid model of ANN and genetic algorithm (ANN-GA) to find the best number of neutrons for the hidden layer and predict GWL in an individual well. Mathur [24] applied hybrid SVM-PSO (particle swarm optimization) model for predicting GWL in Rentachintala region of Andhra Pradesh, India, where optimal parameters of SVM were determined using PSO. Results showed that SVM-PSO was more accurate than the ANN, ANFIS, and ARMA models. Hosseini et al. [25] hybridized ANN and ant colony optimization (ACO) to predict the GWL in Shabestar plain, Iran, and found that the hybrid ANN-ACO model reduced overtraining errors. Zare and Koch [26] demonstrated that the hybridized wavelet-ANFIS model was superior in modelling GWL to other regression models. Balavalikar et al. [27] found that the hybrid ANN-PSO model was better in predicting monthly GWL of Udupi district, India, than the classical ANN model. Malekzadeh et al. [28] evaluated ANN, wavelet extreme machine learning (WEML), SVM, wavelet-SVM, and wavelet-ANN for predicting GWL, and concluded that WEML was more accurate. These studies reveal that hybrid models are more accurate and efficient than single models in predicting GWL and it is inferred from these studies that meta-heuristic optimization algorithms are superior to the classical ones, but require uncertainty analysis for artificial intelligence models.

New hybrid intelligent optimization models can be regarded as appropriate alternative methods with an acceptable range of error for predicting GWL. Among the nature-inspired optimization algorithms, the grasshopper optimization algorithm (GOA) is a novel and robust meta-heuristic method that mimics the swarming behavior of grasshoppers in nature. The GOA is a multi-solution-based algorithm during the optimization process to avoid higher local optima and has high convergence ability toward the optimum [29]. It has different functions than other optimization algorithms that enable it to find the best optimal solution in the search space with high probability. Therefore, this algorithm escapes from local optima and finds the global optimum in the search space. This capability is considered as an advantage of GOA [30] and as reason for the selection of GOA for the current study. Several researchers used GOA for monthly river flow [31], soil compression coefficient [32], coefficients of sediment rating curve [33], and concrete slump [34], but the uncertainty analysis and GWL modeling has not yet been studied.

These models have some drawbacks in the previous studies that are addressed in the current paper. These models are robust tools for modeling many of the nonlinear hydrologic processes such as rainfall-runoff, stream flow, and ground-water level. Despite the wide application of soft computing models, few studies have investigated the capability of novel optimization algorithms, such as GOA integrated with typical predictive methods, for GWL prediction, uncertainty evaluation, and spatial variation modeling. The main problem in developing these models is the using of an appropriate training procedure. Especially, AI tend to be very data intensive in training stage, and there appears to be no established methodology for design and successful implementation of training procedure and error minimizations. Therefore, there are still some questions about AI tools that must be further studied, and important aspects such as local trapping, uncertainty analysis of results, uncertainty due to meta-heuristic optimization algorithms in training, spatial changes modelling with hybrid models must be explored further. Based on the best knowledge of the authors, no published papers exist that evaluate the uncertainty of different meta-heuristic optimizations for groundwater level prediction in hybridization with ANN, ANFIS, and SVM. The main contribution and novelty of the present study is comparative uncertainty analysis of the novel hybrid models, spatial changes modelling by considering PCA as appropriate input selection in regard to uncertainty results. Despite the wide application of soft computing models, few studies have investigated the capability of novel optimization algorithms, such as GOA integrated with typical predictive methods, for GWL prediction, uncertainty evaluation, and spatial variation modeling. The state-of-art models, including ANN, ANFIS, and SVM, have been employed to predict GWL, but these models are easily trapped in local optima and often need longer training times. Hence, the main contribution of this study is to develop and to assess the applicability of hybrid ANFIS-GOA, SVM-GOA, and ANN-GOA models for predicting monthly GWL and uncertainty of results in Ardabil basin in Iran. Application of GOA method integrated with ANN, ANFIS, and SVM models is useful to search the best numerical weights of neurons and bias values. The other objectives of this paper were to (1) compare the GOA with different optimization algorithms of particle swarm (PSO), weed algorithm (WA), cat algorithm (CA), and genetic algorithm (GA); (2) evaluate the uncertainty of the hybridized models for predicting monthly GWL; (3) use principal component analysis to select the appropriate input combinations from time-series data up to 12-month lag; (4) modeling spatial variation of GWL by using hybrid intelligence models results in geospatial analysis.

2. Materials and Methods

2.1. Case Study and Data

The Ardebil plain, with the area of 990 km^2, is located in the northwest of Iran between latitudes 38′3° and 38′27 and the longitudes of 47′55° and 48′20° (Figure 1). The average annual rainfall is 304 mm. The hottest month in this plain is May and the driest month is July. The average annual temperature is 9 °C. In Ardebil plain, groundwater supplies water for drinking, agricultural, and industrial purposes. There is a negative balance of about 550 million m^3 in the Ardebil aquifer. The GWL decreases by 20–30 cm per year, which is the fastest decline. The Ardebil plain has 89 villages, that use groundwater for agricultural uses. The current condition of the GWL in the Ardebil plain has negative impacts on the farmers as its main users. In this study, the following parameters were used as the input to the hybrid ANN, ANFIS, and SVM models. Then, the principal component analysis was used to select the best input combination up to 12-month lag.

$$H(t) = f[H(t-1), H(t-2), H(t-3), \ldots . H(t-12)] \tag{1}$$

where, $H(t)$ is the GWL at month t, $H(t-1)$ is the 1-month lagged H, $H(t-2)$ is the 2-month lagged H, $H(t-3)$ is the 3-month lagged H, and $H(t-12)$ is the 12-month lagged H. The data of 140 months (2000 (January)–2012 (September)) were selected for the current study. A total of 20% of the data set was used for testing, and 80% of the data set was used for the training, that were selected randomly. Nine observed wells (wells 6, 9, 10, 24, 11, 4, 7, 8, and 1) were used to provide the spatiotemporal variation of GWL for different months. Each piezometer had 140 monthly data points. The measurements were made one time during each month.

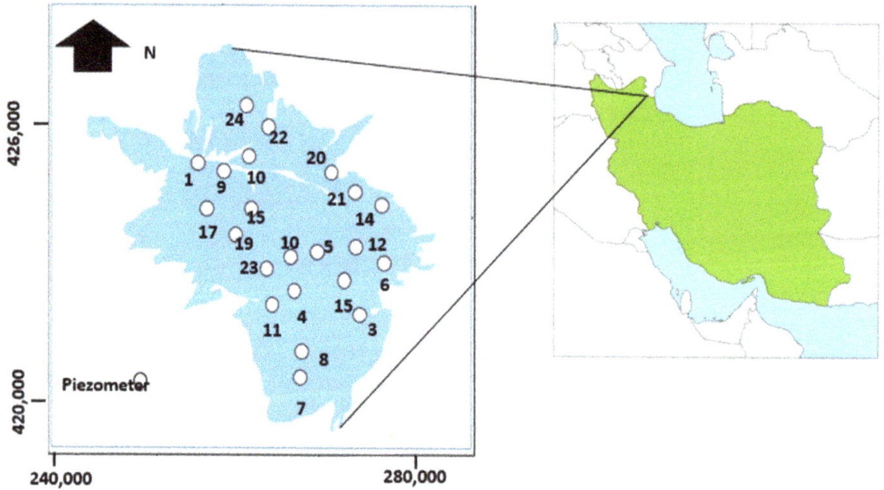

Figure 1. Location of Ardebil Plain as the case study.

2.2. ANFIS Model

The ANFIS model uses fuzzy interface systems which use fuzzy if-then rules to construct a predictive model. The ANFIS model has been widely used for predicting rainfall [33], temperature [34], runoff [35], evaporation [36], and sediment load [37]. Figure 1 shows the structure of the ANFIS model in the framework of the study. The square nodes and circle nodes show the adaptive and fixed nodes, respectively. The ANFIS model has five layers [38]. (1) The inputs are fuzzified in the first layer whose nodes are constant. The membership grade of inputs is the output of the first layer:

$$o_i^1 = u_{Ai}(x), i = 1, 2..$$
$$o_i^1 = u_{B_{i-2}}(y), i = 3, 4, ..$$

(2)

where, o_i^1 is the output of the first layer, $u_{Ai}(x)$ and $u_{B_{i-2}}(y)$ are the fuzzy membership functions for the fuzzy set A_i and B_{i-2}, respectively. The bell-shaped member function is selected for the current study due to its smoothness and concise notation:

$$u_{A_i}(x) = \frac{1}{1 + \left[\left(\frac{x-c_i}{a_i}\right)^2\right]^{b_i}}, i = 1, 2, ..$$

(3)

where a, b, and c are the premise parameters (training algorithms obtain these parameters).

(2) The nodes of the second layer are labelled with M, which shows that they carry out a simple multiplier function. The fuzzy strengths ω_i of each rule are the output of the second layer:

$$o_i^2 = \omega_i = u_{A_i}(x)u_{Bi}(y), i = 1, 2..,$$

(4)

(3) The nodes of the third layer are also fixed. The fuzzy strengths from the previous layer are normalized in the third layer. The sum of weight functions is used to compute the normalization factor. The normalized fuzzy strengths are the output of the third layer:

$$o_i^3 = \omega_i = \frac{\omega_i}{\sum_{i=1}^2 \omega_i}$$

(5)

(4) The nodes of the fourth layer are adaptive and its outputs are computed as:

$$o_i^4 = \omega_i z_i = \omega_i(p_i + q_i y + r_i), i = 1, 2..,$$

(6)

where, p_i, q_i, and r_i are the consequent parameters.

(5) The output in the fifth layer is labelled with S. A fixed node is observed in this layer. This layer computes the total summation of all the incoming signals:

$$o_i^5 = z = \sum_{i=1}^2 \omega_i z_i = \frac{\sum_{i=1}^2 \omega_i z_i}{\sum_{i=1}^2 \omega_i}$$

(7)

In the classical training approach, a combination of the least square and gradient descent methods is commonly used as a hybrid learning algorithm to adjust the parameters of the ANFIS model. The consequent parameters of ANFIS model are updated by applying the least square method in the forward pass. Additionally, in the backward pass, the gradient descent method is used for updating the premise parameters. In the hybridized schemes, tuning and adjusting the consequent and premise parameters are determined by the optimization algorithms as the hybrid training scheme.

2.3. ANN Model

The artificial neural network uses behavioral patterns to provide a framework for modeling mechanisms. It consists of three layers: input, hidden, and output layers, and includes the processing units named neurons which are arranged in several layers [39]. The connection weights link the neurons of preceding layers to the neurons of the following layers. The output of the middle layer (hidden layer) is used as the input to the following layer. The input data is received by the input layer, while the last layer generates the final output of the ANN model. The middle layers receive and transmit the input data to the connected nodes in the following layers. The weighted sum of inputs is used by the hidden neurons to produce the intermediate output. The ANN model uses the activation functions to compute the outputs of the hidden and output neurons. It uses the bias values to set the output along with the weighted sum of inputs to the neuron. The process of ANN modelling has two major levels: (1) preparing the network structure, and (2) adjustment of the weights of connections. The literature review indicates that the backpropagation training algorithm is wildly used in different fields, such as water engineering [40]. First, the output of the ANN model is obtained as a response of the ANN model. In the next level, the error between observed and estimated values is minimized to find the weights of the model. If the output is different from the observed value, the modification of weights and biases will start to decrease the error values. However, the backpropagation algorithm has a slow convergence rate and to overcome its inherent weakness the meta-heuristic optimization algorithms are used in the present study. Figure 1 shows the structure of the ANN model and its hybridization with intelligence algorithms.

2.4. SVM Model

The SVM model has been widely used for predicting solar radiation [41], rainfall [42], landslides [43], and drought [44]. In the SVM model, the input data are divided into testing and training samples. The selected input vector (training sample) is mapped into a high-dimensional feature space. Then, the optimal decision function is generated [44]. Equation (7) shows the regression estimation function of the SVM model:

$$f(x) = W^T \phi(x) + b \tag{8}$$

where, $\phi(x)$ is the nonlinear mapping function for mapping sample data (x) into an m-dimensional feature vector, b is the bias, and W^T is the weight vector of the independent function. W^T and b are computed by minimizing the following function:

$$D(f) = \frac{1}{2}\|w\|^2 + \frac{C}{n}\sum_{j=1}^{n} R_\varepsilon\left[y_j, f(x_j)\right] \tag{9}$$

where, $D(f)$ is the generalized optimal function, $\|w\|^2$ is the complexity of the model, C is the penalty parameter, and R_ε is the error control function of ε. Thus, the optimization problem is defined as follows:

$$\begin{aligned}
&\min Q(W, \xi) = \tfrac{1}{2}\|w\|^2 + C\sum_{j=1}^{n} \xi_j + \xi_j^n \\
&W^T\phi(x_j) + b - y_j \le \varepsilon + \xi_j \\
&y_j - W^T\phi(x_j) - b \le \varepsilon + \xi^*_j \\
&\xi_j \ge 0, \xi^*_j \ge 0, j = 1, 2, .., n
\end{aligned} \tag{10}$$

where, ξ_j and ξ^*_j are the relation factors. Adjusting the partial derivatives of W, b, ξ_j, and ξ^*_j to 0 and using the Lagrangian equation, an optimization problem can be formulated as follows:

$$
\begin{aligned}
L(W, a, b, \varepsilon, y) &= min\tfrac{1}{2} \sum_{j=1}^{n} (a_r - a_r^*)^T H_{r,j*} \\
&\quad a_r - a_r^* + \varepsilon \sum_{j=1}^{n} (a_r - a_r^*) + \sum_{j=1}^{n} y_r(a_r - a_r^*) \\
&\quad \sum_{r=1}^{n} (a_r - a_r^*) = 0, \, (0 \leq a_r, a_r^* \leq C) \\
&\quad H_{r,j} = K(x, x_j) = \phi(x_r)^T \phi(x_j), \, (r = 1, 2, .., n)
\end{aligned}
\tag{11}
$$

where, $K(x, x_j)$ is the kernel function. The most popular kernel function is the radial basis function:

$$
K(x, x_j) = exp\left(-\frac{|x - x_j|^2}{2\gamma^2}\right)
\tag{12}
$$

where, γ is the radial basis function parameter. The SVM based model uses the grid search algorithm (GS) to find the optimal value of parameters C and γ. Specifically, a set of initial values is chosen for both parameters γ and C. To select γ and C using cross-validation, the available data are divided into k subsets. One subset is regarded as testing data and then assessed using the remaining k-1 training subsets. Then, the cross-validation error is computed using the split error for the SVM model using different values of C and γ. Various combination of parameters C and γ are evaluated and the one yielding the lowest cross-validation error is chosen and used to train the SVM model for the whole dataset. The structure of the SVM model is shown in Figure 2.

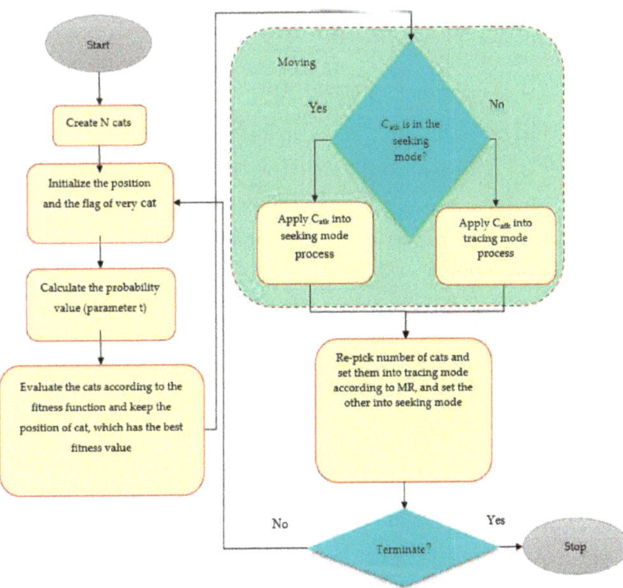

Figure 2. Developed methodology framework for modeling groundwater level time series.

2.5. Optimization Algorithms

2.5.1. Grasshoppers Optimization Algorithm (GOA)

Grasshoppers are regarded as pests because they damage agricultural crops. They are a group of insects that can generate large insect swarms. The mathematical function to investigate the swarming behavior of grasshoppers is demonstrated with the following equation [45]:

$$X_i = S_i + G_i + A_i \tag{13}$$

where, X_i is the position of the ith grasshopper, S_i is the classical interaction, G_i is the gravity force on the ith grasshopper, and A_i is the wind advection. The classical interaction is simulated as follows:

$$S_i = \sum_{j=1}^{N} s(d_{ij})\hat{d}_{ij} \tag{14}$$

where, d_{ij} is the distance between the ith and jth grasshoppers, and s is a function for the definition of the strength of social forces.

$$d_{ij} = |x_i - x_j| \\ \hat{d}_{ij} = \frac{x_j - x_i}{d_{ij}} \tag{15}$$

The function s is computed as follows:

$$s(r) = f e^{-\frac{r}{l}} - e^{-r} \tag{16}$$

where, f is the intensity of attraction, and l is the attractive length scale. The distance between grasshoppers ranges between 0 and 15. Repulsion is observed in the interval [0 2.079]. The grasshoppers enter the comfort zone if they are far from 2.079 units from other grasshoppers. G component is computed as follows:

$$G_i = -g\hat{e}_g \tag{17}$$

where, g is the gravitational constant and $\hat{e}_g$ is a unity vector towards the center of the earth. The A parameter is computed as follows:

$$A_i = u\hat{e}_w \tag{18}$$

where, u is a constant drift and $\hat{e}_w$ is a unit vector in the direction of the wind. Finally, the new position of a grasshopper is computed using its common position, the food source position, and the position of all other grasshoppers:

$$X_i = \sum_{\substack{j=1 \\ j \neq i}}^{N} s(|x_j - x_i|)\frac{x_j - x_i}{d_{ij}} - g\hat{e}_g + u\hat{e}_w \tag{19}$$

where, N is the number of grasshoppers. However, Equation (18) cannot be directly used for optimization because grasshoppers do not converge to a specified point. Thus, a corrected equation is used to update the grasshopper's position:

$$X_i^d = c\left[\sum_{\substack{j=1 \\ j \neq i}}^{N} c\frac{ub_d - lb_d}{2}s(|x_j^d - x_i^d|)\frac{x_j - x_i}{d_{ij}}\right] + \hat{T}_d \tag{20}$$

where, ub is the upper bound; lb_d is the lower bound; $\hat{T}_d$ is the value of the D_{th} dimension in the target space (optimal solution found so far); and c is a decreasing coefficient to shrink the comfort zone, repulsion zone, and attraction zone. Figure 2 shows the flowchart of GOA.

2.5.2. Weed Algorithm (WA)

Weeds have a very adaptive nature that converts them to undesirable plants in agriculture. Figure 3 shows the flowchart of the WA algorithm [46]. The WA starts with initializing a random population of weeds in the search space. A predefined number of weeds are randomly distributed over the entire dimensional space, indicated as a solution space. The fitness of weeds is assessed by considering its fitness function to optimize the problem. Each agent of the current population can produce some seeds via a predefined region considering its own location. In this way, the number of produced seeds relies on its fitness function in the population regarding the best and worst solutions, as observed in Figure 3. The number of seeds is computed as follows [46]:

$$Number(of)seed(around)weed_i = \frac{F_i - F_{worst}}{F_{best} - F_{worst}}(Smin_{max} + S_{min}) \tag{21}$$

where, F_{worst} is the worst fitness function, F_{best} is the best fitness function, S_{min} is the minimum number of seeds, S_{max} is the maximum number of seeds, and F_i is ith fitness function. The distribution of seeds is random over the search space and is based on the standard deviation σ_i and zero mean. The standard deviation of the distribution of seeds varies as follows:

$$\sigma_{cur} = \frac{\left(iter_{max}()^n\right)}{\left(iter_{max}()^n\right)\left(\sigma_{init} - \sigma_{final}\right) + \sigma_{final}} \tag{22}$$

where, $iter_{max}$ is the maximum number of iterations, σ_{cur} is the standard deviation at the current iteration, σ_{final} is the final value of standard deviation, σ_{init} is the predefined initial value of standard deviation, and n is the nonlinear modulation index. Seeds are produced by each weed and then are distributed over the space. The competitive exclusion is the final level in the WA. If a weed does not generate seeds, it will be extinct. If all the weeds generate seeds, the number of weeds increases exponentially. Therefore, the number of seeds is limited to the maximum value (P_{max}). The weeds with better fitness function are allowed to reproduce. Weeds with worse fitness function are removed (see Figure 4).

1: Objective function f(x), x = (x1, x2, ..., xdim), dim= no. of dimensions
2: Generate initial population of n grasshoppers xi= (i=1, 2, ..., n)
3: Calculate fitness of each grasshopper
4: T = the best search agent
5: while stopping criteria not met do
6: Update c1 using equation (20)
7: for each grasshopper gh in population do
8: Normalize the distances between grasshoppers in [1,4]
9: Update the position of the grosshopers by Eq. (19)
10: If required, update bounds of gh
11: end for
12: If there is a better solution, update T
13: end while
14: Output the T.

Figure 3. The flowchart of grasshopper optimization algorithm (GOA) [30].

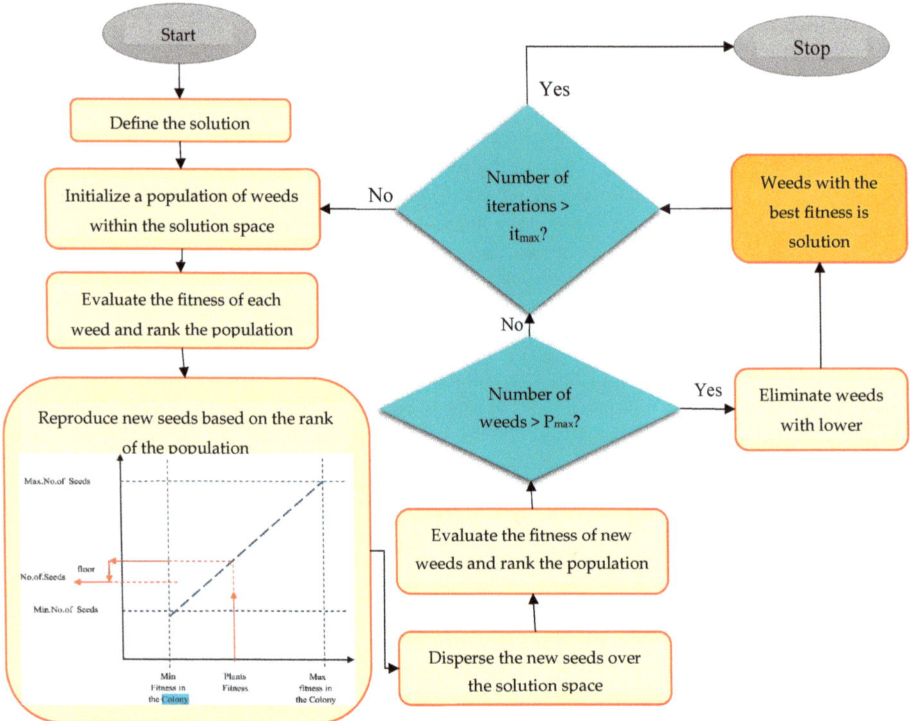

Figure 4. The flowchart of weed algorithm (WA) [46].

2.5.3. Cat Swarm Optimization (CSO)

Recently, CSO has gained popularity among other optimization algorithms because of its exploration ability and is widely used in different fields, such as wireless sensor networks [47], robotics [48], data clustering [49], and dynamic multi-objective algorithms [50]. Chu et al. (2006) introduced the cat swarm algorithm [51]. Figure 5 shows the flowchart of CSO. The CSO uses hunting and resting skills for optimization. First, the initial population of cats is initialized randomly. The seeking mode and tracing mode are two important operation modes in the CSO model. The seeking mode demonstrates the resting ability of cats which change their position and remain alert. This mode is regarded as a local search for the solutions. The seeking memory pool (SMP), the seeking range of selected dimension (SRD), and counts of dimension to change (CDS) affect the cat's behavior. The number of duplicate cats is denoted by SMP. CDC shows that the dimensions are to be mutated and SRD denotes change value of chosen dimensions. In the seeking mode, most of the cat's time is in the resting time, even though they remain alert [52]. The seeking mode includes the following levels:

- Generate replicas of the cats as per SMP.
- The position of each copy is updated as follows:

$$x_{k,d} = \left[\begin{array}{c} (1 + (2 \times rand - 1) * SRD) * x_{j,d} \leftarrow if(D) \in N \\ x_{j,d} \leftarrow otherwise \end{array} \right] \tag{23}$$

where, $x_{k,d}$ is the position of the kth cat in the dth dimension (new position of the cat), *rand* is the random number, N is the number of cats, D is the number of dimensions, and $x_{j,d}$ is the position of jth cat in the d dimension (old position of the cat).

- Compute the objective function for all copies and choose the best objective function value (x_{best}) of the cat.
- Substitute $x_{j,p}$ with the best cat if the x_{best} is better than $x_{j,p}$ in terms of the objective function value.

The hunting skill of cats is represented by the tracing mode. Cats trace the objectives with high energy by changing their locations with their own velocities. The velocity is updated as follows:

$$v_{j,d,new} = \omega \times v_{j,d} + r_1 \times c_1 \times \left(x_{best,d} - x_{j,d} \right) \tag{24}$$

where, ω is the inertia weight, c_1 is a constant, and $v_{j,d}$ is the velocity of jth cat in the d dimension, and $v_{j,d,new}$ is the new velocity of the jth cat. The position of cats in the tracing mode is updated as follows:

$$x_{j,d} = x_{j,d} + v_{jd} \tag{25}$$

where, $x_{j,d,new}$ is the jth position of the kth cat in the dth dimension (new position of the cat).

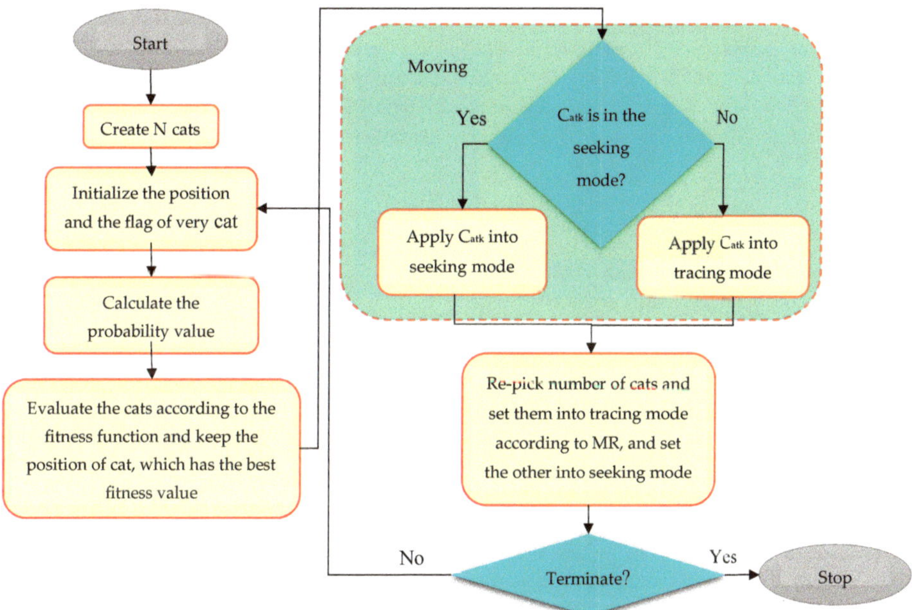

Figure 5. Flowchart of cat swarm optimization (CSO) for the optimization problems [49].

2.5.4. Particle Swarm Optimization (PSO)

In PSO, a set of particles that are generated randomly search the best adjacent solutions for optimization. The updating equations for the new position and velocity of particles are written as [53]:

$$x_{id}(t+1) = x_{id}(t) + v_{id}(t+1) \tag{26}$$

$$v_{id}(t+1) = \psi * v_{id}(t) + r_1 * c_1 * [p_{id}(t) - x_{id}(t)] + r_2 * c_2 * [g_d(t) - x_{id}(t)] \tag{27}$$

where, d is the number of dominions; ψ is the inertia weight; r_1 and r_2 are the random values; c_1 and c_2 are the acceleration coefficients; $g_d(t)$ is the global best position obtained by neighbors; and p_{id} is the personal best position.

The particles find the solutions of optimization problems by adjusting the position and velocity of particles. The main advantages of PSO are easy implementation and computational efficiency.

2.5.5. Genetic Algorithm (GA)

Genetic algorithm is one of the most popular algorithms that is extensively applied for optimization problems. Each chromosome in GA is a candidate solution [19]. The genes of chromosomes simulate the variables of optimization. First, the initial population of chromosomes is randomly initialized for optimization and the selection operator is used to select the best chromosomes for the production of the next generation. The chromosomes with better fitness values have a great chance of being chosen by the selection operator. The crossover operator is used to exchange genes between two chromosomes for producing new solutions. Finally, the mutation operator is used to cause changes in the genes. The mutation operator is applied to the chromosomes of new genes to generate different solutions with new genes. If the convergence criteria are satisfied, the algorithm stops; otherwise, the algorithm runs again. The drawback of GA shows that GA requires a high number of iterations [20].

2.5.6. Krill Herd Algorithm (KHA)

Gandomi and Alavi [54] introduced the KHA using the krill's behavior in nature [54]. The KHA is widely used in different fields, such as text document clustering analysis [55] and structural seismic reliability [56]. The KHA acts, based on three main concepts: (1) mutation-induced, (2) foraging mutation, and (3) physical diffusion. The following formulation uses the three behaviors mentioned above [54]:

$$\frac{dY_i}{dt} = N_i + F_i + D_i \tag{28}$$

where, Y is the location of the ith krill, N_i is the motion induced by another krill, F_i is the foraging motion, and D_i is the physical diffusion of the ith krill. Equation (28) describes the motion-induced by another individual krill.

$$N_{new,i} = N\left(\alpha_{local,i} + \alpha_{t\,arg\,et,i}\right)_{n\,old,i\,max} \tag{29}$$

where, N_{max} is the maximum induced speed, $\alpha_{local,i}$ is the neighbor's local effect, $\alpha_{t\,arg\,et,i}$ is the krill's target direction, ω_n is the inertia weight of induced motion, and $N_{old,i}$ is the old motion-induced for the ith individual krill. The foraging motion can be formulated as:

$$F_i = V_f\left(\beta_{food,i} + \beta_{best,i}\right) + \omega_f F_{old,i} \tag{30}$$

where, V_f is the foraging speed, $\beta_{food,i}$ is the food attractive, $\beta_{best,i}$ is the effect of the best fitness of the ith krill, and $F_{old,i}$ is the last foraging motion. The diffusion can be computed as:

$$D_i = D_{max} \tag{31}$$

where, D_{max} is the maximum diffusion speed, and δ is the random direction.

Finally, the position of a krill is computed as follows:

$$X_{new,i} = X_{old,i} + \Delta t \frac{dX_i}{dt} \tag{32}$$

where, $X_{new,i}$ is the value of the next individual krill location, and $X_{old,i}$ represents the current position of solution number I, and Δt is the essential constant. Figure 5 shows the flowchart of the krill algorithm.

2.6. Principal Component Analysis (PCA)

PCA is a statistical orthogonal transformation to obtain a set of values of linearly uncorrelated (principal components) from a set of observations. When the user has the number of inputs but he cannot identify the appropriate inputs, the PCA is used to reduce the number of inputs. The final data set should be able to demonstrate most of the variance of the original input data by creating a variable reduction [57]. PCA can be explained, based on the following equation [57]:

$$Z_i = a_{i1} + a_{i2} + \ldots + a_{ip}x_p \tag{33}$$

where, Z_i shows the principal component, a_{ip} is the related eigenvector, and x_i is the input variable. The information is obtained by solving Equation (34):

$$|R - \lambda I| = 0 \tag{34}$$

where, R is the variance-covariance matrix, I is the unit matrix, and λ is the eigenvalues.

2.7. Taguchi Model

The random parameters of optimization algorithms are the most important parameters affecting the outputs of the optimization algorithms. Thus, determining the appropriate values of random parameters is necessary to construct the optimization models. The Taguchi model is widely used to design different parameters of different experiments or experimental models. First, the initial level is determined for each of the random parameters in the optimization algorithms. In the Taguchi method, parameters are classified into two groups: (1) controllable, and (2) uncontrollable (noise). In the Taguchi model, each parameter combination that has a higher S (signal)/N (noise) ratio is regarded as the best combination [58].

$$S/N = -10\,log\left(\frac{1}{n}\sum_{i=1}^{n} Y_i^2\right) \tag{35}$$

where, n is the number of data, and Y_i is the fitness function that is obtained by the Taguchi model. For example, consider the PSO algorithm with four parameters and three levels. When the population size is at level 1, the acceleration coefficient is tested at levels 1, 2, 3, and 4. Similarly, the inertia coefficient is tested at levels 1, 2, 3, and 4.

2.8. Hybrid ANN, ANFIS, and SVM Models with Optimization Algorithms

The optimization algorithms can be used as a robust training algorithm for the ANN models. The process starts with the initialization of a group of random agents (particles, chromosomes, krill, grasshoppers, weeds, or cats). The position of agents represents the ANN weights and biases. Following this level, using the initial biases and weights (i.e., the initial position of agents), the hybrid ANN-optimization algorithms are trained, and the error between the observed and estimated value is calculated. At each iteration, the calculated error is decreased by the updating of agent locations.

The model procedure in ANFIS-optimization algorithm models starts with the initialization of a set of agents (particles, chromosomes, krill, grasshoppers, weeds, or cats) and continues with the random choice of agents and finally adjusts a location for each agent. First, the ANFIS model is trained. Then, the consequent and premise parameters are optimized by the optimization algorithms. The root mean square error (RMSE) is defined as an objective function. The aim of optimization algorithms is to minimize the objective function value with finding the appropriate values of consequent and premise parameters.

In SVM, the C parameter and kernel function parameters have significant effects on the accuracy of the SVM. The random population of agents (particles, chromosomes, krill, grasshoppers, weeds, or cats) are initialized for training the SVM parameters. The RMSE is defined as an objective function. The aim of hybrid SVM-optimization algorithm models is to minimize model errors. Figure 2 shows the developed framework of hybrid ANN, ANFIS, and SVM-optimization models for modeling groundwater level.

Thus, the model parameters are considered as decision variables for optimization algorithms. The optimization algorithms aim to minimize the error function to find the optimal value of model parameters. The PCA selects the appropriate input combinations. Then the hybrid and standalone models uses the input combinations to forecast GEWL. The models uses the optimized model parameters to accurately forecast monthly GWL.

2.9. Uncertainty Analysis of Soft Computing Models

The input data and the inability of model structure are the sources of uncertainty. In this research, an integrated framework is developed to simultaneously evaluate the input data and model structure.

Input Data Uncertainty

The combined Bayesian uncertainty was used to compute the uncertainty contributed by input data. The input error model was used to account for the uncertainty of input data [59]:

$$H_{a,t} = KH_t, K \sim N\left(m, \sigma_m^2\right) \tag{36}$$

where, $H_{a,t}$: the adjusted groundwater level (GWL), H_t: the observed GWL, t: the given month, K: the normally distributed random, m: mean, and σ_m: variance. For each soft computing model, m and σ_m were added to the system. A dynamically dimensioned search was used to find the value of m: mean and σ_m: variance as defined by [59].

Mode Structure Uncertainty

Bayesian model average (BMA) is used for model uncertainty. The posterior model probability and averaging over the best models were used to estimate the uncertainty of the models. The weighted average prediction of quantity of target variable is computed as follows [59]:

$$H_j = \sum_{k=1}^{k} \beta_k F_{jk} + e_j \tag{37}$$

where, F_j: the point prediction of each model, e_j: noise, β_k: the weight vector of model, H: n observation of GWL, k: number of models, and j: number of observations. For accurate application of BMA model, the standard deviation of normal probability distribution functions and weights should be estimated accurately. The log-likelihood function is used to calculate the weights and standard deviation as follows [59]:

$$L(\beta_{BMA}, \sigma_{BMA}|F, H) = \sum_{i=1}^{n} \log\left\{\sum_{k=1}^{k} \beta_k \frac{1}{\sqrt{2\pi\sigma_k^2}} \exp\left[-\frac{1}{2}\sigma_k^{-2}\left(H_j - F_{jk}\right)^2\right]\right\} \tag{38}$$

where, β_{BMA}: maximum likelihood Bayesian weight. Markov Chain Monte Carlo (MCMC) simulations are used to compute the log-likelihood function. The integrated framework is defined as follows:

1. A number of models are selected to simulate the GWL.
2. The prior probability is assigned to each model.
3. An error input model is defined.
4. The posterior distribution of input error models and model parameters are obtained.
5. A predetermined number of GWLs for each model is provided using probabilistic parameter estimations obtained from level 2 to level 4.
6. The variance and weight of models are estimated.
7. The weights for ensemble members of models are summed to compute the weight models.
8. To the experimental soft computing models. The following indices were used to quantify the uncertainty of models:

$$p = \frac{1}{n}count[H|X_L \leq H \leq X_U] \tag{39}$$

$$d = \frac{d_x}{\sigma_x}$$

$$d_x = \frac{1}{k} \sum_{l=1}^{k} (X_U - X_L) \tag{40}$$

9. where k is the number of observed data, X_U is the upper bound of data, X_L is the lower bound of data, σ_x is standard deviation, p is bracketed by 95% of predicted uncertainties, d is the distance between the upper and lower bounds, and d_x is the average distance between the upper and lower bounds [59,60].

2.10. Statistical Indices for Evaluation of Different Models

In this study, the following indices were used to evaluate the performance of models:
Root mean square error:

$$RMSE = \sqrt{\frac{1}{N} \sum_{t=1}^{n} ((H_0(t)) - (H_s(t)))^2} \tag{41}$$

Mean absolute error:

$$MAE = \frac{1}{N} \sum_{t=1}^{1} \left| H_0(t) - H_s(t) \right|^2 \tag{42}$$

Nash Sutcliffe efficiency:

$$NSE = 1 - \frac{\sum_{i=1}^{n} \left| H_s(t) - H_0(t) \right|^2}{\sum_{i=1}^{n} \left| H_s - H_0(t) \right|^2} \tag{43}$$

Percent bias (PBIAS):

$$PBIAS = \left[\frac{\sum_{i=1}^{n} (H_s(t) - H_0(t))^2}{\sum_{i=1}^{n} (H_o^t)^2} \right] \tag{44}$$

where, N is the number of data, H_0 is the observed value, and P_s is the predicted value.

RMSE and MAE show a good match between observed data and estimated values when it equals 0. The NSE shows a good match between the observed values and estimated values when it equals 1. The best value of PBIAS is zero.

3. Results and Discussion

3.1. Inputs Selection by PCA

In this study, 12 input variables (H(t-1), ..., H(t-12)) were considered to select the input lag times of monthly GWL. As presented in the flowchart and framework of the current study in Figure 1, the first step of the model developments is the appropriate selection of time lags for GWL modelling by PCA analysis. Table 1 shows the variance contribution rate for PCAs as the principal component loadings. There are the loadings of 12 principal components versus 12 input lag times of GWL. The first four PCs variance summed up a contribution of 91%, among which the first PC variance had a contribution of 48% loadings. It was observed that the inputs H(t-1), H (t-2), H (t-3), H (t-4), and H (t-5) had higher factor loading in comparison with other inputs of the PCs. Thus, the first four PCs were selected for the hybrid soft computing models which included inputs H(t-1), H (t-2), H (t-3), H (t-4), and H (t-5) because of their higher loading factor. This loading analysis of variables reduced the raw initial input parameter numbers from 12 to 5, that decrease the model development efforts. The coefficients of more 0.75 are significant for Eigen value verifications [60].

Table 1. Principal component loadings.

PC	1	2	3	4	5	6	7	8	9	10	11	12
H (t-1)	**0.98**	**0.95**	**0.93**	**0.90**	0.89	0.88	0.88	0.86	0.75	0.62	0.52	0.45
H (t-2)	**0.84**	**0.82**	**0.88**	**0.86**	0.85	0.84	0.85	0.82	0.62	0.60	0.51	0.44
H (t-3)	**0.83**	**0.81**	**0.80**	**0.77**	0.74	0.72	0.84	0.80	0.61	0.55	0.43	0.40
H (t-4)	**0.82**	**0.80**	**0.78**	**0.76**	0.75	074	0.83	0.78	0.60	0.54	0.39	0.37
H (t-5)	**0.81**	**0.79**	**0.76**	**0.75**	0.72	0.71	0.82	0.79	0.55	0.51	0.38	0.35
H (t-6)	0.73	0.67	0.74	0.73	0.71	0.70	0.80	0.77	0.54	0.50	0.37	0.34
H (t-7)	0.62	0.55	0.72	0.70	0.65	0.64	0.76	0.75	0.53	0.47	0.33	0.30
H (t-8)	0.61	0.50	0.71	0.69	0.54	0.52	0.65	0.64	0.51	0.46	0.30	0.29
H (t-9)	0.54	0.54	0.70	0.65	0.42	0.64	0.54	0.52	0.50	0.45	0.29	0.25
H (t-10)	0.42	0.42	0.69	0.66	0.41	0.62	0.45	0.44	0.49	0.42	0.28	0.26
H (t-11)	0.42	0.42	0.55	0.54	0.40	0.55	0.42	0.40	0.47	0.41	0.27	0.24
H (t-12)	0.40	0.40	0.45	0.43	0.38	0.52	0.40	0.38	0.46	0.38	0.25	0.23
Eigen value	5.78	3.22	1.12	0.90	0.6	0.27	0.05	0.03	0.02	0.003	0.003	0.03
Cumulative variance	48%	74%	84%	91%	96%	99	99.5	99.7	99.99	99.99	99.99	100%

3.2. Selection of Random Parameters by the Taguchi Model

The Taguchi model was used to find the value of random parameters rather than the classical trial and error methods. Table 2 shows the computed signal-to-noise (S/N) ratio for each random parameter in the optimization module of the hybrid training of ANFIS. Each parameter had four levels and the best level of each parameter is selected based on the S/N values. The S/N ratio was computed for each level of parameters. The best value of parameters had the highest S/N rate. For example, sensitivity analysis for different values of GOA parameters was done, as shown in Table 2. The results indicated that the population size = 300 had the highest value of S/N. Thus, the optimal size of population was 300. The maximum S/N ratio for parameter l was 1.23. Thus, the optimal value of parameter l was 1.5. The maximum S/N ratio for parameter f was 1.14. Thus, the optimal value of parameter f was 0.5.

Table 2. Results of Taguchi model for a: GOA, b: particle swarm optimization (PSO), c: genetic algorithm (GA), d: WA, e: CSO, and f: krill algorithm.

(a)						
Population size	S/N	l	S/N	f	S/N	
100	1.05	0.5	1.07	0.1	1.09	
200	1.15	1	1.19	0.3	1.12	
300	**1.20**	**1.5**	**1.23**	**0.5**	**1.14**	
400	1.02	2	1.18	0.7	1.10	

(b)							
Population size	S/N	c_1	S/N	c_2	S/N	ω	S/N
100	1.25	1.6	1.20	1.6	1.21	0.3	1.19
200	1.29	1.8	1.27	1.8	1.25	0.50	1.18
300	1.23	2.0	1.26	2.0	1.23	0.70	1.17
400	1.20	2.2	1.22	2.2	1.25	0.90	1.24

Table 2. *Cont.*

(c)					
Population size	**S/N**	**Mutation probability**	**S/N**	**Crossover rate**	**S/N**
100	1.18	0.01	1.16	1.6	1.21
200	1.20	0.03	1.17	1.8	1.25
300	1.21	0.05	1.20	2.0	1.23
400	1.17	0.07	1.19	2.2	1.25

(d)			
P_{max}	**S/N**	**n**	**S/N**
50	1.12	1	1.14
100	1.23	2	1.17
150	1.19	3	1.18
200	1.17	4	1.19

(e)					
Population size	**S/N**	**SMP**	**S/N**	**MR**	**S/N**
100	1.11	5	1.10	0.10	1.12
200	1.24	10	1.15	0.30	1.16
300	1.17	15	1.17	0.50	1.18
400	1.15	20	1.21	0.70	1.20

(f)					
Population size	**S/N**	V_f	**S/N**	N_{max}	**S/N**
100	1.10	0.005	1.12	0.02	1.14
200	1.12	0.010	1.15	0.04	1.17
300	1.14	0.015	1.17	0.06	1.12
400	1.16	0.020	1.14	0.08	1.21

3.3. Results of Hybrid ANN, ANFIS, and SVM Models

In this section, the results of developed hybrid models are presented and compared with each other and with the usual ANFIS, ANN, and SVM models. These models are hybridized with GOA, CSO, KA, WA, PSO, and GA meta-heuristic optimization algorithms. The results of models in three piezometers of 6, 9, and 10 as shown in Figure 6, are presented and discussed. These piezometers were selected as samples to evaluate the ability of new hybrid models.

Begin

Step 1: Initialization. Set the generation counter G=1; initialize the population P of NP krill

individuals randomly; set the foraging speed Vf, the maximum diffusion speed Dmax,

and the maximum induced speed Nmax.

Step 2: While the termination criteria is not satisfied or G<Gmax do

Sort the population/krill from best to worst.

for i=1:NP (all krill) do

Perform the following motion calculation.

Motion induced by the presence of other individuals

Foraging motion

Physical diffusion

Implement the genetic operators.

Update the krill individual position in the search space.

Evaluate each krill individual according to its position.

end for i

Sort the population/krill from best to worst and find the current best.

G=G+1.

Figure 6. The flowchart of the krill algorithm [54].

- piezometer 6

Table 3 and Figure 7a show the results of hybrid optimized and standalone soft computing models for piezometer 6. Results indicated that ANFIS-GOA was the most accurate model and is selected as the optimum model that was verified by a value of RMSE = 1.12 m, MAE = 0.812 m, NSE = 0.95, and PBIAS = 0.12 for the training level. For the testing phase assessed with the ANFIS-GOA, results indicated a value of RMSE: 1.21 m, MAE: 0.878 m, NSE: 0.93, and PBIAS: 0.15 which reflected better performance in comparison to other models. From Table 3, results indicated that the SVM model with the higher values of RMSE, MAE, and PBIAS and lower values of NSE was the worst model among other models. Among the hybrid ANN models, the ANN-GOA outperformed the ANN-CSO, ANN-GA, ANN-PSO, ANN-WA, and ANN-KA models with the best values for RMSE = 1.21 m, MAE = 0.878 m, NSE = 0.93, PBIAS = 0.15 in the test stage. The ability of GA was lower than that of CSO, PSO, WA, and KA because of higher values of RMSE, MAE, and PBIAS and lower values of NSE in train and test steps as presented in Table 3. Among SVM models, the hybrid SVM-GOA was observed to have the lowest value of NSE and the highest values of RMSE, MAE, and PBIAS. It was important to mention that the standalone SVM, ANN, and ANFIS had worse performance than hybrid ANN, SVM, and ANFIS models that indicates the superiority of hybridization in model developments. Among PSO, CSO, GA, KA, and WA, the CSO had better results than the other optimization algorithms. The general results showed that ANFSI model was superior to the SVM and ANN models. Additionally, the ANN model had lower values of RMSE and MAE than did the SVM model. Additionally, the results of ANFIS-GOA as the best model in piezometer 6 in comparison with standalone ANFIS shows that meta-heuristic hybridizations improved the model performances in train and test steps. The percent of RMSE, MAE, NSE, and PBIAS improvements by ANFIS-GOA in train step were 15%, 4%, 13%, and 208% and these values for the test steps of ANFIS-GOA are 33%, 44.6%, 16.3%, and 173%, respectively, that clearly confirm the superiority of developed hybridization schemes in GWL modelling. Additionally, in Figure 7a, the scatter plots of training and testing steps visualize the performance of ANFIS-GOA compared to the other models. Furthermore, simulations coincide very well with the observed values and all of the data points concentrated over the y = x line with R^2 = 0.93. Furthermore, this figure shows that other hybridized models such as CSO, PSO, KA, WA, GA, and standalone ANFIS have less accuracy in high and low values of GWL, while the ANFIS-GOA over all of low to high values of GWL performed accurately in regard to the observations.

Table 3. Statistical characteristics of applied hybrid models for piezometer 6.

Model	Training					Testing				
	RMSE	MAE	NSE	PBIAS	R^2	RMSE	MAE	NSE	PBIAS	R^2
ANFIS-GOA	**1.12**	**0.812**	**0.95**	**0.12**	**0.95**	**1.21**	**0.878**	**0.93**	**0.15**	**0.93**
ANN-GOA	1.24	0.815	0.92	0.14	**0.94**	1.25	0.897	0.91	0.16	0.92
SVM-GOA	1.25	0.817	0.91	0.17	**0.91**	1.29	0.901	0.90	0.18	0.90
ANFIS-CSO	1.14	0.819	0.94	0.15	**0.94**	1.30	0.899	0.92	0.17	0.92
ANN-CSO	1.28	0.821	0.93	0.18	**0.93**	1.34	0.935	0.90	0.19	0.90
SVM-CSO	1.32	0.823	0.90	0.20	**0.89**	1.38	0.939	0.89	0.22	0.87
ANFIS-KA	1.19	0.825	0.93	0.16	**0.93**	1.41	1.01	0.91	0.24	0.90
ANN-KA	1.30	0.829	0.91	0.22	**0.90**	1.42	1.09	0.89	0.25	0.88
SVM-KA	1.33	0.832	0.89	0.24	**0.88**	1.43	1.12	0.87	0.26	0.85
ANFIS-WA	1.21	0.827	0.92	0.27	**0.92**	1.45	1.10	0.89	0.28	0.93
ANN-WA	1.32	0.832	0.90	0.29	**0.90**	1.47	1.14	0.86	0.31	0.87
SVM-WA	1.35	0.833	0.88	0.33	**0.84**	1.51	1.16	0.85	0.35	0.83
ANFIS-PSO	1.24	0.829	0.88	0.35	**0.90**	1.53	1.12	0.84	0.37	0.89
ANN-PSO	1.35	0.835	0.87	0.37	**0.89**	1.55	1.17	0.85	0.39	0.86
SVM-PSO	1.37	0.839	0.86	0.39	**0.83**	1.52	1.19	0.83	0.43	0.82
ANFIS-GA	1.28	0.835	0.87	0.35	**0.88**	1.59	1.21	0.82	0.37	0.87
ANN-GA	1.32	0.839	0.85	0.39	**0.87**	1.62	1.23	0.81	0.40	0.84
SVM-GA	1.35	0.842	0.83	0.41	**0.82**	1.71	1.25	0.80	0.42	0.81
ANFIS	1.30	0.844	0.84	0.37	**0.85**	1.61	1.27	0.80	0.41	0.83
ANN	1.38	0.849	0.82	0.43	**0.87**	1.73	1.29	0.78	0.45	0.84
SVM	1.40	0.851	0.81	0.45	**0.80**	1.75	1.32	0.77	0.47	0.79

- piezometer 9

Results of hybrid models for piezometer 9 in Figure 7b and Table 4 indicated that the hybrid ANN, ANFIS, and SVM models had better performance than the standalone ANN, SVM, and ANFIS models, the same as the results for piezometer 6 in the previous subsection. Among ANFIS hybrid models, the hybrid ANFIS-GOA was confirmed to have the best performance with the smallest values of RMSE = 1.16 m, MAE = 0.818 m, and PBIAS = 0.14 and the highest values of NSE = 0.94 in the training stage and in testing stage these values were 1.22 m, 0.881 m, 0.17, and 0.92 respectively. The ANFIS model provided the best RMSE, PBIAS, MAE, and NSE among other models. The best values of RMSE, MAE, PBIAS, and NSE for ANN-GOA in the training phase were 1.25 m, 0.819 m, 0.91, and 0.19, respectively. Results indicated that the SVM model had the worst performance among other models. For the testing phase assessed with SVM-GOA, the results indicated a value of RMSE: 1.31 m, MAE: 0.903 m, NSE: 0.89, and PBIAS: 0.20 which reflected better performance than the SVM model and indicates the improvements when SVM is hybridized with the GOA. Results of Table 4 indicated that GOA and GA were the best and worst algorithms among other algorithms. As observed in Table 4 and Figure 7b, the evolutionary ANN models had more accuracy than the evolutionary SVM model because of lower values of RMSE, MAE, and PBIAS and higher values of NSE. However, no major differences were observed in GWL predictions of piezometers 6 and 9 predictions by ANFIS-GOA.

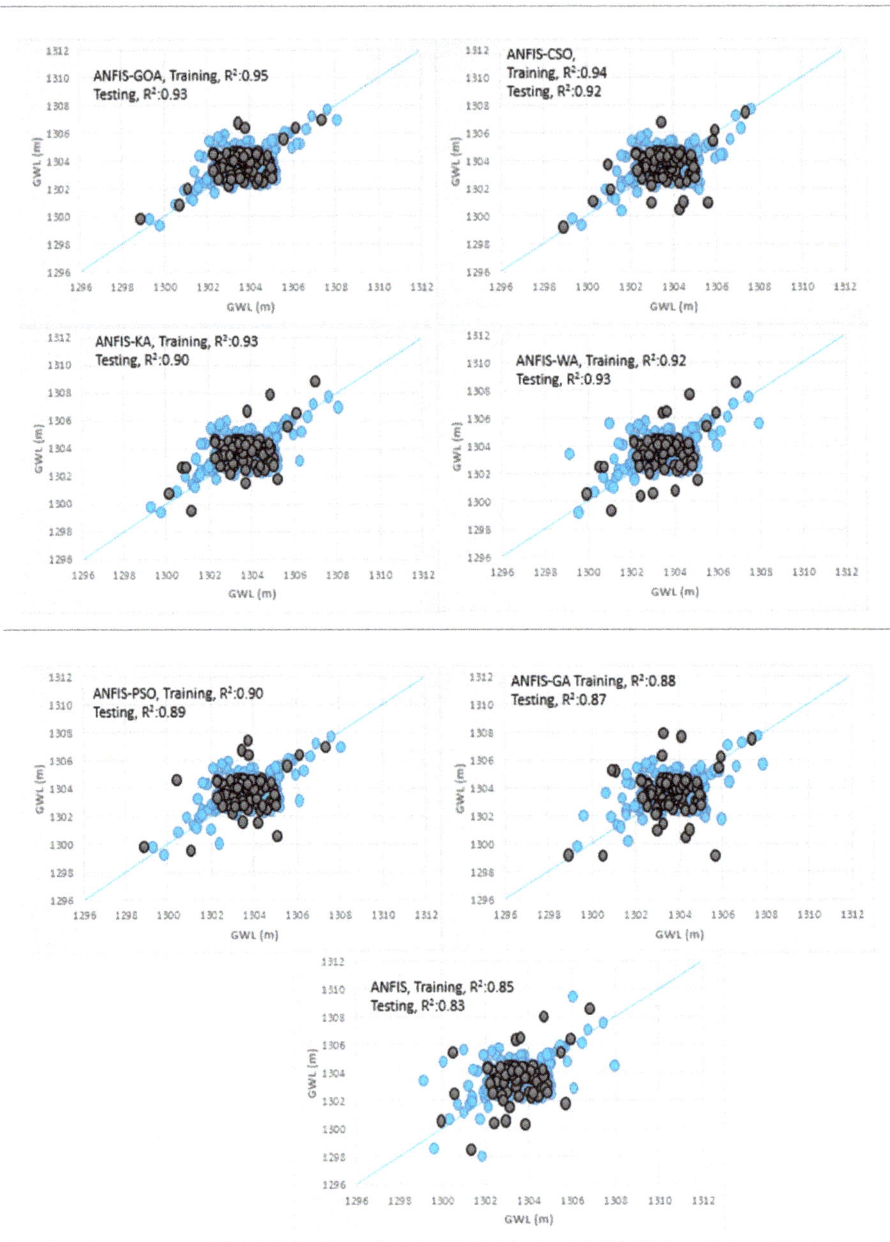

Figure 7. *Cont.*

Figure 7. *Cont.*

(a)

Figure 7. *Cont.*

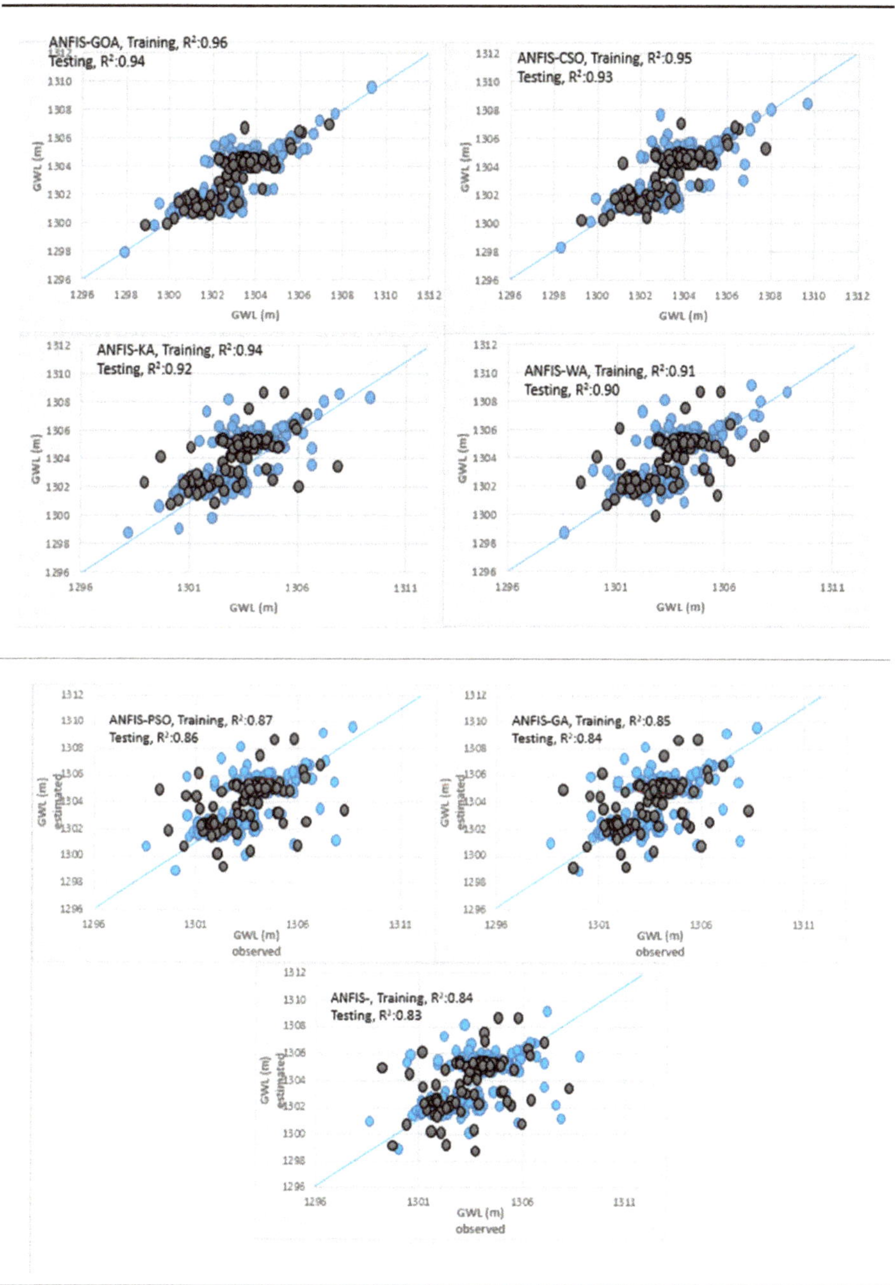

Figure 7. *Cont.*

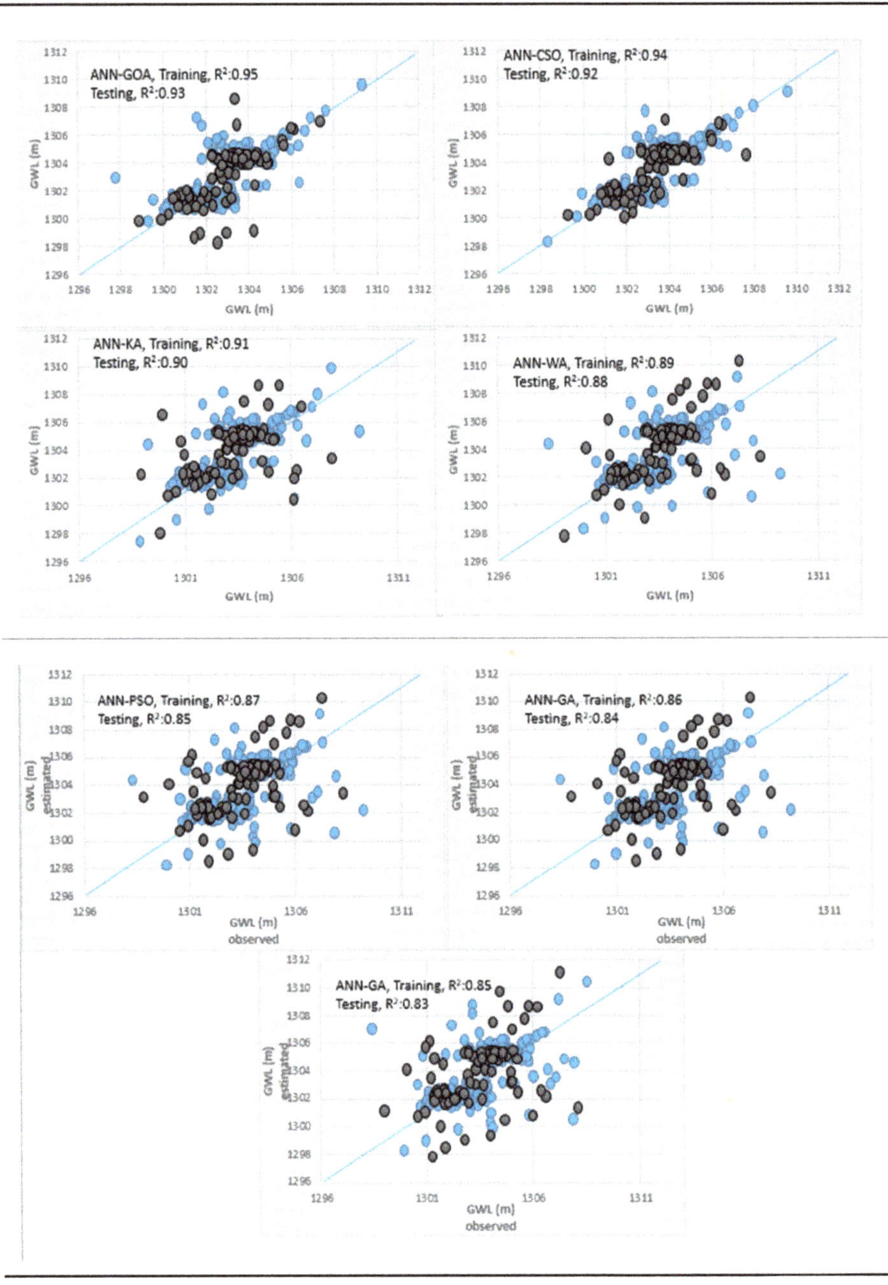

Figure 7. *Cont.*

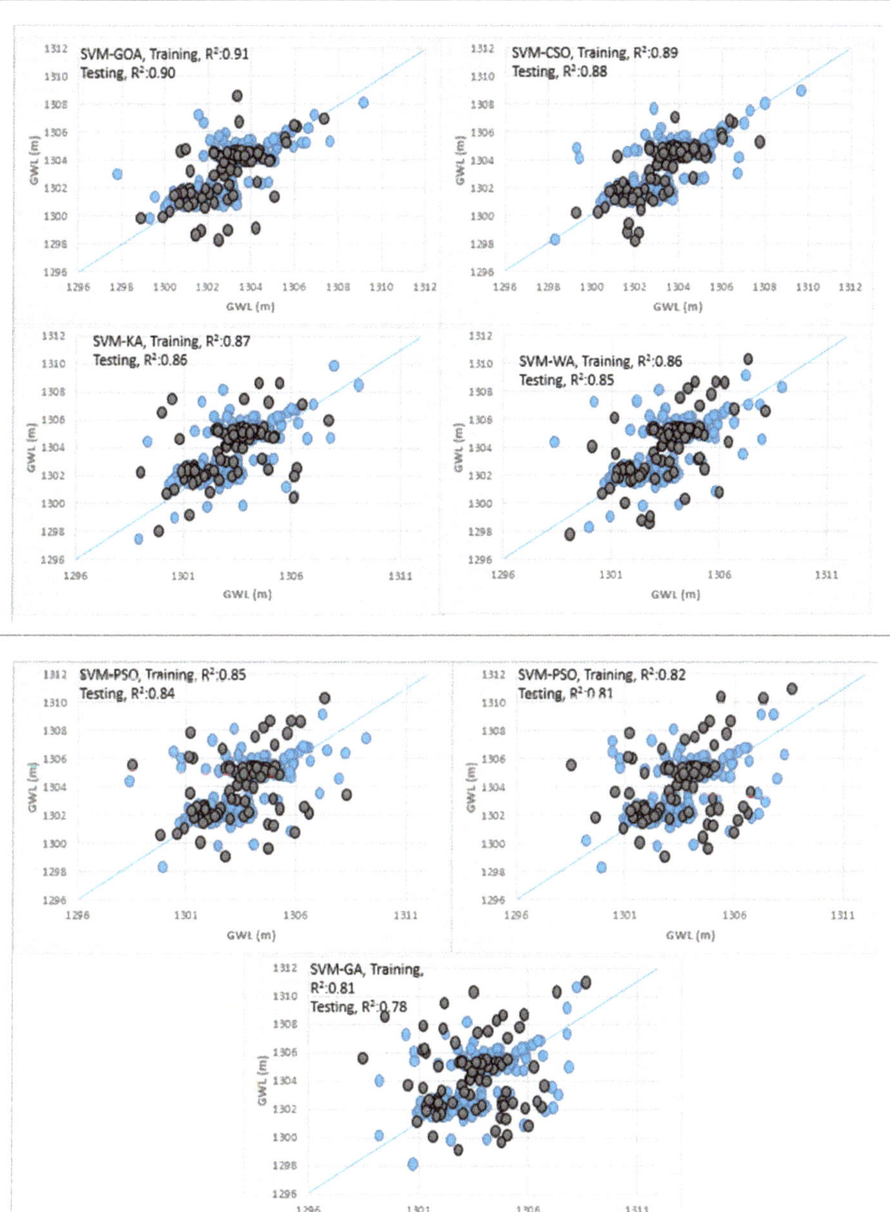

(b)

Figure 7. *Cont.*

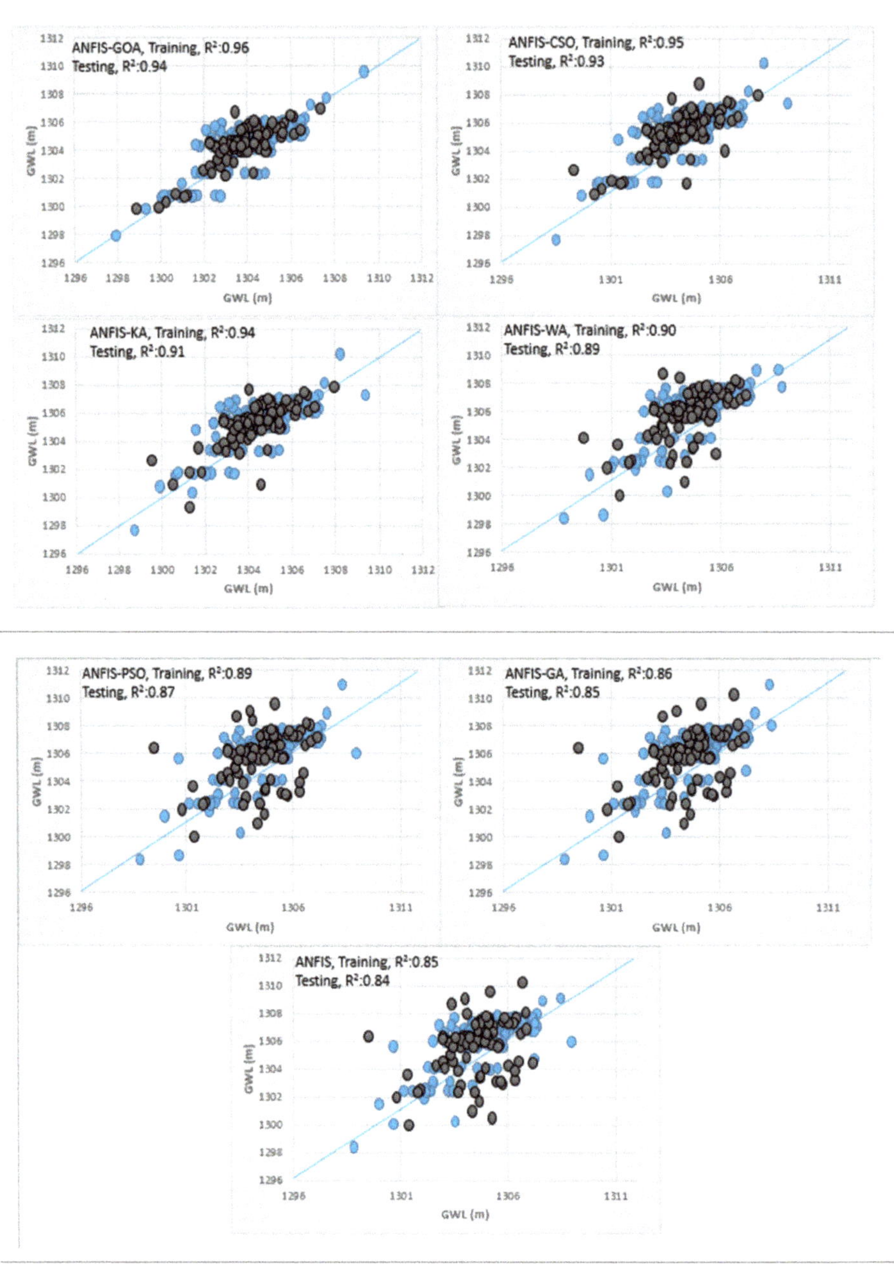

Figure 7. *Cont.*

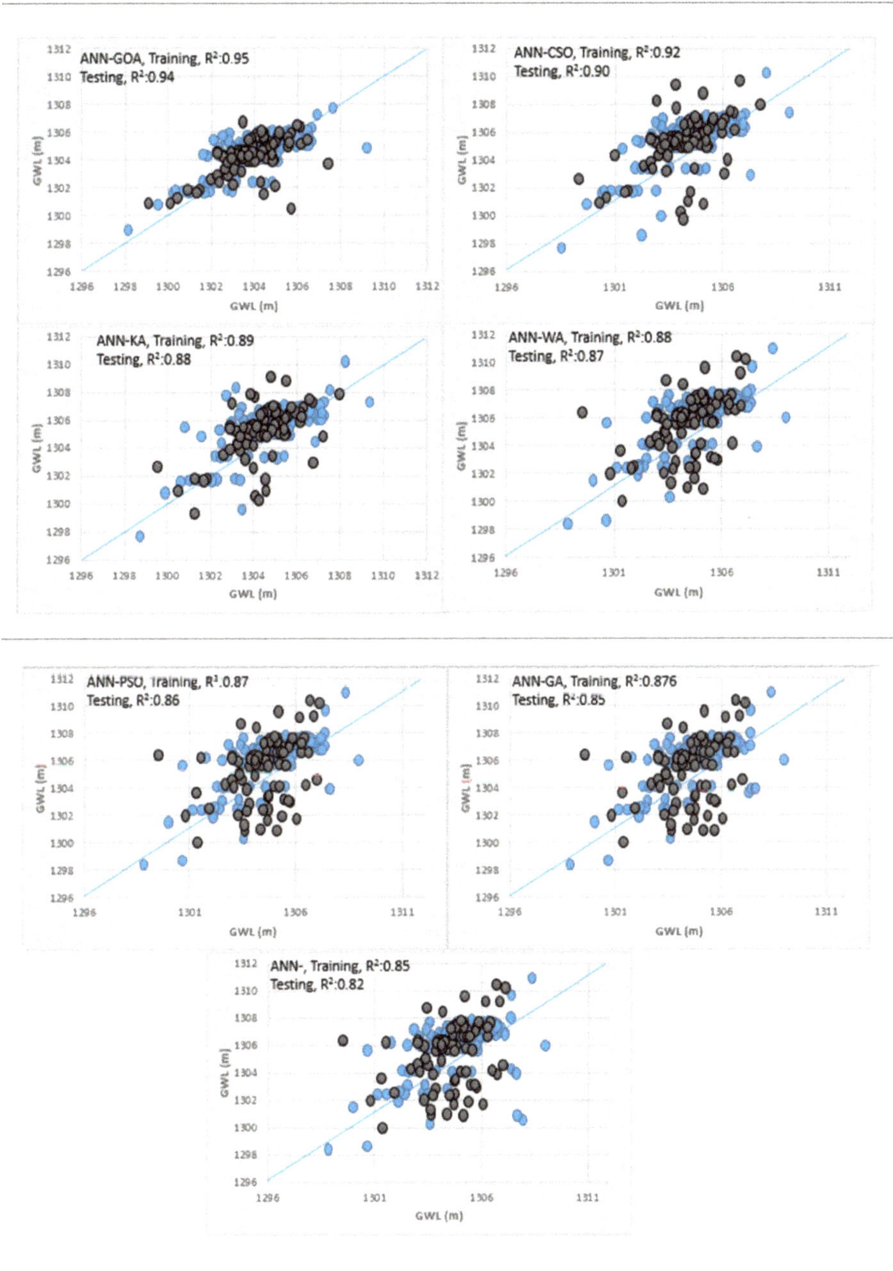

Figure 7. *Cont.*

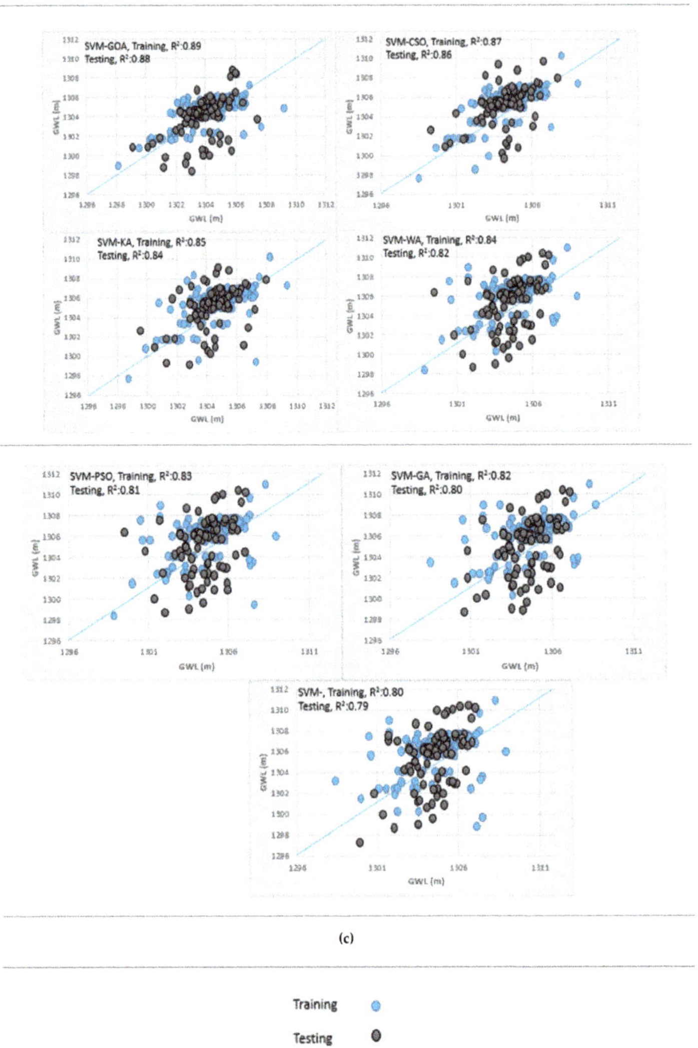

Figure 7. The scatter plots of exanimated soft computing models for predicting groundwater level (GWL), (**a**) piezometer 6, (**b**) piezometer 9, and (**c**) piezometer 10.

Table 4. Statistical characteristics of applied hybrid models for piezometer 9.

Model	Training					Testing				
	RMSE	MAE	NSE	PBIAS	R^2	RMSE	MAE	NSE	PBIAS	R^2
ANFIS-GOA	**1.16**	**0.818**	**0.94**	**0.14**	**0.96**	**1.22**	**0.881**	**0.92**	**0.17**	**0.94**
ANN-GOA	1.25	0.819	0.91	0.15	**0.95**	1.27	0.899	0.90	0.18	0.93
SVM-GOA	1.27	0.821	0.90	0.19	**0.91**	1.31	0.903	0.89	0.20	0.90
ANFIS-CSO	1.18	0.820	0.93	0.16	**0.95**	1.32	0.901	0.91	0.19	0.93
ANN-CSO	1.29	0.823	0.92	0.19	**0.94**	1.36	0.938	0.88	0.18	0.92
SVM-CSO	1.33	0.825	0.91	0.22	**0.89**	1.39	0.940	0.87	0.20	0.88
ANFIS-KA	1.20	0.827	0.92	0.18	**0.94**	1.34	1.05	0.90	0.22	0.92
ANN-KA	1.31	0.831	0.90	0.23	**0.91**	1.44	1.10	0.86	0.23	0.90
SVM-KA	1.35	0.833	0.88	0.25	**0.87**	1.45	1.14	0.85	0.27	0.86
ANFIS-WA	1.22	0.829	0.91	0.28	**0.91**	1.49	1.12	0.83	0.29	0.90
ANN-WA	1.36	0.834	0.89	0.30	**0.89**	1.51	1.15	0.82	0.32	0.88
SVM-WA	1.38	0.835	0.87	0.34	**0.86**	1.53	1.17	0.83	0.37	0.85
ANFIS-PSO	1.27	0.831	0.86	0.36	**0.87**	1.55	1.19	0.81	0.39	0.86
ANN-PSO	1.39	0.837	0.85	0.38	**0.87**	1.57	1.23	0.80	0.40	0.85
SVM-PSO	1.40	0.840	0.84	0.40	**0.85**	1.59	1.25	0.83	0.45	0.84
ANFIS-GA	1.29	0.839	0.83	0.39	**0.85**	1.61	1.28	0.80	0.39	0.84
ANN-GA	1.42	0.840	0.82	0.40	**0.86**	1.63	1.29	0.79	0.42	0.84
SVM-GA	1.43	0.843	0.81	0.42	**0.82**	1.69	1.32	0.77	0.43	0.81
ANFIS	1.33	0.845	0.82	0.39	**0.84**	1.71	1.39	0.79	0.42	0.83
ANN	1.44	0.851	0.80	0.44	**0.85**	1.76	1.40	0.77	0.47	0.83
SVM	1.45	0.852	0.79	0.47	**0.81**	1.77	1.43	0.76	0.49	0.78

- piezometer 10

Here the results of models in piezometer 10 are evaluated. As observed in Table 5, results indicated that the ANFIS-GOA was better in terms of minimizing RMSE, MAE, and PBIAS than the other models. ANFIS-GOA reduced RMSE error by 7.01% and 7.04% compared to ANN-GOA and SVM-GOA, respectively. The standalone ANFIS, ANN, and SVM models provided worse results than the hybrid models. The SVM model provided the worst performance among other models. The NSE of ANFIS-GOA, ANFIS-CSO, ANFIS-KA, ANFIS-WA, ANFIS-PSO, and ANFIS-GA was 0.91, 0.90, 0.89, 0.79, and 0.75, respectively. GA had the worst performance among other algorithms. As is shown in Table 5, the error in the estimated GWL by using GA was more than that of PSO, KA, WA, GA, CSO, and GOA. Overall, the percent of improvements in the ANFIS-GOA versus standalone ANFIS in piezometer 6 were 14.4%, 3%, 17.8%, and 181% for RMSE, MAE, NSE, and PBIAS in training stage and 40.7%, 55%, 25%, and 132% in testing stage, respectively. These values again confirm that all of the hybridized models performed more accurately than the stand-alone models and indicate the generality of hybridizing Taguchi with training procedure compared to the classical standalone models.

Table 5. Statistical characteristics of applied hybrid models for piezometer 10.

Model	Training					Testing				
	RMSE	MAE	NSE	PBIAS	R^2	RMSE	MAE	NSE	PBIAS	R^2
ANFIS-GOA	**1.18**	**0.819**	**0.93**	**0.16**	**0.96**	**1.23**	**0.911**	**0.91**	**0.19**	**0.94**
ANN-GOA	1.27	0.821	0.90	0.17	**0.95**	1.28	0.921	0.90	0.20	0.94
SVM-GOA	1.29	0.823	0.89	0.20	**0.89**	1.32	0.925	0.87	0.21	0.88
ANFIS-CSO	1.20	0.822	0.92	0.17	**0.95**	1.34	0.914	0.90	0.22	0.93
ANN-CSO	1.31	0.824	0.91	0.20	**0.92**	1.37	0.926	0.87	0.23	0.90
SVM-CSO	1.35	0.827	0.90	0.23	**0.87**	1.40	0.930	0.86	0.25	0.86
ANFIS-KA	1.22	0.829	0.89	0.19	**0.94**	1.41	1.10	0.89	0.24	0.91
ANN-KA	1.33	0.833	0.87	0.24	**0.89**	1.43	1.12	0.85	0.26	0.88
SVM-KA	1.37	0.835	0.86	0.27	**0.85**	1.47	1.17	0.84	0.28	0.84
ANFIS-WA	1.24	0.837	0.90	0.29	**0.90**	1.50	1.14	0.82	0.30	0.89
ANN-WA	1.37	0.839	0.88	0.31	**0.88**	1.52	1.16	0.81	0.33	0.87
SVM-WA	1.39	0.840	0.86	0.35	**0.84**	1.54	1.18	0.80	0.38	0.82
ANFIS-PSO	1.29	0.838	0.85	0.37	**0.89**	1.56	1.20	0.79	0.40	0.87
ANN-PSO	1.40	0.842	0.84	0.39	**0.87**	1.58	1.25	0.78	0.41	0.86
SVM-PSO	1.41	0.844	0.83	0.41	**0.83**	1.60	1.27	0.77	0.43	0.81
ANFIS-GA	1.31	0.839	0.82	0.42	**0.86**	1.62	1.29	0.76	0.42	0.85
ANN-GA	1.44	0.845	0.81	0.43	**0.88**	1.65	1.32	0.75	0.44	0.85
SVM-GA	1.45	0.847	0.80	0.44	**0.82**	1.71	1.33	0.74	0.45	0.80
ANFIS	1.35	0.849	0.79	0.45	**0.85**	1.73	1.41	0.73	0.44	0.84
ANN	1.45	0.853	0.78	0.47	**0.85**	1.77	1.42	0.72	0.49	0.82
SVM	1.47	0.855	0.77	0.49	**0.8**	1.78	1.45	0.70	0.50	0.79

3.4. Analysis of Scatterplots of Soft Computing Models

- piezometer 6

Scatterplots for the soft computing models are provided in Figure 7a for the training and testing phases. It is clear that the hybrid ANFIS-GOA predictions were much closer to the measured data in the testing and training phases with a higher coefficient of determination. This result indicated a better correlation and a larger degree of statistical match between measured and predicted data of ANFIS-GOA relative to the other hybrid ANN and SVM models. The R^2 values were found to vary in the range of 0.84–0.94 and 0.79–0.91 for the ANN (hybrid ANN models and based ANN model) and SVM models (hybrid SVM models and based SVM model), respectively. The SVM model had the lowest R^2 among other models. Additionally, the ANFIS-GA, ANN-GA, and SVM-GA models had the lowest R^2 among other hybrid ANFIS, ANN, and SVM models. There is a weak agreement between the lower and higher values of the actual and estimated GWLs in this scatter plots of piezometer 6, unlike the ANFIS-GOA results.

- piezometer 9

As observed in Figure 7b, the R^2 values of testing phase were 0.94, 0.93, 0.92, 0.90, 0.86, 0.84, and 0.83 for ANFIS-GOA, ANFIS-CSO, ANFIS-KA, ANFIS-WA, ANFIS-PSO, ANFIS-GA, and ANFIS model, respectively. GOA had a better performance than other optimization algorithms. The outputs indicated that all hybrid optimized ANFIS, ANN, and SVM models outperformed the standalone ANFIS, ANN, and SVM models. As the results in Table 4 show incorporating the Taguchi and GOA in ANFIS training enhanced the R^2 values 13% in comparison with the standalone ANFIS and in all of the developed models the hybridized meta-heuristic models outperformed the single standalone models.

- Piezometer 10

The results of Figure 7c indicated that the ANFIS-GOA and SVM models produced the best and the worst results, respectively. It is clear that developed hybrid ANFIS-GOA model forecasting of GWL was less scattered and closer to the straight line of 1:1 than those the other models and it shows impressive results in regard to the other models. For training and testing phases, GA had a worse performance than CSO, PSO, KA, WA, and GOA because of the lower values of R^2. The standalone ANFIS model had the worst performance among the ANFIS-GOA, ANFIS-CSO, ANFIS-WA, ANFIS-PSO, ANFIS-GA, and ANFIS-KA models. The ANFIS-GOA model with $R^2 = 0.94$ as is presented in Table 5, the values of GWL simulated by the ANFIS-GOA are almost equal to the observed values of GWL. The linear fit of the forecasted GWL and measured GWL results have a high correlation coefficient that is very close to 1.00 ($R^2 = 0.97$) and a perfect correlation coefficient (R^2 value) of 0.94, confirmed that the simulation model has provided a very good prediction of the observed values of GWL. Additionally, 94% of the observed GWL values accurately fit the hybrid ANFIS-GOA model predictions.

3.5. Uncertainty Analysis of Soft Computing Models

As stated in the aims of the current study, the uncertainty analysis of hybrid intelligence models is another major contribution and novelty of the present study. The same as the previous subsections, in this section the results of uncertainty analysis of hybrid models in selected three piezometers are provided and comparative evaluation between different hybrid models are presented. The hybrids of ANFIS, SVM, and ANN models with GOA, WA, KA, PSO, and GA are joined with the non-parametric Monte-Carlo Simulations (MCSs) to quantify the uncertainty of developed models in GWL simulations. The probability of model predictions in MCSs is considered as a degree of uncertainty of model results and demonstrates the probabilities in the GWL forecasting bands that enclosed the observed GWL inside these bounds of probability.

- Piezometer 6

In the trained hybrid models, the uncertainty in the model trained parameters and weights is the major source of uncertainty in model results. Here the effects of uncertainty in trained, optimization, and determination of parameters, and weights of intelligence developed hybrid models for piezometer 6 are presented. For training and testing stage, the uncertainty of the models results in piezometer 6 are provided in Figure 8a and in Table 6. The uncertainty results are quantified by the two indices of p and d and visualized by the uncertainty bounds of 95%. At first, the values of p show how many of the observed GWL values in the training and testing stages are positioned inside the 95% confidence bounds. Secondly, the d-factor as the measure of deviations should be small also. Figure 8 indicated that the highest and lowest d was obtained for SVM and ANFIS-GOA, respectively. Based on p and d indices, CSO had better performance than PSO, GA, KA, and WA. Results indicated that the standalone ANN, ANFIS, and SVM models had higher d and lower p than hybrid ANFIS, ANN, and SVM models that indicate higher uncertainty in the standalone model results. The overall comparison of the results indicated that the ANN model outperformed the SVM model. The d values of uncertainties of models in Table 6 show that in all of developed models for GWL the d value is lower than 1, that proves the superior tight bounds of developed models. The best results are derived by the ANFIS-GOA with d = 012 and p = 0.94 indicates that developed model 95% of observations are covered by the uncertainty bounds. The desired values for p in model uncertainty analysis have values greater than 80% [49].

- Piezometer 9

As presented in Table 6 and in Figure 8b, SVM-GOA and SVM had the lowest and highest d among SVM models. According to Table 6, GOA outperformed CSO and KA, but both algorithms were better than GA, PSO, and WA. The *p*-value of the standalone ANFIS model was increased by the optimization algorithms. GA provided lower performance in the optimization of ANN with p equal to 0.83 and d equal to 0.24, compared to WA, GOA, PSO, KA, CSO, and WA.

Again, the comparisons confirm the superiority of ANFIS-GOA in uncertainty verifications that have p = 0.94 and d = 0.16. As confirmed by these values of p in all of the developed models, all of them are satisfactory and the major part of GWL simulations are enclosed by the 95% prediction interval based on model prediction in Monte Carlo simulations. However, the d values that measure the average distance from upper and lower limits of prediction interval, for the ANFIS-GOA models are significantly and considerably smaller than those of all of the other models. In general, the benefits of ANFIS-GOA models over the other models is two-fold. At first, the GOA based models provide a more accurate prediction of GWL with fewer errors. Secondly, the confidence interval of ANFIS-GOA model results is much narrower and yet encloses almost the greatest percent of observation in MCSs.

Table 6. The results of uncertainty of soft computing models.

Model	Piezometer 6		Piezometer 9		Piezometer 10	
	p	d	p	d	p	d
ANFIS-GOA	0.94	0.14	0.94	0.16	0.95	0.17
ANN-GOA	0.93	0.16	0.91	0.17	0.93	0.19
SVM-GOA	0.86	0.23	0.86	0.20	0.89	0.27
ANFIS-CSO	0.93	0.15	0.93	0.15	0.92	0.17
ANN-CSO	0.91	0.19	0.92	0.21	0.91	0.20
SVM-CSO	0.84	0.21	0.88	0.23	0.88	0.29
ANFIS-KA	0.90	0.15	0.92	0.17	0.89	0.18
ANN-KA	0.89	0.20	0.87	0.19	0.87	0.21
SVM-KA	0.86	0.21	0.89	0.19	0.86	0.29
ANFIS-WA	0.90	0.19	0.89	0.19	0.85	0.18
ANN-WA	0.86	0.23	0.84	0.24	0.84	0.19
SVM-WA	0.89	0.27	0.85	0.25	0.83	0.29
ANFIS-PSO	0.89	0.21	0.86	0.19	0.84	0.19
ANN-PSO	0.84	0.25	0.85	0.20	0.82	0.21
SVM-PSO	0.84	0.27	0.84	0.24	0.81	0.31
ANFIS-GA	0.87	0.20	0.83	0.24	0.82	0.20
ANN-GA	0.84	0.27	0.86	0.25	0.80	0.25
SVM-GA	0.80	0.32	0.89	0.29	0.79	0.30
ANFIS	0.85	0.20	0.87	0.24	0.78	0.20
ANN	0.82	0.28	0.83	0.24	0.77	0.27
SVM	0.80	0.35	0.82	0.29	0.76	0.33

- Piezometer 10

From Table 6, it was observed that ANFIS-GOA yielded the most dominant performance among other models. The weakest model in the optimization of the ANFIS model was ANFIS-GA with a p of 0.82 and d of 0.20. The ANN model provided better performance than the SVM model. The corresponding performance values of the SVM-GA model had p of 0.79 and d of 0.30. The standalone SVM model had the worst performance among other models.

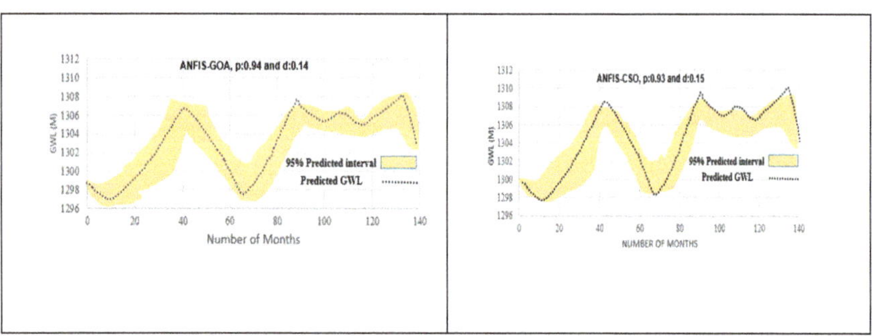

Figure 8. *Cont.*

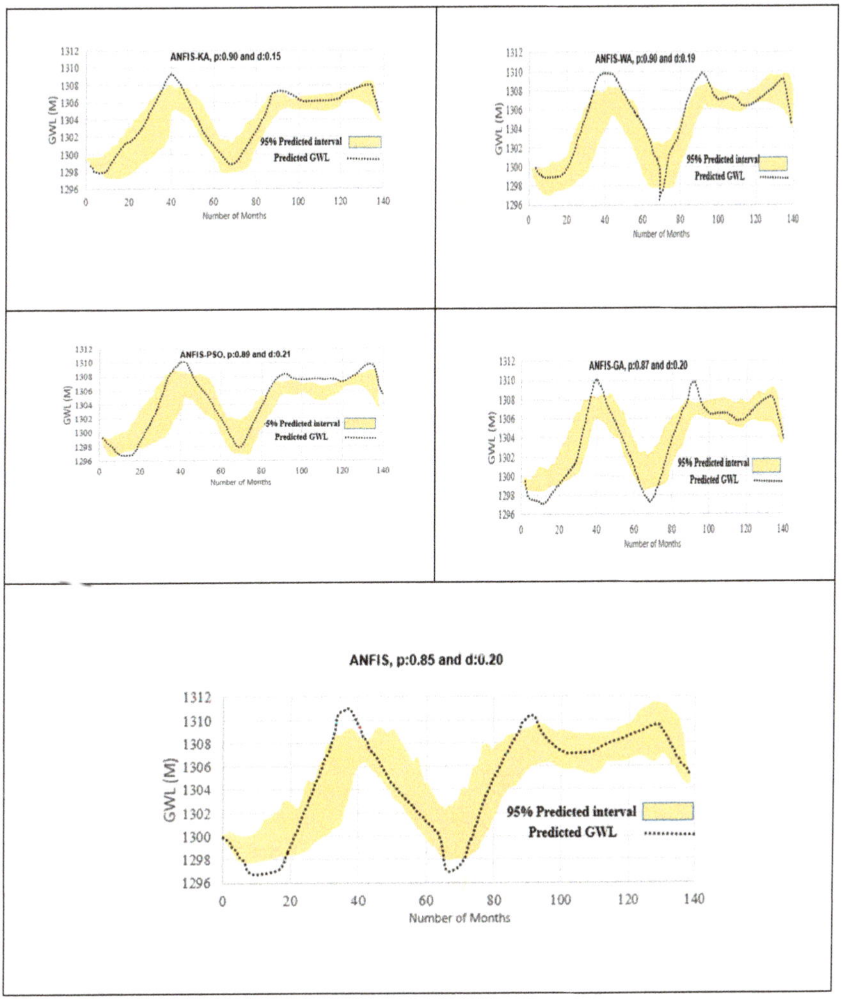

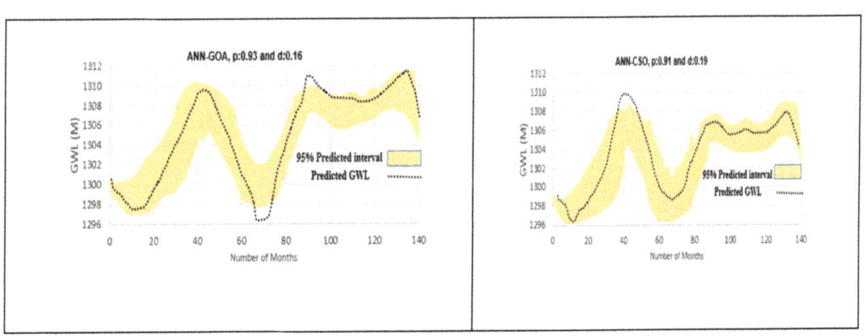

(a)

Figure 8. *Cont.*

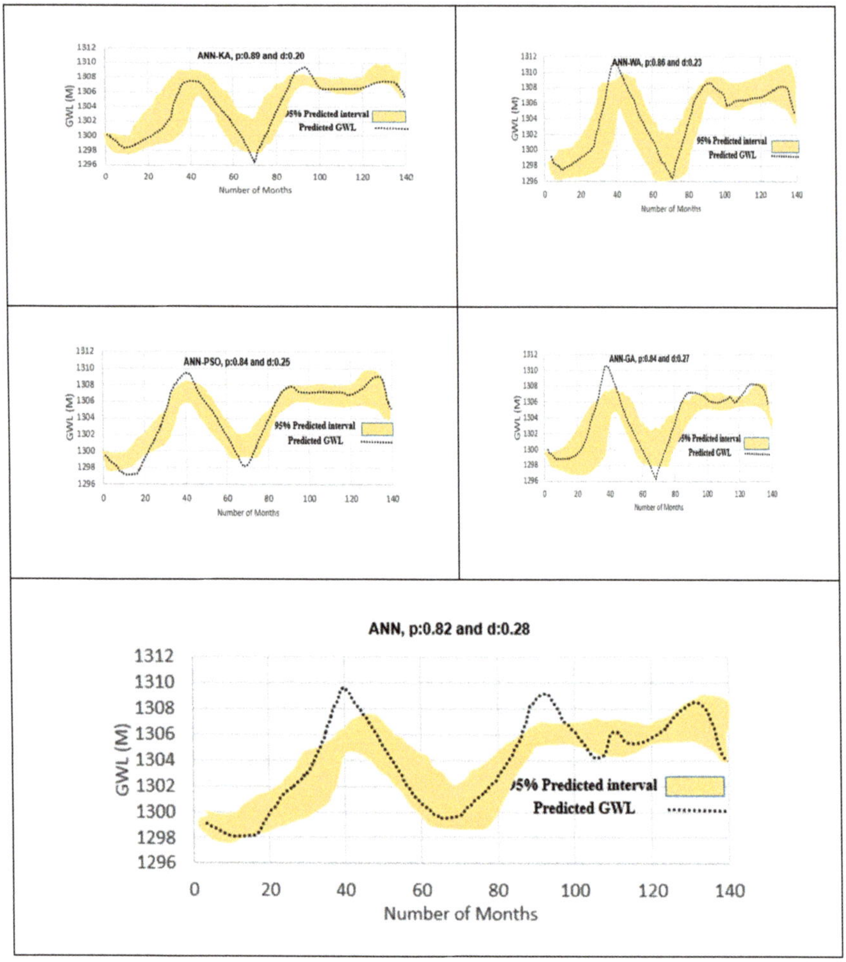

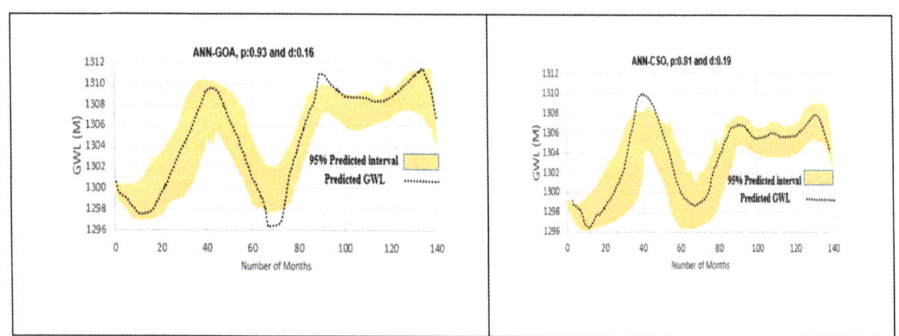

Figure 8. *Cont.*

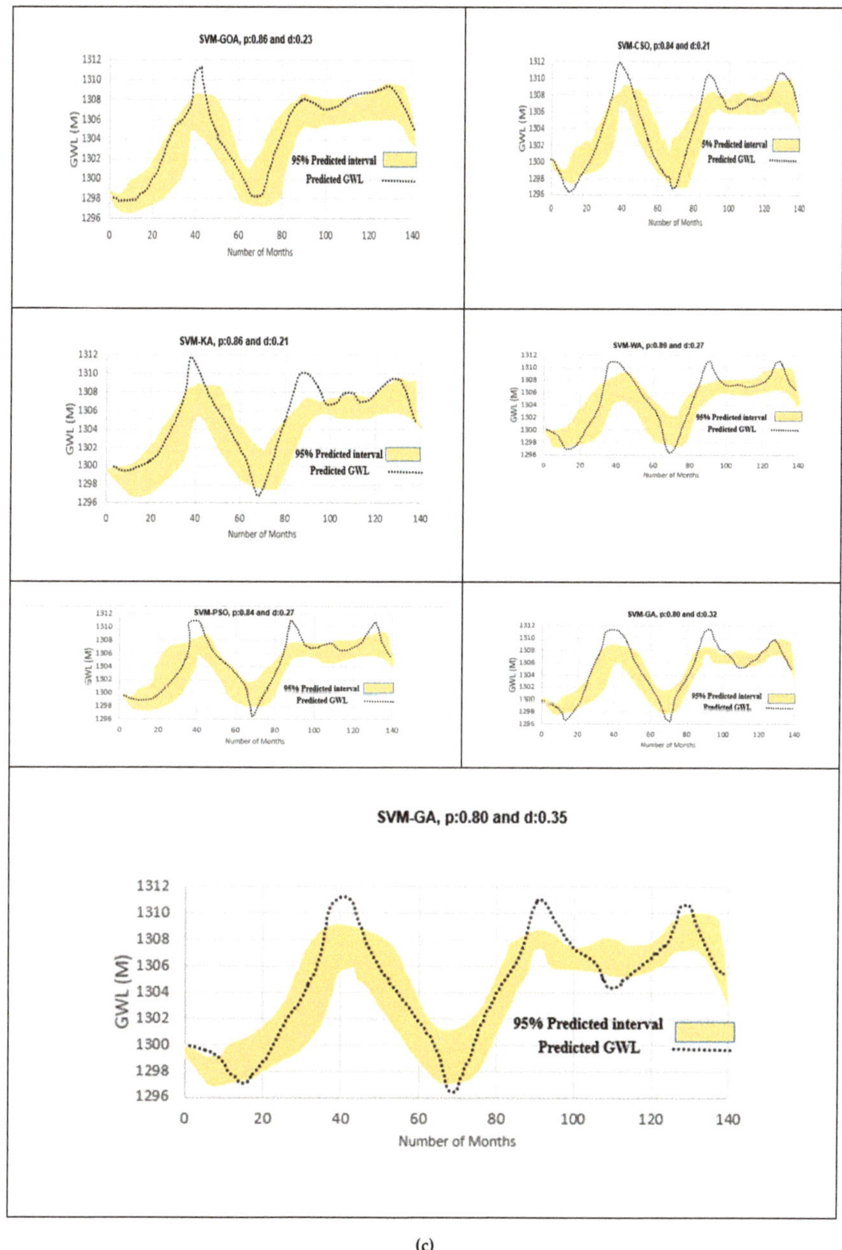

(c)

Figure 8. Computed uncertainty bound for piezometer 6; (**a**) ANFIS, (**b**) ANN, (**c**) SVM.

However, general results indicated that the ANFIS-GOA has the best performance among other models. Figure 9 shows the coefficient of variation for different optimization algorithms. ANFIS-GOA had a lower coefficient of variation than other models and optimization algorithms. The worst results were for GA. In general, there are three main sources that generate the uncertainty of model outputs: the first one is the data and knowledge uncertainty, the second one is the parametric uncertainty due to unknown model parameters, and the third one is the structural uncertainty due to physical complexity of phenomena. The main contribution of the current paper is the uncertainty analysis of hybrid models prediction of GWL in the form of parametric uncertainty due to regulatory parameters and weights produced in the training stage of models.

3.6. Spatiotemporal Variation of GWL

The previous section indicated that the GOA improved the performance of ANN, ANFIS, and SVM models. The results indicated that the GOA had better performance than other optimization algorithms. As shown in Figure 9, the hybrid GOA models (ANFIS-GOA, ANN-GOA, and SVM-GOA) have low variation coefficients in modeling.

Most literature reviews revealed only a few quantity comparisons. Furthermore, they did not include the spatiotemporal variation of GWL. In this section, the latitude, longitude, H(t-1), H (t-2), H (t-3), H (t-4), and H (t-5), hydraulic conductivity (HC), and specific yield of nine observed wells (well 6, 9, 10, 24, 11, 4, 7, 8, and 1) were used to provide the spatiotemporal variation of GWL for different months. The Ardebil plain is a heterogeneous aquifer. Thus, the hydraulic conductivity and specific yield spatially vary in the Ardebil plain. HC is a measure of a material's capacity to transmit water. The specific yield is defined as the ratio of the volume of water that an aquifer will yield by gravity to the total volume of the aquifer. A pumping test method was used to obtain the value of the hydraulic conductivity and specific yield. Figure 10 shows the measured hydraulic conductivity and specific yield for the Ardebil plain. In this section, the ANFIS, ANN, and SVM models with the best algorithm (GOA) were used to provide the spatiotemporal variation of GWL. The difference between estimated GWL models and observed GWL was computed for all months of years. The RMSE was used as an error function to compare the estimated data with the observed data. From Figure 11, it was clear that the ANFIS-GOA provided more accurate estimation than ANN-GOA and SVM-GOA. It was clear that the RMSE of ANFIS-GOA varied from white (1.2 m) to dark blue (2.2), while the RMSE of ANN-GOA and SVM-GOA varied from 1.7 (yellow) to 2.7 m (light green). Thus, results indicated that ANFIS-GOA has higher accuracy for the heterogeneous aquifers. The heterogeneous aquifers are considered as complex hydraulic systems because their hydraulic parameters vary spatially and temporally. Additionally, the climate parameters, such as temperature and rainfall, can increase the complexity of prediction of GWL for heterogeneous aquifers.

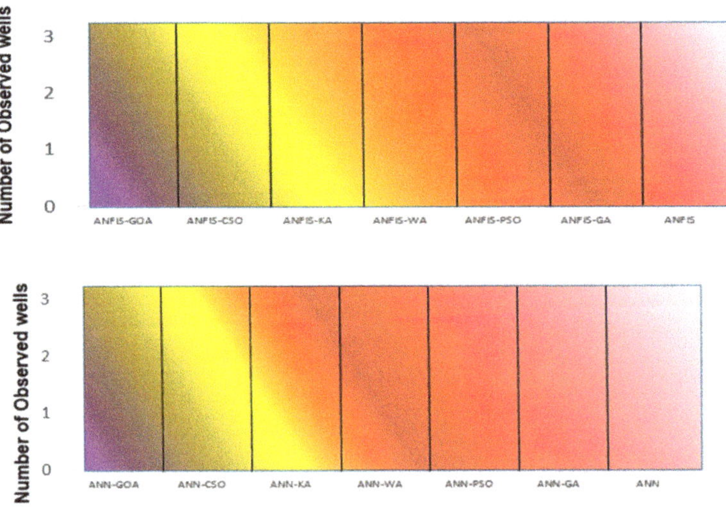

Figure 9. *Cont.*

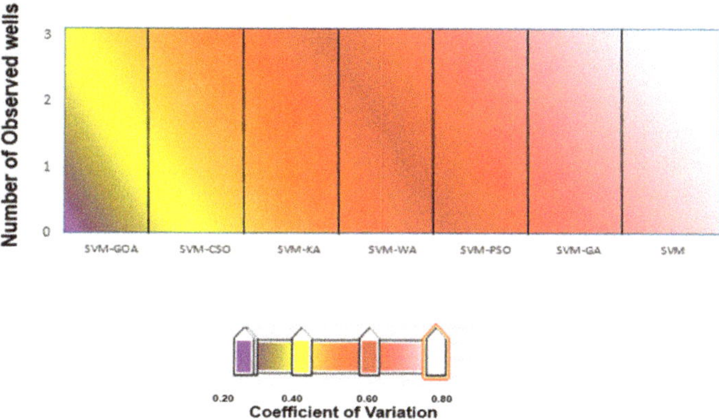

Figure 9. The map of variations coefficient of different models for 100 random runs of objective function.

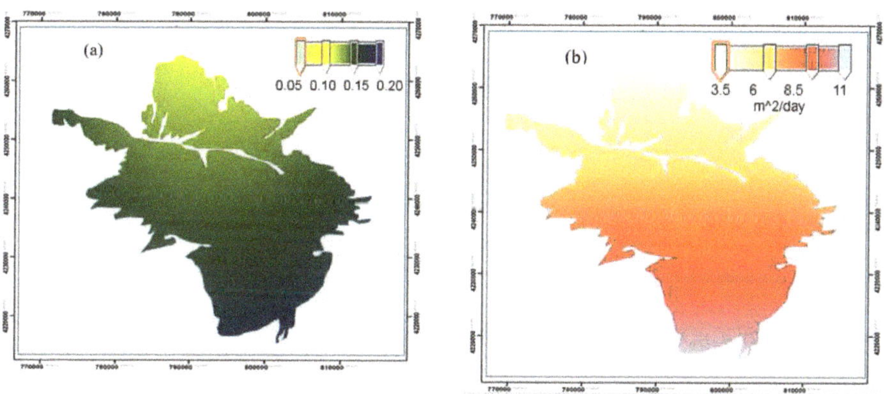

Figure 10. (**a**) Spatial specific yield and (**b**) hydraulic conductivity.

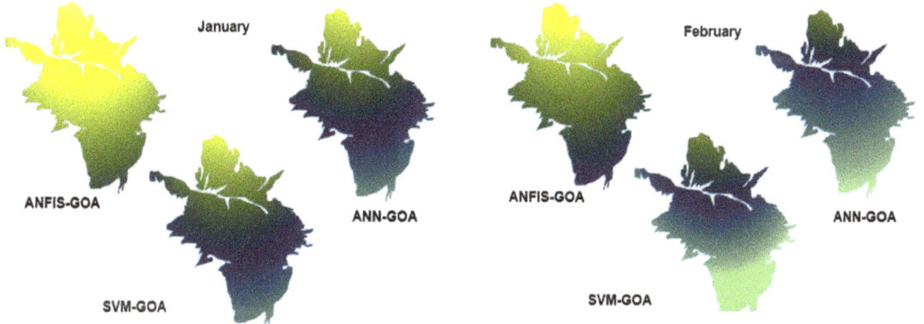

Figure 11. *Cont.*

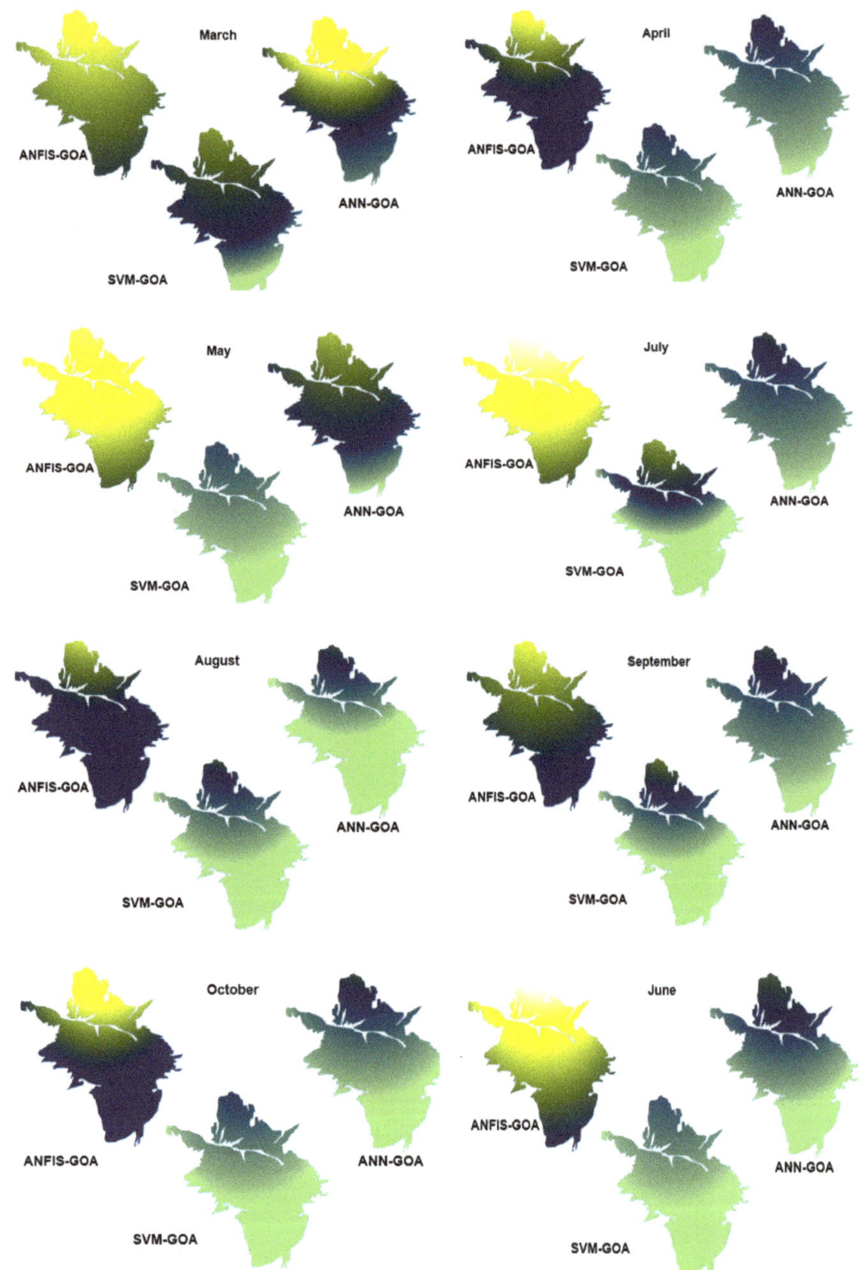

Figure 11. *Cont.*

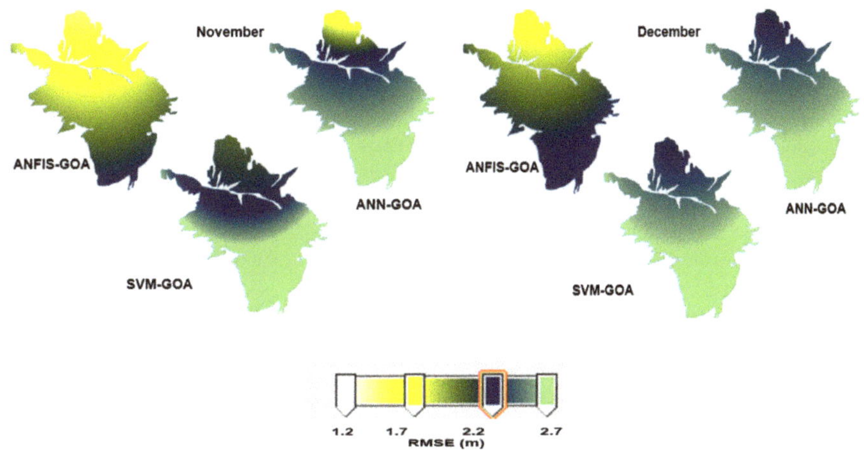

Figure 11. The spatial and temporal variation of GWL.

4. Conclusions

In this study, the ANFIS, ANN, and SVM models were used to predict groundwater level. The GOA, CSO, GA, PSO, WA, and KA were used to fine-tune and integrate with the ANN, SVM, and ANFIS models. Three piezometers (6, 9, and 10) in the Ardebil plain were considered as a case study for the GWL investigation. The input combinations of time series (up to 12-month lag) were reduced using principal component analysis (PCA). For the testing phase and piezometer 6 ANFIS-GOA indicated a value of RMSE: 1.21, MAE: 0.878, NSE: 0.93, and PBIAS: 0.15 which reflected better performance than the other models. The R^2 values were found to vary in the range of 0.84–0.94 and 0.79–0.91 for the ANN (hybrid ANN models and based ANN model) and SVM models (hybrid SVM models and based SVM model), respectively. The results indicated that the SVM model had the lowest R^2 among other models. It was observed that the ANFIS-GOA yielded the most dominant performance among other models. From uncertainty analysis, the weakest model in the optimization of the ANFIS model was ANFIS-GA with a p = 0.87 and d = 0.21. However, general results indicated that the ANFIS-GOA had better performance than other models. Additionally, the results of spatiotemporal variations maps of GWL showed that ANFIS-GOA has high accuracy for the heterogeneous Ardebil aquifer. Future studies can evaluate the accuracy of these models under climate change conditions. The climate parameters such as temperature and rainfall can be simulated for future periods. Then, these parameters can be used as input to the models to simulate GWL for the future periods.

Author Contributions: Conceptualization, A.S., and M.E.; methodology, A.S., and M.E.; writing—review and editing, A.S., M.E., V.P.S., and A.M.; validation; A.M., A.S., and M.E.; supervision, V.P.S.; Funding acquisition, A.M. All authors have read and agreed to the published version of the manuscript.

Funding: This work is supported by the Hungarian State and the European Union under the EFOP-3.6.1-16-2016-00010 project and the 2017-1.3.1-VKE-2017-00025 project.

Acknowledgments: Support of the Hungarian State and the European Union under the EFOP-3.6.1-16-2016-00010 project and the 2017-1.3.1-VKE-2017-00025 project is acknowledged. We also acknowledge the support of the German Research Foundation (DFG) and the Bauhaus-Universität Weimar within the Open-Access Publishing Programme.

Conflicts of Interest: The authors declare no conflict of interest.

References

1.   Sattari, M.T.; Mirabbasi, R.; Sushab, R.S.; Abraham, J. Prediction of Groundwater Level in Ardebil Plain Using Support Vector Regression and M5 Tree Model. *Groundwater* **2018**. [CrossRef] [PubMed]

2. Jeong, J.; Park, E. Comparative applications of data-driven models representing water table fluctuations. *J. Hydrol.* **2019**, *572*, 261–273. [CrossRef]

3. Alizamir, M.; Kisi, O.; Zounemat-Kermani, M. Modelling long-term groundwater fluctuations by extreme learning machine using hydro-climatic data. *Hydrol. Sci. J.* **2018**, *63*, 63–73. [CrossRef]

4. Yoon, H.; Kim, Y.; Lee, S.H.; Ha, K. Influence of the range of data on the performance of ANN-and SVM-based time series models for reproducing groundwater level observations. *Acque Sotter. Ital. J. Groundwater.* **2019**. [CrossRef]

5. Mohanty, S.; Jha, M.K.; Kumar, A.; Sudheer, K.P. Artificial neural network modeling for groundwater level forecasting in a river island of eastern India. *Water Resour. Manag.* **2010**, *24*, 1845–1865. [CrossRef]

6. Natarajan, N.; Sudheer, C. Groundwater level forecasting using soft computing techniques. *Neural Comput. Appl.* **2019**, 1–18. [CrossRef]

7. Lee, S.; Lee, K.K.; Yoon, H. Using artificial neural network models for groundwater level forecasting and assessment of the relative impacts of influencing factors. *Hydrogeol. J.* **2019**, *27*, 567–579. [CrossRef]

8. Khan, U.T.; Valeo, C. Dissolved oxygen prediction using a possibility theory based fuzzy neural network. *Hydrol. Earth Syst. Sci.* **2016**, *20*, 2267–2293. [CrossRef]

9. Jeihouni, E.; Eslamian, S.; Mohammadi, M.; Zareian, M.J. Simulation of groundwater level fluctuations in response to main climate parameters using a wavelet–ANN hybrid technique for the Shabestar Plain, Iran. *Environ. Earth Sci.* **2019**, *78*, 293. [CrossRef]

10. Alian, S.; Mayer, A.; Maclean, A.; Watkins, D.; Mirchi, A. Spatiotemporal Dimensions of Water Stress Accounting: Incorporating Groundwater–Surface Water Interactions and Ecological Thresholds. *Environ. Sci. Technol.* **2019**, *53*, 2316–2323. [CrossRef]

11. Jalalkamali, A.; Sedghi, H.; Manshouri, M. Monthly groundwater level prediction using ANN and neuro-fuzzy models: A case study on Kerman plain, Iran. *J. Hydroinformatics.* **2010**, *13*, 867–876. [CrossRef]

12. Trichakis, I.C.; Nikolos, I.K.; Karatzas, G.P. Artificial neural network (ANN) based modeling for karstic groundwater level simulation. *Water Resour. Manag.* **2011**, *25*, 1143–1152. [CrossRef]

13. Fallah-Mehdipour, E.; Haddad, O.B.; Mariño, M.A. Prediction and simulation of monthly groundwater levels by genetic programming. *J Hydro-Environment. Res.* **2013**, *7*, 253–260. [CrossRef]

14. Moosavi, V.; Vafakhah, M.; Shirmohammadi, B.; Behnia, N. A wavelet-ANFIS hybrid model for groundwater level forecasting for different prediction periods. *Water Resour. Manag.* **2013**, *27*, 1301–1321. [CrossRef]

15. Emamgholizadeh, S.; Moslemi, K.; Karami, G. Prediction the Groundwater Level of Bastam Plain (Iran) by Artificial Neural Network (ANN) and Adaptive Neuro-Fuzzy Inference System (ANFIS). *Water Resour. Manag.* **2014**, *28*, 5433–5446. [CrossRef]

16. Suryanarayana, C.; Sudheer, C.; Mahammood, V.; Panigrahi, B.K. An integrated wavelet-support vector machine for groundwater level prediction in Visakhapatnam, India. *Neurocomputing* **2014**, *145*, 324–335. [CrossRef]

17. Mohanty, S.; Jha, M.K.; Raul, S.K.; Panda, R.K.; Sudheer, K.P. Using artificial neural network approach for simultaneous forecasting of weekly groundwater levels at multiple sites. *Water Resour. Manag.* **2015**, *29*, 5521–5532. [CrossRef]

18. Yoon, H.; Hyun, Y.; Ha, K.; Lee, K.K.; Kim, G.B. A method to improve the stability and accuracy of ANN-and SVM-based time series models for long-term groundwater level predictions. *Comput. Geosci.* **2016**, *90*, 144–155. [CrossRef]

19. Zhou, T.; Wang, F.; Yang, Z. Comparative analysis of ANN and SVM models combined with wavelet preprocess for groundwater depth prediction. *Water* **2017**, *9*, 781. [CrossRef]

20. Choubin, B.; Malekian, A. Combined gamma and M-test-based ANN and ARIMA models for groundwater fluctuation forecasting in semiarid regions. *Environ. Earth Sci.* **2017**, *76*, 538. [CrossRef]

21. Das, U.K.; Roy, P.; Ghose, D.K. Modeling water table depth using adaptive Neuro-Fuzzy Inference System. *ISH J. Hydraul. Eng.* **2019**, *25*, 291–297. [CrossRef]

22. Hadipour, A.; Khoshand, A.; Rahimi, K.; Kamalan, H.R. Groundwater Level Forecasting by Application of Artificial Neural Network Approach: A Case Study in Qom Plain, Iran. *J. Hydrosci. Environ.* **2019**, *3*, 30–34. [CrossRef]

23. Jalalkamali, A.; Jalalkamali, N. Groundwater modeling using hybrid of artificial neural network with genetic algorithm. *Afr. J. Agric. Res.* **2011**, *6*, 5775–5784. [CrossRef]

24. Mathur, S. Groundwater level forecasting using SVM-PSO. *Int. J. Hydrol. Sci. Technol.* **2012**, *2*, 202–218. [CrossRef]

25. Hosseini, Z.; Gharechelou, S.; Nakhaei, M.; Gharechelou, S. Optimal design of BP algorithm by ACO R model for groundwater-level forecasting: A case study on Shabestar plain, Iran. *Arab. J. Geosci.* **2016**, *9*, 436. [CrossRef]

26. Zare, M.; Koch, M. Groundwater level fluctuations simulation and prediction by ANFIS-and hybrid Wavelet-ANFIS/Fuzzy C-Means (FCM) clustering models: Application to the Miandarband plain. *J. Hydro-Environ. Res.* **2018**, *18*, 63–76. [CrossRef]

27. Balavalikar, S.; Nayak, P.; Shenoy, N.; Nayak, K. Particle swarm optimization based artificial neural network model for forecasting groundwater level in Udupi district. In *Proceedings of the AIP Conference*; AIP Elsevier: New York, NY, USA, 2018. [CrossRef]

28. Malekzadeh, M.; Kardar, S.; Saeb, K.; Shabanlou, S.; Taghavi, L. A Novel Approach for Prediction of Monthly Ground Water Level Using a Hybrid Wavelet and Non-Tuned Self-Adaptive Machine Learning Model. *Water Resour. Manag.* **2019**, *33*, 1609–1628. [CrossRef]

29. Mirjalili, S.Z.; Mirjalili, S.; Saremi, S.; Faris, H.; Aljarah, I. Grasshopper optimization algorithm for multi-objective optimization problems. *Appl. Intell.* **2018**, *48*, 805–820. [CrossRef]

30. Zeynali, M.J.; Shahidi, A. Performance Assessment of Grasshopper Optimization Algorithm for Optimizing Coefficients of Sediment Rating Curve. *AUT J. Civ. Eng.* **2018**, *2*, 39–48.

31. Alizadeh, Z.; Yazdi, J.; Kim, J.; Al-Shamiri, A. Assessment of Machine Learning Techniques for Monthly Flow Prediction. *Water* **2018**, *10*, 1676. [CrossRef]

32. Moayedi, H.; Gör, M.; Lyu, Z.; Bui, D.T. Herding Behaviors of Grasshopper and Harris hawk for Hybridizing the Neural Network in Predicting the Soil Compression Coefficient. *Meas. J. Int. Meas. Confed.* **2019**, 107389. [CrossRef]

33. Gampa, S.R.; Jasthi, K.; Goli, P.; Das, D.; Bansal, R.C. Grasshopper optimization algorithm based two stage fuzzy multiobjective approach for optimum sizing and placement of distributed generations, shunt capacitors and electric vehicle charging stations. *J. Energy Storage.* **2020**, *27*, 101117. [CrossRef]

34. Moayedi, H.; Kalantar, B.; Foong, L.K.; Tien Bui, D.; Motevalli, A. Application of three metaheuristic techniques in simulation of concrete slump. *Appl. Sci.* **2019**, *9*, 4340. [CrossRef]

35. Kumar, A.; Kumar, P.; Singh, V.K. Evaluating Different Machine Learning Models for Runoff and Suspended Sediment Simulation. *Water Resour. Manag.* **2019**, *33*, 1217–1231. [CrossRef]

36. Khosravi, K.; Daggupati, P.; Alami, M.T.; Awadh, S.M.; Ghareb, M.I.; Panahi, M.; ThaiPham, B.; Rezaei, F.; Qi, C.; Yaseen, Z.M. Meteorological data mining and hybrid data-intelligence models for reference evaporation simulation: A case study in Iraq. *Comput. Electron. Agric.* **2019**, *167*, 105041. [CrossRef]

37. Kisi, O.; Yaseen, Z.M. The potential of hybrid evolutionary fuzzy intelligence model for suspended sediment concentration prediction. *Catena* **2019**, *174*, 11–23. [CrossRef]

38. Dubdub, I.; Rushd, S.; AlYaari, M.; Ahmed, E. Application of Artificial Neural Network to Model the Pressure Losses in the Water-Assisted Pipeline Transportation of Heavy Oil. Proceedings of the SPE Middle East Oil and Gas Show and Conference. *Soc. Pet. Eng.* **2019**. [CrossRef]

39. Moghaddam, H.K.; Moghaddam, H.K.; Kivi, Z.R.; Bahreinimotlagh, M.; Alizadeh, M.J. Developing comparative mathematic models, BN and ANN for forecasting of groundwater levels. *Groundw. Sustain. Dev.* **2019**, 100237. [CrossRef]

40. Fan, J.; Wang, X.; Wu, L.; Zhou, H.; Zhang, F.; Yu, X.; Lu, X.; Xiang, Y. Comparison of Support Vector Machine and Extreme Gradient Boosting for predicting daily global solar radiation using temperature and precipitation in humid subtropical climates: A case study in China. *Energy Convers. Manag.* **2018**, *164*, 102–111. [CrossRef]

41. Pour, S.H.; Shahid, S.; Chung, E.S.; Wang, X.J. Model output statistics downscaling using support vector machine for the projection of spatial and temporal changes in rainfall of Bangladesh. *Atmos. Res.* **2018**, *213*, 149–162. [CrossRef]

42. Pham, B.T.; Bui, D.T.; Prakash, I. Bagging based Support Vector Machines for spatial prediction of landslides. *Environ. Earth Sci.* **2018**, *77*, 146. [CrossRef]

43. Deo, R.C.; Salcedo-Sanz, S.; Carro-Calvo, L.; Saavedra-Moreno, B. Drought prediction with standardized precipitation and evapotranspiration index and support vector regression models. In *Integrating Disaster Science and Management*, 1st ed.; Elsevier: Amsterdam, The Netherlands, 2018; pp. 151–174. [CrossRef]

44. Mafarja, M.; Aljarah, I.; Faris, H.; Hammouri, A.I.; Ala'M, A.Z.; Mirjalili, S. Binary grasshopper optimisation algorithm approaches for feature selection problems. *Expert Syst. Appl.* **2019**, *117*, 267–286. [CrossRef]
45. Mehrabian, A.R.; Lucas, C. A novel numerical optimization algorithm inspired from weed colonization. *Ecol. Inform.* **2006**, *1*, 355–366. [CrossRef]
46. Chandirasekaran, D.; Jayabarathi, T. Cat swarm algorithm in wireless sensor networks for optimized cluster head selection: A real time approach. *Cluster Comput.* **2019**, *22*, 11351–11361. [CrossRef]
47. Karpenko, A.P.; Leshchev, I.A. Advanced Cat Swarm Optimization Algorithm in Group Robotics Problem. *Procedia Comput. Sci.* **2019**, *150*, 95–101. [CrossRef]
48. Ramezani, F. Solving Data Clustering Problems using Chaos Embedded Cat Swarm Optimization. *J. Adv. Comput. Res.* **2019**, *10*, 1–10.
49. Orouskhani, M.; Shi, D. Fuzzy adaptive cat swarm algorithm and Borda method for solving dynamic multi-objective problems. *Expert Syst.* **2018**, *35*, e12286. [CrossRef]
50. Pradhan, P.M.; Panda, G. Solving multiobjective problems using cat swarm optimization. *Expert Syst. Appl.* **2012**, *39*, 2956–2964. [CrossRef]
51. Chu, S.C.; Tsai, P.W.; Pan, J.S. Cat swarm optimization. In Proceedings of the Pacific Rim International Conference on Artificial Intelligence; Springer: Berlin/Heidelberg, Germany, 2006; pp. 854–858. [CrossRef]
52. Saha, S.K.; Ghoshal, S.P.; Kar, R.; Mandal, D. Cat swarm optimization algorithm for optimal linear phase FIR filter design. *ISA Trans.* **2013**, *52*, 781–794. [CrossRef]
53. Kennedy, J.; Eberhart, R.C. A discrete binary version of the particle swarm algorithm. Proceedings of 1997 IEEE International conference on systems, man, and cybernetics. *Comput. Cybern. Simul. IEEE* **1997**, 4104–4108. [CrossRef]
54. Gandomi, A.H.; Alavi, A.H. Krill herd: A new bio-inspired optimization algorithm. *Commun. Nonlinear Sci. Numer. Simul.* **2012**, *17*, 4831–4845. [CrossRef]
55. Abualigah, L.M.; Khader, A.T.; Hanandeh, E.S. A combination of objective functions and hybrid Krill herd algorithm for text document clustering analysis. *Eng. Appl. Artif. Intell.* **2018**, *73*, 111–125. [CrossRef]
56. Asteris, P.G.; Nozhati, S.; Nikoo, M.; Cavaleri, L.; Nikoo, M. Krill herd algorithm-based neural network in structural seismic reliability evaluation. *Mech. Adv. Mater. Struct.* **2019**, *26*, 1146–1153. [CrossRef]
57. Lu, C.; Feng, J.; Liu, W.; Lin, Z.; Yan, S. Tensor robust principal component analysis with a new tensor nuclear norm. *IEEE Trans. Pattern Anal. Mach. Intell.* **2019**. [CrossRef]
58. Priyadarshi, D.; Paul, K.K. Optimisation of biodiesel production using Taguchi model. *Waste Biomass Valori.* **2019**, *10*, 1547–1559. [CrossRef]
59. Yen, H.; Wang, X.; Fontane, D.G.; Harmel, R.D.; Arabi, M. A Framework for Propagation of Uncertainty Contributed by Parameterization, Input Data, Model Structure, and Calibration/Validation Data in Watershed Modeling. *Environ. Model. Softw.* **2014**. [CrossRef]
60. Youcai, Z.; Sheng, H. Pollution Characteristics of Industrial Construction and Demolition Waste. In *Pollution Control and Resource Recovery*; Elsevier Inc.: New York, NY, USA, 2017. [CrossRef]

 *sustainability*

Article

Bayesian Model Averaging: A Unique Model Enhancing Forecasting Accuracy for Daily Streamflow Based on Different Antecedent Time Series

Sungwon Kim [1,*] **, Meysam Alizamir [2], Nam Won Kim [3,*] and Ozgur Kisi [4,5]**

[1]   Department of Railroad Construction and Safety Engineering, Dongyang University, Yeongju 36040, Korea
[2]   Department of Civil Engineering, Hamedan Branch, Islamic Azad University, Hamedan 65181-15743, Iran; meysamalizamir@gmail.com
[3]   Department of Land, Water and Environment Research, Korea Institute of Civil Engineering and Building Technology, Goyang-si 10223, Korea
[4]   Department of Civil Engineering, School of Technology, Ilia State University, Tbilisi 0162, Georgia; ozgur.kisi@iliauni.edu.ge
[5]   Institute of Research and Development, Duy Tan University, Da Nang 550000, Vietnam
*   Correspondence: swkim1968@dyu.ac.kr (S.K.); nwkim@kict.re.kr (N.W.K.)

Received: 19 September 2020; Accepted: 19 November 2020; Published: 21 November 2020

Abstract: Streamflow forecasting is a vital task for hydrology and water resources engineering, and the different artificial intelligence (AI) approaches have been employed for this purposes until now. Additionally, the forecasting accuracy and uncertainty estimation are the meaningful assignments that need to be recognized. The addressed research investigates the potential of novel ensemble approach, Bayesian model averaging (BMA), in streamflow forecasting using daily time series data from two stations (i.e., Hongcheon and Jucheon), South Korea. Six categories (i.e., M1–M6) of input combination using different antecedent times were employed for streamflow forecasting. The outcomes of BMA model were compared with those of multivariate adaptive regression spline (MARS), M5 model tree (M5Tree), and Kernel extreme learning machines (KELM) models considering four assessment indexes, root mean square error (RMSE), Nash-Sutcliffe efficiency (NSE), correlation coefficient (R), and mean absolute error (MAE). The results revealed the superior accuracy of BMA model over three machine learning models in daily streamflow forecasting. Considering RMSE values among the best models during testing phase, the best BMA model (i.e., BMA2) enhanced the forecasting accuracy of MARS1, M5Tree4, and KELM3 models by 5.2%, 5.8%, and 3.4% in Hongcheon station. Additionally, the best BMA model (i.e., BMA1) improved the forecasting accuracy of MARS1, M5Tree1, and KELM1 models by 6.7%, 9.5%, and 3.7% in Jucheon station. In addition, the best BMA models in both stations allowed the uncertainty estimation, and produced higher uncertainty of peak flows compared to that of low flows. As one of the most robust and effective tools, therefore, the BMA model can be successfully employed for streamflow forecasting with different antecedent times.

Keywords: streamflow forecasting; Bayesian model averaging; multivariate adaptive regression spline; M5 model tree; Kernel extreme learning machines; South Korea

---

1. Introduction

Implementing a stable model to forecast streamflow can be influential for the fields of hydrology and water resources researches [1–4]. Streamflow forecasting, however, is an intricate project because of nonstationary time series and reliance on temporal and spatial parameters which have unclear and complicated components [5–7]. Increasing issue complications often depend on long antecedent times (or lead times) such as days and months [8–11]. Therefore, streamflow forecasting using different antecedent times can be categorized as universal assignment for hydrology and water resources researches [12–16].

Machine learning (ML) models have popular and flexible approaches for simulating and catching the nonlinear phenomena for science and engineering during three decades including multivariate adaptive regression spline (MARS), M5 model tree (M5Tree), and Kernel extreme learning model (KELM), etc. The MARS model has been successfully applied and employed for solving streamflow forecasting problems until now. Al-Sudani et al. [17] surveyed the ability of MARS incorporated with differential evolution (MARS-DE) model to forecast streamflow in Tigris River, Iraq. They investigated that the MARS-DE model provided a reliable forecasted accuracy for semi-arid streamflow. Adamowski et al. [18] managed the MARS model to forecast streamflow in Himalayan watershed, Uttaranchal State, India. They found that the MARS model preformed a superior forecasted accuracy compared to the artificial neural network (ANN) model. Tyralis et al. [19] utilized the MARS model for daily streamflow forecasting in 511 basins, USA. The MARS model, however, did not improve the performance of linear regression model obviously compared to the other models (e.g., extremely randomized trees, XGBoost, and polyMARS).

M5Tree model has also been utilized for perceiving the pros and cons of streamflow forecasting. Solomatine and Xue [20] applied the M5Tree model for flood forecasting in the Huai River, China. They provided that the forecasted accuracy of M5Tree model were similar with that of ANN models, and the hybrid model covering M5Tree and ANN indicated the best forecasted accuracy. Štravs and Brilly [21] developed the M5Tree model for low streamflow forecasting in the Sava River basin, Slovenia. They employed the recession streamflow data based on 7-day lead time for forecasting and showed the reliable accuracy. Sattari et al. [22] hired the M5Tree model for daily streamflow forecasting in the Sohu River, Turkey. They demonstrated that the M5Tree model forecasted 7-day lead time streamflow accurately. Adnan et al. [23] worked using the M5Tree model for monthly and daily streamflow forecasting in the Hunza River, Pakistan. This experiment said that the M5Tree model could not forecast monthly and daily streamflow effectively compared to the least square support vector machine (LSSVM) model.

However, the diverse researches using multiple machine learning models can be found for streamflow forecasting including the MARS and M5Tree models from the published articles and reports. Yaseen et al. [24] evaluated the MARS and M5Tree models for monthly streamflow forecasting in Turkey and Iraq. This document explained that the LSSVM model, however, forecasted the monthly streamflow accurately compared to the MARS and M5Tree models. Yin et al. [25] utilized the MARS and M5Tree models for streamflow forecasting in a semiarid and mountainous region, Northwestern China. They evaluated that the performance of M5Tree model was superior to the support vector regression (SVR) and MARS models for 1-, 2-, and 3-day lead times forecasting. Kisi et al. [26] investigated the MARS and M5Tree models for streamflow forecasting in the Mediterranean region, Turkey. This article showed that the MARS and M5Tree model did not accomplish the outstanding performance compared to the LSSVM model. Rezaie-Balf et al. [27] handled the MARS and M5Tree models to forecast daily streamflow in Iran and South Korea. They indicated that the MARS model combined ensemble empirical mode decomposition (EEMD) was the effective method to forecast streamflow based on 1-, 2-, 3-, and 4-day lead times. Additionally, Rezaie-Balf et al. [28] explored the MARS and M5Tree models to forecast the reservoir inflow for Aswan High Dam, Egypt. They discovered that the MARS model embedded with complete ensemble empirical mode decomposition with adaptive noise (CEEMDAN) suggested the reliable accuracy to forecast dam inflow up to 6-month lead time.

Extreme learning machines (ELM) has also been accomplished to understand the nonlinear behavior of streamflow forecasting. Lima et al. [29] forecasted the daily streamflow using the ELM model in British Columbia, Canada. This research explained that the online sequential extreme learning machine (OSELM) model was trained utilizing abundant dataset to choose the optimal parameters, and generated the effective performance to forecast streamflow based on 1-, 2-, and 3-day lead times. Yadav et al. [30] verified the ELM model for streamflow forecasting in the Neckar River, Germany.

They illustrated that the OSELM model forecasted streamflow up to 6 h lead time accurately compared to the ANN, support vector machine (SVM), and genetic programming. Yaseen et al. [2] investigated the ELM model for forecasting monthly streamflow in the Tigris River, Iraq. They concluded that the ELM model surpassed the SVR and the generalized regression neural network (GRNN) models to forecast the monthly streamflow. Rezaie-Balf and Kisi [13] applied the ELM model for daily streamflow forecasting in the Tajan River, Iran. This study revealed that the evolutionary polynomial regression (EPR) model outperformed the multilayer perceptron neural network (MLPNN) and optimally pruned extreme learning machine (OPELM) models to forecast daily streamflow. Niu et al. [31] developed the ELM model for forecasting daily streamflow in Xinfengjiang Reservoir, China. They demonstrated that the ELM integrated with quantum particle swarm optimization (ELM-QPSO) model enhanced the performance accuracy of ELM model to forecast daily streamflow. Under addressed research, Kernel extreme learning machine (KELM), a special type of ELM model, has been considered.

BMA model is a unique approach to implement a mechanism and clarify the model uncertainty [32]. However, the limited researchers have developed and applied the BMA model for fields of hydrology and water resources engineering including streamflow, rainfall, and water stage, etc. Duan et al. [33] employed the BMA model to develop the stable hydrologic predictions. They surveyed that the BMA model carried out the effective probabilistic prediction compared to the original ensemble model. Jiang et al. [34] investigated the BMA model for evaluating the multi-satellite precipitation using simulated hydrological streamflows, South China. This research showed that the satellite streamflow was merged by the BMA model, and the simulated streamflow was improved effectively. Wang et al. [35] developed the BMA model for rainfall forecasting based on seasonal concept, Australia. They inspected that the BMA model outperformed the specific model with two fixed predictors to forecast the merging seasonal rainfall. Rathinasamy et al. [36] developed the BMA model for forecasting streamflow at different time-scales (i.e., daily, weekly, and monthly) in two stations, USA. They produced several wavelet Volterra to obtain ensemble BMA model. The BMA model coupling ensemble multi wavelet Volterra outperformed the single wavelet Volterra and the mean averaged ensemble wavelet Volterra clearly to forecast daily, weekly, and monthly streamflow. Liu and Merwade [37] developed the BMA model to operate the system of water stage prediction in the Black River watershed, Missouri and Arkansas, USA. They reported that the BMA model provided the accurate prediction for flood water stage. In addition, flood inundation range estimated from BMA flood map was more effective than the probabilistic flood inundation range.

It can be considered from literature reviews of the BMA model that there have been no previously published the articles using the BMA model to compare the performance accuracy of MARS, M5Tree, and KELM models for streamflow forecasting until now. The purposes of this article can be arranged as follows: (1) to evaluate various input category of streamflow data with different antecedent times, (2) to compare and assess the performance accuracy of multivariate adaptive regression spline (MARS), M5 model tree (M5Tree), Kernel extreme learning model (KELM), and Bayesian model averaging (BMA) models for streamflow forecasting, and (3) to map the uncertainty ranges utilizing the performance of novel BMA model.

2. Methodology

2.1. Multivariate Adaptive Regression Spline (MARS)

MARS model (see Figure 1) does not require the particular presumptions of practical relationships between input and output indicators [38]. The performance of MARS model using spline functions gives larger flexibility than linear ones based on curvature and thresholds. The basic functions (BFs), which are assigned as smooth polynomials (e.g., splines), are built using two step approaches. In the first approach, the model performance is enhanced until probabilistic nodes are identified. The second includes the elimination of lowest real terms. Imagine y is an output indicator and $X = (X_1, \ldots, X_p)$ is an input indicator. Therefore, the actual response can be expressed using following Equation (1) [27,39].

$$y = f(X_1, \ldots, X_p) + e = f(x) + e \tag{1}$$

where e = the error distribution. The MARS model achieves the approximate function (f) using the BFs. The Equation (2) provides a linear combination of BFs and shared relation for the MARS model.

$$f(x) = \beta_0 + \sum_{m=1}^{M} \beta_m \lambda_m(x) \tag{2}$$

where individual $\lambda_m(x)$ = a spline function or output of two (or more) spline functions. The least squares method (LSM) can evaluate the coefficients β_0 (i.e., constant value). A model, therefore, can form the training error (e.g., having maximum reduction) using separating β_0 and basis pair. The following pair is boosted to the addressed model based on the M BFs as [27,39]:

$$\hat{\beta}_{M+1} \lambda_1(X) max(0, X_j - t) + \hat{\beta}_{M+2} \lambda_1(X) max(0, t - X_j) \tag{3}$$

where LSM can be applied for estimating β. When a novel BF is boosted to the model space, the associated interactions are recognized among the BFs. BFs are accumulated to the model for acquiring the maximum number of terms that deliver a sufficient fitness model. Then, a backward technique is applied to reduce the numbers of terms effectively. In the backward technique, BFs with the lowest accuracy are deleted to determine the best alternate model. Generalized cross validation (GCV), a method for comparing alternative models, can be represented as [27,39].

$$GCV = \frac{MSE}{\left[1 - \frac{N + dN}{M}\right]^2} \tag{4}$$

where M and N = the number of observations and BFs, respectively, MSE = mean squared error, and d = the penalty of each BF. To broaden the knowledge of MARS model, [27] provided the detailed theory and application using MARS models for streamflow forecasting.

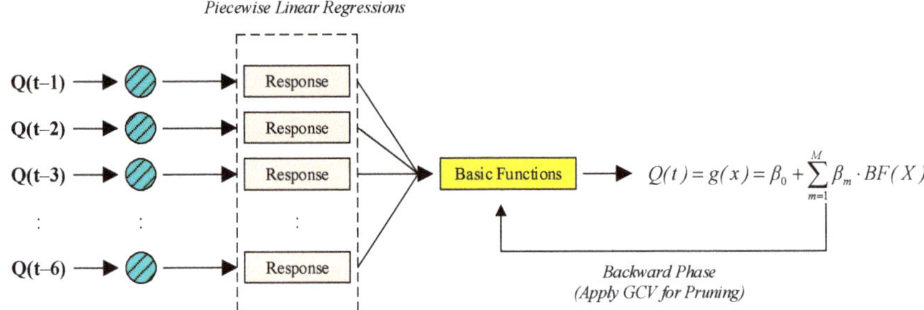

Figure 1. Architecture of multivariate adaptive regression spline (MARS) model (M6 category).

2.2. M5 Model Tree (M5Tree)

M5Tree model (see Figure 2) is a layered algorithm to judge the connection between input and output indicators [27,40]. The classification and regression trees (CART) is the basic algorithm for developing M5Tree model [41]. The M5Tree model assembles a linear-based model to the specific division which calculates the class properties of data portion leading to the leaf [27]. The standard deviation reduction (SDR) can influence to build the tree of M5Tree model. Additionally, it can reinforce the expected error reduction for specific points using Equation (5).

$$SDR = sd(E) - \sum_i \frac{|E_i|}{|E|} sd(E_i)$$ (5)

where E = a group of demonstrations that reach the leaf and E_i = a sub-group of input data to antecedent leaf. The pruning method was employed to suppress the overfitting burden and attain the accurate formation [27]. To apply the M5Tree model for streamflow forecasting, the previous articles (e.g., [20,27]) furnished the core approach to solve the addressed problems of streamflow forecasting.

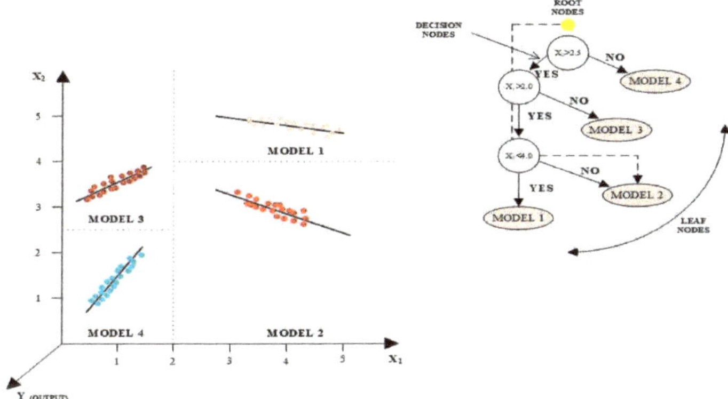

Figure 2. Architecture of M5Tree model.

2.3. Kernel Extreme Learning Machines (KELM)

ELM model (see Figure 3), one of novel training algorithms for feedforward neural networks (FFNN) with single hidden layer, was recommended by the previous article of Huang et al. [42] to lessen the handicaps of conventional training algorithm and enhance the model accuracy [43,44]. The training speed of ELM model, which generates the connection weights randomly in the hidden layer, is faster than that of other models. Additionally, the performance of ELM model shows robust generalizations with accurate control [42]. The aforementioned specification evaluates the ELM model as a superior model compared to other models with conventional training algorithm. The conventional version of the ELM model meets the disadvantages of providing diverse accuracies in various trials because of randomly assigned connection weights. To solve the weak point of standard ELM model, Huang et al. [45] supplied the Kernel ELM (i.e., KELM) model by improving the process of allocating random connection weights between the input and hidden layers, which explains briefly the theory of KELM model. Detailed demonstration can be found in published article of Huang et al. [45]. The conventional FFNN model (i.e., having single hidden layer) with N hidden nodes can be shown using Equation (6).

$$\sum_{i=1}^{N} \beta_i g(W_i x_i + b_i) = y_k, \ k = 1, 2, \ldots, M$$ (6)

where $g(\cdot)$, b_i, W_i, and β_i = transfer function, specified bias randomly, connection weights from hidden to output layer, and connection weights from hidden and output layer, respectively. Equation (6), therefore, can be re-written as [43].

$$H\beta = Y$$ (7)

where $Y = N$ target values and H = the matrix of hidden layer.

$$H = \begin{bmatrix} g(W_1 \cdot x_1 + b_1) & \cdots & g(W_M \cdot x_1 + b_M) \\ \vdots & \ddots & \vdots \\ g(W_1 \cdot x_N + b_1) & \cdots & g(W_M \cdot x_N + b_M) \end{bmatrix}_{N \times M} \tag{8}$$

where M = the number of nodes in the hidden layer. The connection weights in the output layer can be generated applying the Moore-Penrose generalized inverse (H^+) of hidden layer matrix.

$$\beta = H^+ Y \tag{9}$$

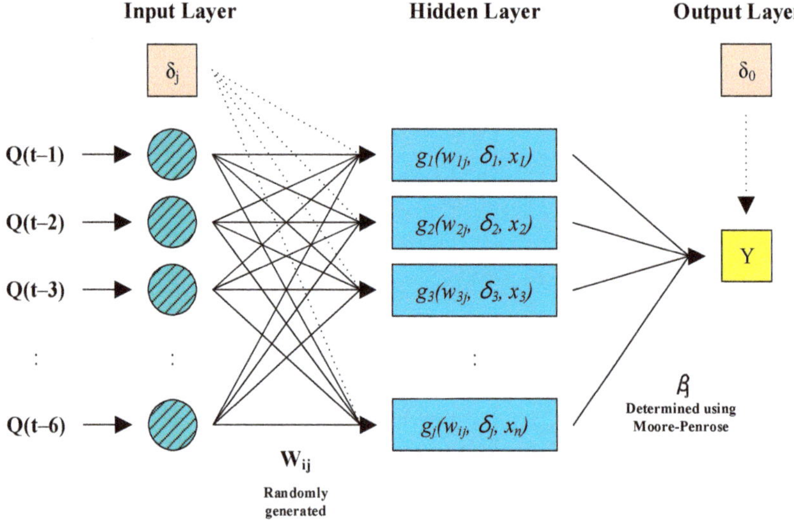

Figure 3. Architecture of ELM model (M6 category).

As one of accurate nonlinear regression models, the ELM model has been employed widely in the fields of hydrology and water resources engineering (e.g., [13,46]). In this study, 12 neurons and polynomial kernel were applied in hidden layer by applying trial and error process. Additionally, the regularization coefficient of KELM model was set to 10 to minimize difference between observed and forecasted streamflow values in both stations.

2.4. Bayesian Model Averaging (BMA)

BMA model, as a Bayesian inference, is implemented for model selection, forecasting, prediction, and estimation, and is also developed to combine the interferences and predictions from statistical models [32]. It can provide a criteria for simple model selection with limited simulations. One can simulate the parameter uncertainty utilizing a prior distribution as well as posterior parameter when requesting BMA model. This approach causes the brash inferences by neglecting uncertainty of candidate models [47]. This unique characteristics can provide a way to forecast natural behavior utilizing statistical post-processing approach [48–50]. Dismissing process, therefore, to acquire the posterior densities on BMA model parameters can be found from predictive probability density function (PDF) of x.

$$p(x|f_1, f_2, \ldots, f_M, \theta_1, \theta_2, \ldots, \theta_M) = \sum_{k=1}^{M} \omega_k g_k(x|f_k \theta_k) \tag{10}$$

where $f_1, f_2, \ldots, f_M$ = the group for candidate models of a specific quantity x based on temporal and spatial scale, θ_k = estimated parameter, and ω_k = connection weights for the relative performance of every ensemble member (f_k). Therefore, the connection weights form the probability density and $\sum_{k=1}^{M} \omega_k = 1$ [51]. In BMA model' category, f_k demonstrates a component PDF ($g_k(x|f_k\theta_k)$) [52].

2.5. Assessment of Models Performance

To figure out the performance of MARS, M5Tree, KELM, and BMA models, four assessment indexes were handled.

2.5.1. Root Mean Square Error (RMSE)

The error discrepancy between observed and forecasted streamflow values can be assessed by root mean square error (RMSE) function [53]. Perfect forecasting can be judged by RMSE = 0. In case of the highest error caused by the peak and higher values, RMSE can be distorted [54] and exploited for model evaluation with absolute units [55].

2.5.2. Nash-Sutcliffe Efficiency (NSE)

Evaluating the models' capability can be accomplished by the Nash-Sutcliffe efficiency (NSE) function [56]. NSE = 0 when the squared difference between observed and forecasted streamflow values is immense to approve the variance in observed streamflow values. If NSE < 0, this indicates that the observed mean is better than forecasted one by the model [57]. If NSE = 1, all points are ideal category [58].

2.5.3. Correlation Coefficient (R)

The correlation coefficient (R) is defined as the ratio of dependent indicator from the independent one. If R = 0, it implies that streamflow cannot be forecasted using developed models, whereas if R = 1, it demonstrates that the observed and forecasted streamflows have a strong correlation.

2.5.4. Mean Absolute Error (MAE)

The mean absolute error (MAE) can supply better knowledge for a model's forecasting, and cannot be contemplated in the vicinity of higher or lower significance. However, it assesses all derivations from observed streamflow values in the same manner [59]. If MAE = 0, it defines that the employed models can forecast streamflow absolutely, while if MAE = 1, it describes that the observed and forecasted streamflows do not have any relationship for forecasting category.

Four assessment indexes (i.e., RMSE, NSE, R, and MAE) can be implemented as Equations (11)–(14), respectively.

$$RMSE = \sqrt{\frac{1}{n}\sum_{i=1}^{n}[S_{obs} - S_{for}]^2} \tag{11}$$

$$NSE = 1 - \frac{\sum_{i=1}^{n}[S_{obs} - S_{for}]^2}{\sum_{i=1}^{n}[S_{obs} - \overline{S}_{for}]^2} \tag{12}$$

$$R = \frac{\sum_{i=1}^{n}(S_{obs} - \overline{S}_{obs})(S_{for} - \overline{S}_{for})}{\sqrt{\sum_{i=1}^{n}(S_{obs} - \overline{S}_{obs})^2 \sum_{i=1}^{n}(S_{for} - \overline{S}_{for})^2}} \tag{13}$$

$$MAE = \frac{1}{N}\sum_{i=1}^{N} \left| S_{for} - S_{obs} \right| \tag{14}$$

where S_{obs} and S_{for} are the observed and forecasted streamflow values; $\overline{S}_{obs}$ and $\overline{S}_{for}$ are the observed and forecasted mean streamflow values; and n is total number of employed data.

3. Study Area and Data

Under the addressed research, two stations (i.e., Hongcheon and Jucheon) were appointed for forecasting streamflow of the Hongcheon and Jucheon Streams (e.g., branches of the Han River), South Korea. Hongcheon station is located at Hongcheon Bridge with a latitude of 37°41′ N and a longitude of 127°52′ E, and Jucheon station is located at Jucheon Bridge with a latitude of 37°16′ N and a longitude of 128°15′ E, respectively.

Streamflow data in both stations have been collected and managed in the Water Resources Management Information System (WAMIS) of South Korea. The data available were divided into two phases: 80% (1 October 2003–30 September 2011) of the whole data was employed for the training phase and the remainder (i.e., 20%) (1 October 2011–30 September 2013) of the dataset was kept for the testing phase. The schematic diagrams of Hongcheon and Jucheon stations are provided in Figure 4. The properties of the used data for models' development are summed up in Table 1. The statistical evidence that the streamflow data have highly skewed distributions indicates the chaotic behavior of the studied data. Since the streamflow follows a complicated transformation of excess rainfall, surface, and subsurface flows, Salas et al. [60] reported that the streamflow bounced off chaotic behavior with low dimension for the outlet of specific watershed.

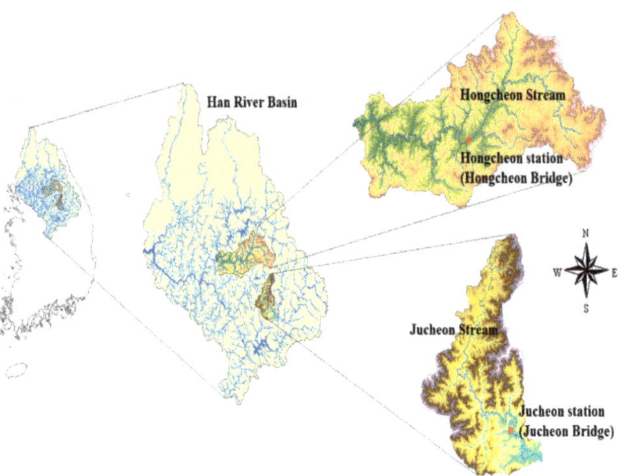

Figure 4. Schematic diagram of research area.

Table 1. Statistical properties of the streamflow data.

	Hongcheon		Jucheon	
	Training	**Testing**	**Training**	**Testing**
Number	**2922**	**731**	**2922**	**731**
Maximum	1951.5	1362	2720.4	515.5
Minimum	2.92	0.92	0.01	1.12
Average	67.245	32.552	27.519	16.203
Standard Deviation	111.949	83.321	105.677	39.099
Skewness	4.844	9.310	11.966	7.207

One of the important projects for streamflow forecasting is determination of the appropriate input variables [1,27]. For the application of forecasting model, the optimal selection of the best input combination based on the antecedent times was suggested using six different combinations. The detection of appropriate antecedent times for streamflow forecasting can be clarified as identification of catchment characteristics including area, shape, length, and slope, etc. The previous research demonstrated that the recent antecedent times (e.g., $(t-1)$, $(t-2)$, and $(t-3)$) were better associated than the ancient ones [27,61]. Rezaie-Balf et al. [27] accomplished that the antecedent times for daily streamflow forecasting were determined as $(t-1)\sim(t-4)$ days in Tajan (Iran) and Hongcheon (South Korea) rivers. Under the addressed study, the antecedent times were increased to verify the effective outcomes (e.g., forecasting accuracy) of input combinations based on the article of [27]. Thus, the six input combinations (i.e., six categories from M1 to M6) using six antecedent times values which are provided in Table 2 were employed for the implemented methods.

Table 2. Different input combinations for streamflow forecasting.

Types	Input Combinations	Functions
M1	$t-1$	$Q(t) = f(Q(t-1))$
M2	$t-1, t-2$	$Q(t) = f(Q(t-1), Q(t-2))$
M3	$t-1, t-2, t-3$	$Q(t) = f(Q(t-1), Q(t-2), Q(t-3))$
M4	$t-1, t-3, t-5$	$Q(t) = f(Q(t-1), Q(t-3), Q(t-5))$
M5	$t-2, t-4, t-6$	$Q(t) = f(Q(t-2), Q(t-4), Q(t-6))$
M6	$t-1, t-2, t-3, t-4, t-5, t-6$	$Q(t) = f(Q(t-1), Q(t-2), Q(t-3),$ $Q(t-4), Q(t-5), Q(t-6))$

4. Application and Results

4.1. Hongcheon Station

Forecasting accuracy of developed models during testing phase are provided in Table 3 for Hongcheon station. Bold values indicate the best category of each model (i.e., MARS, M5Tree, KELM, and BMA). The MARS1 model (RMSE = 52.214 m^3/s and NSE = 0.609) suggested the best performance among all the MARS models. Additionally, the M5Tree4 model (RMSE = 52.528 m^3/s and NSE = 0.605) provided the best achievement among all the M5Tree models. Besides, the KELM3 model (RMSE = 51.242 m^3/s and NSE = 0.624) supported the best accomplishment among all the KELM models. Finally, the BMA2 model (RMSE = 49.507 m^3/s and NSE = 0.649) furnished the best accuracy among all the BMA models for streamflow forecasting. Additionally, it can be found from Table 3 that the BMA models provided better performance than the MARS, M5Tree, and KELM models based on each category (i.e., Categories M1–M6) during testing phase. Therefore, the BMA2 (i.e., having $t-1$ and $t-2$ antecedent times) model supplied the best forecasting accuracy compared to the other models, whereas the M5-based models (i.e., MARS5, M5Tree5, KELM5, and BMA5) showed the worst performance considering all models and categories.

Table 3. Performance of MARS, M5Tree, KELM, and Bayesian model averaging (BMA) models in terms of root mean square error (RMSE), Nash-Sutcliffe Efficiency (NSE), correlation coefficient (R), and mean absolute error (MAE) values during testing phase (Hongcheon station).

Category	Assessment Indexes	MARS1	M5Tree1	KELM1	BMA1
M1	RMSE (m^3/s)	**52.214**	52.866	51.541	50.887
	NSE	**0.609**	0.600	0.619	0.629
	R	**0.780**	0.780	0.789	0.798
	MAE (m^3/s)	**15.510**	14.860	15.890	19.280
		MARS2	**M5Tree2**	**KELM2**	**BMA2**
M2	RMSE (m^3/s)	56.026	55.442	55.160	**49.507**
	NSE	0.550	0.560	0.564	**0.649**
	R	0.743	0.749	0.751	**0.812**
	MAE (m^3/s)	16.210	15.160	14.740	**17.200**
		MARS3	**M5Tree3**	**KELM3**	**BMA3**
M3	RMSE (m^3/s)	54.280	57.704	**51.242**	50.212
	NSE	0.578	0.523	**0.624**	0.639
	R	0.760	0.732	**0.789**	0.805
	MAE (m^3/s)	15.500	14.320	**13.470**	15.010
		MARS4	**M5Tree4**	**KELM4**	**BMA4**
M4	RMSE (m^3/s)	58.394	**52.528**	53.611	49.933
	NSE	0.511	**0.605**	0.588	0.643
	R	0.715	**0.779**	0.767	0.808
	MAE (m^3/s)	16.110	**15.720**	14.080	14.720
		MARS5	**M5Tree5**	**KELM5**	**BMA5**
M5	RMSE (m^3/s)	71.612	72.264	69.496	69.283
	NSE	0.266	0.252	0.308	0.313
	R	0.524	0.505	0.555	0.562
	MAE (m^3/s)	26.860	23.910	21.360	25.040
		MARS6	**M5Tree6**	**KELM6**	**BMA6**
M6	RMSE (m^3/s)	54.154	58.372	51.473	51.212
	NSE	0.580	0.512	0.620	0.624
	R	0.762	0.715	0.794	0.790
	MAE (m^3/s)	15.570	16.310	15.690	14.410

The scatter diagrams between the observed and forecasted streamflow values using the best models (i.e., MARS1, M5Tree4, KELM3, and BMA2) based on each model (i.e., MARS, M5Tree, KELM, and BMA) and category for Hongcheon station are illustrated in Figure 5a–d including the exact ($y = x$) line, fitted line, and R value, respectively. It can be judged that the forecasted streamflow values based on the BMA2 model were more adjacent to the equivalent observed values during testing phase. Figure 6 supports the RMSE values for each model during testing phase in Hongcheon station. It can be observed from Figure 6 that the BMA models based on M1–M6 categories provided lower RMSE compared to other models during testing phase.

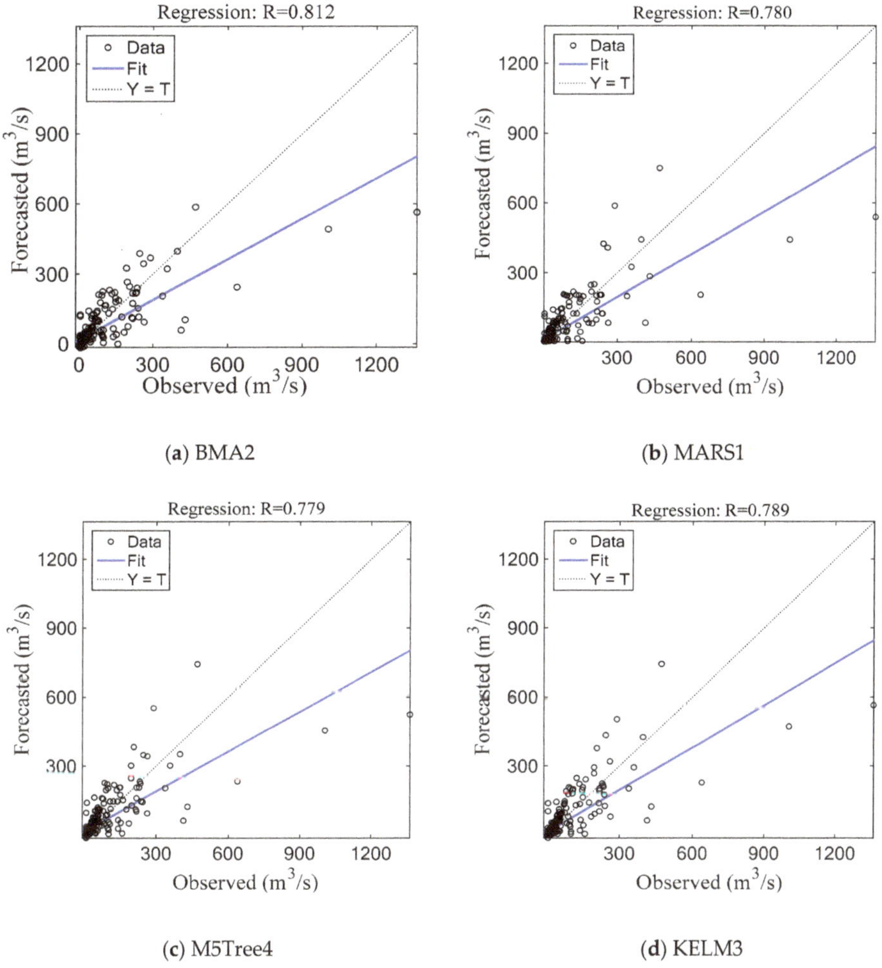

(**a**) BMA2   (**b**) MARS1

(**c**) M5Tree4   (**d**) KELM3

Figure 5. Scatter diagrams for the best models during testing phase (Hongcheon station).

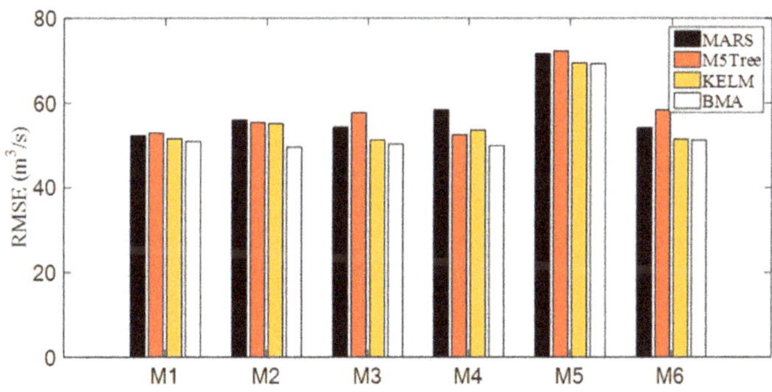

Figure 6. Comparison of RMSE values for each model during testing phase (Hongcheon station).

Figure 7a,b present the comparison of methods in streamflow forecasting based on M2 category during testing phase in Hongcheon station. It can be judged from Figure 7a that the BMA2 model forecasted the observed streamflow closely compared to other models (i.e., MARS2, M5Tree2, and KELM2). Additionally, Figure 7b explains the uncertainty estimation using 95% prediction interval for the BMA2 model. It can be found from Figure 7b that the BMA2 model provided higher uncertainty of peak flows compared to that of low flows.

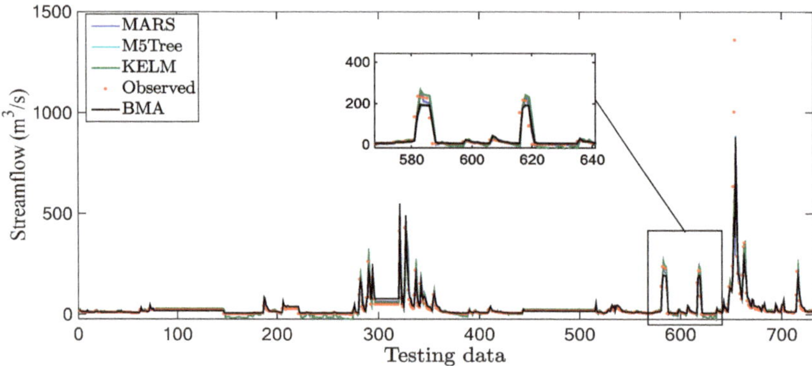

(**a**) Relationship between observed and forecasted daily streamflow

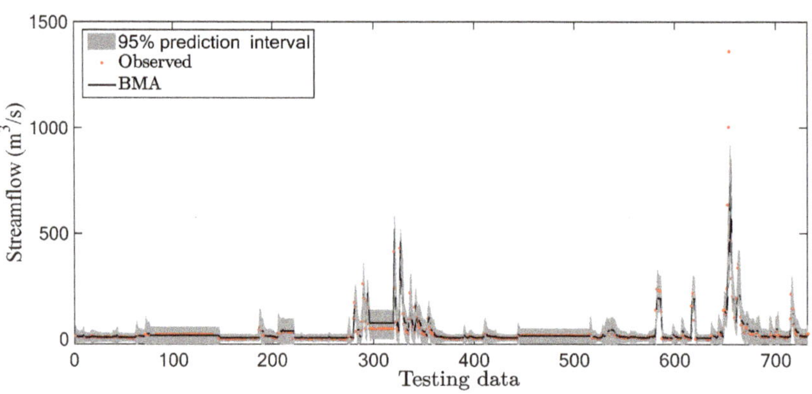

(**b**) 95% prediction interval for uncertainty estimation

Figure 7. Comparison of streamflow based on M2 category during testing phase (Hongcheon station).

In addition, Figure 8a,b provide that the comparison of streamflows based on M4 category during testing phase in Hongcheon station. Furthermore, it can be seen from Figure 8a that the BMA4 model better forecasted the observed streamflow than the alternative models (i.e., MARS4, M5Tree4, and KELM4). Additionally, Figure 8b represents the uncertainty estimation using 95% prediction interval for the BMA4 model. It can be seen from Figure 8b that the BMA4 model has also higher uncertainty in catching peak flows compared to that of low flows.

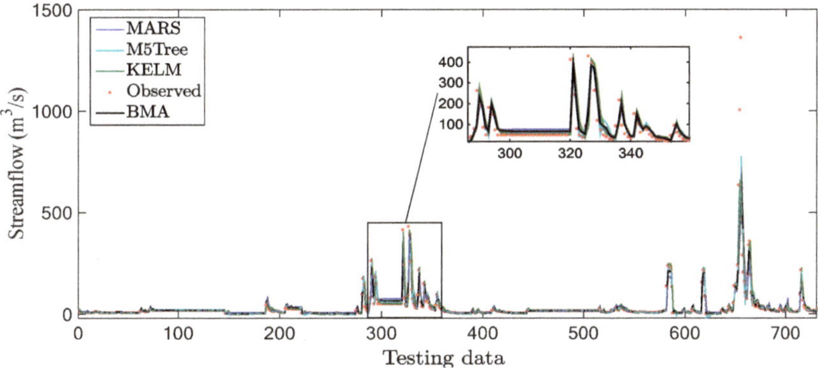

(a) Relationship between observed and forecasted daily streamflow

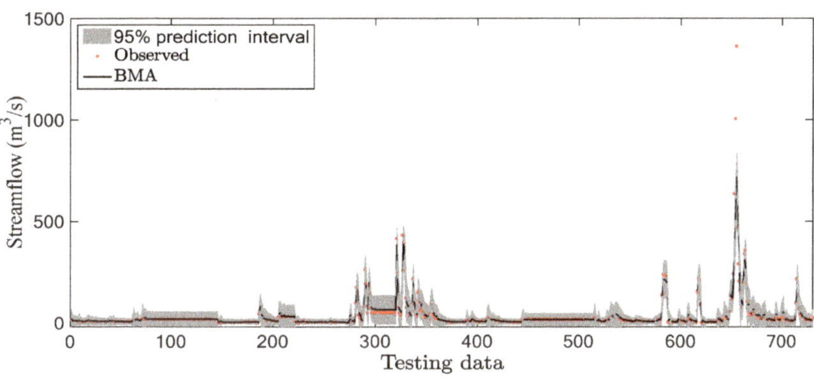

(b) 95% prediction interval for uncertainty estimation

Figure 8. Comparison of streamflow based on M4 category during testing phase (Hongcheon station).

4.2. Jucheon Station

Table 4 supplies the forecasted accuracy of employed models during testing phase in Jucheon station. Bold values display the best category of each model (i.e., MARS, M5Tree, KELM, and BMA). M1-based models (RMSE = 30.429 m^3/s and NSE = 0.397 in MARS1; RMSE = 31.367 m^3/s and NSE = 0.360 in M5Tree1; RMSE = 29.498 m^3/s and NSE = 0.434 in KELM1; RMSE = 28.396 m^3/s and NSE = 0.475 in BMA1) suggested the best accuracy based on each category (i.e., Categories M1–M6) for streamflow forecasting. Additionally, it can be seen from Table 4 that the BMA model provided better forecasting ability than the MARS, M5Tree, and KELM models considering each category. Based on all models and categories, the BMA1 (i.e., having $t-1$ antecedent time) model furnished the best forecasting compared to the other models, while the worst accuracy was accomplished by the M5-based models during testing phase in Jucheon station.

Table 4. Performance of MARS, M5Tree, KELM, and BMA models in terms of RMSE, NSE, and R values during testing phase (Jucheon station).

Category	Assessment Indexes	MARS1	M5Tree1	KELM1	BMA1
		MARS1	**M5Tree1**	**KELM1**	**BMA1**
M1	RMSE (m^3/s)	**30.429**	**31.367**	**29.498**	**28.396**
	NSE	**0.397**	**0.360**	**0.434**	**0.475**
	R	**0.688**	**0.670**	**0.664**	**0.689**
	MAE (m^3/s)	**9.910**	**10.390**	**7.380**	**7.850**
		MARS2	**M5Tree2**	**KELM2**	**BMA2**
M2	RMSE (m^3/s)	34.026	31.972	29.730	29.083
	NSE	0.247	0.335	0.425	0.449
	R	0.642	0.654	0.667	0.670
	MAE (m^3/s)	10.760	9.900	7.600	7.840
		MARS3	**M5Tree3**	**KELM3**	**BMA3**
M3	RMSE (m^3/s)	32.925	32.554	30.346	29.321
	NSE	0.294	0.310	0.401	0.440
	R	0.656	0.658	0.648	0.664
	MAE (m^3/s)	10.490	13.180	7.700	7.940
		MARS4	**M5Tree4**	**KELM4**	**BMA4**
M4	RMSE (m^3/s)	31.896	31.648	29.923	28.972
	NSE	0.337	0.347	0.416	0.454
	R	0.654	0.665	0.651	0.674
	MAE (m^3/s)	10.540	11.720	7.550	7.990
		MARS5	**M5Tree5**	**KELM5**	**BMA5**
M5	RMSE (m^3/s)	37.917	38.960	35.990	34.657
	NSE	0.063	0.011	0.156	0.219
	R	0.465	0.415	0.445	0.469
	MAE (m^3/s)	15.410	19.110	11.960	11.870
		MARS6	**M5Tree6**	**KELM6**	**BMA6**
M6	RMSE (m^3/s)	30.690	31.417	29.723	29.092
	NSE	0.386	0.357	0.424	0.449
	R	0.676	0.651	0.656	0.670
	MAE (m^3/s)	9.730	11.800	7.610	8.190

Considering all models and categories, the observed and forecasted streamflow values using the best models (i.e., MARS1, M5Tree1, KELM1, and BMA1) are illustrated in Figure 9a–d including the exact ($y = x$) line, fitted line, and R value for Jucheon station, respectively. It can be seen that the forecasted streamflow values based on BMA1 model were more neighboring to the corresponding observed values during testing phase. Figure 10 explains the RMSE values for each model during testing phase in Jucheon station. It can be seen from the figure that the BMA models based on M1–M6 categories provided lower RMSE compared to other models.

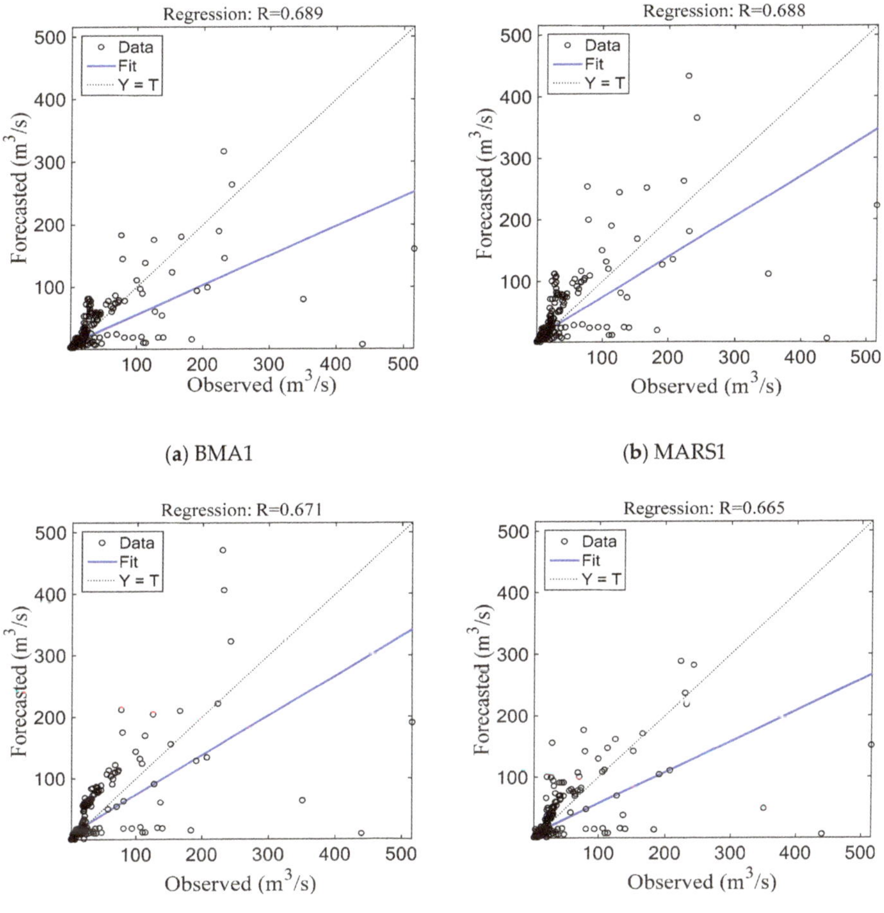

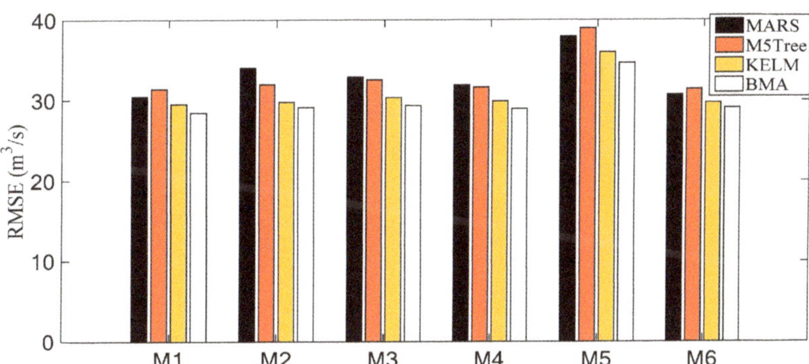

(**a**) BMA1   (**b**) MARS1

(**c**) M5Tree1   (**d**) KELM1

Figure 9. Scatter diagrams for the best models during testing phase (Jucheon station).

Figure 10. Comparison of RMSE values for each model during testing phase (Jucheon station).

Figure 11a,b present the comparison of methods in streamflow forecasting based on M1 category during testing phase in Jucheon station. It can be seen in Figure 11a that the BMA1 model provided better forecasting performance compared to other models (i.e., MARS1, M5Tree1, and KELM1). Additionally, Figure 11b explains the uncertainty estimation using 95% prediction interval for the BMA1 model. It can be considered from Figure 11b that the BMA1 model provided higher uncertainty of peak flows compared to that of low flows. Besides, Figure 12a,b yields the comparison of methods based on M4 category during testing phase in Jucheon station. It can be seen from Figure 12a that the BMA4 model produced better forecasting compared to other models (i.e., MARS4, M5Tree4, and KELM4). Additionally, Figure 12b describes the uncertainty estimation using 95% prediction interval for the BMA4 model. It can be considered from Figure 12b that the BMA4 model provided higher uncertainty for catching peak flows compared to that of low flows.

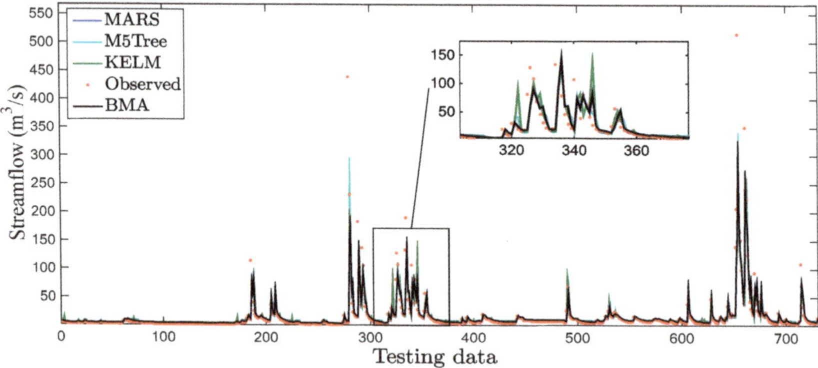

(**a**) Relationship between observed and forecasted daily streamflow

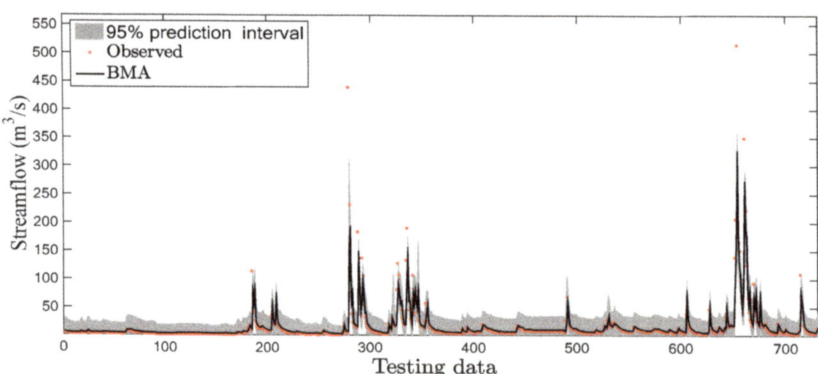

(**b**) 95% prediction interval for uncertainty estimation

Figure 11. Comparison of streamflow based on M1 category during testing phase (Jucheon station).

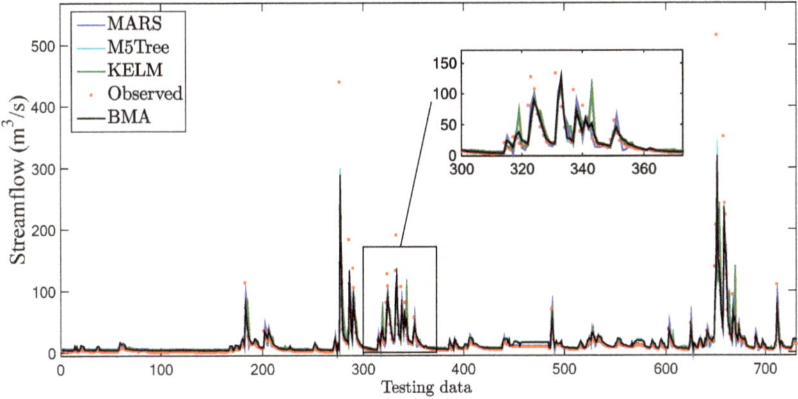

(**a**) Relationship between observed and forecasted daily streamflow

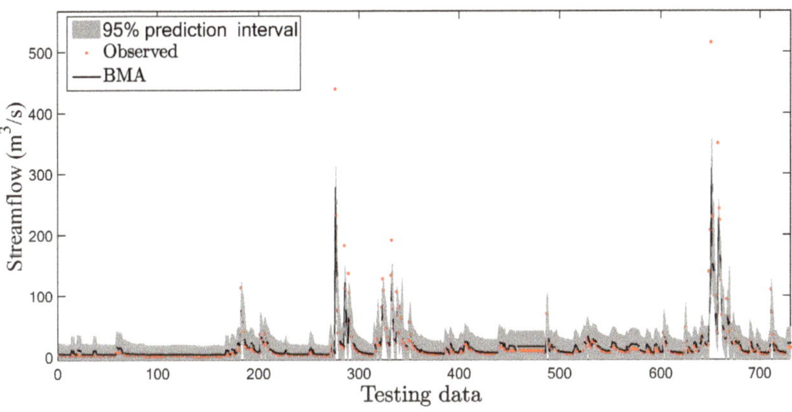

(**b**) 95% prediction interval for uncertainty estimation

Figure 12. Comparison of streamflow based on M4 category during testing phase (Jucheon station).

4.3. Discussion

The addressed research boosted that the BMA models based on each category correctly captured the nonlinear time series of streamflow and could carry out the accurate forecasting in both stations. The comparison of individual RMSE values among the best models in Hongcheon station supplied that the BMA2 model enhanced the accomplishment by 5.2%, 5.8%, and 3.4% compared to MARS1, M5Tree4, and KELM3 models during testing phase, respectively. Additionally, the comparison of individual RMSE values among the best models in Jucheon station furnished that the BMA1 model increased an efficiency by 6.7% (MARS1 model), 9.5% (M5Tree1 model), and 3.7% (KELM1 model) during testing phase.

Based on the category of the best models, the forecasted accuracy using the BMA model in Hongcheon and Jucheon stations was found to be slightly better than the other models. In addition, the best models in the Hongcheon station could be found considering the different category (i.e., MARS1, M5Tree4, KELM3, and BMA2 models), whereas the best models in the Jucheon station were discovered based on the M1 category (i.e., MARS1, M5Tree1, KELM1, and BMA1) during testing phase, respectively. The improvement difference between both stations might be derived from the characteristics (e.g., maximum and minimum values) of data available. The similar results can be found from the previous documents [2,62,63]. Additionally, comparison of two stations revealed that the employed models were more successful in forecasting streamflows of Jucheon station compared to Hongcheon station. This can be explained by the different properties of the data sets, for example, training data of Jucheon station have much more skewed distribution (skewness = 11.966, Table 1) than those of the other station. If the different models produced the best accuracy using the same data, the additional statistical skills (e.g., null hypothesis [64] and Akaike's information criterion [65]) are proposed to determine the best model for the undergoing project. Considering the previous researches for Bayesian approaches, Rasouli et al. [62] proposed that three machine learning models (i.e., Bayesian neural network (BNN), support vector regression (SVR), and Gaussian process (GP)) were utilized to forecast the daily streamflow using from 1- to 7-day lead time, British Columbia, Canada. The BNN model outperformed other models slightly. Wang et al. [35] proved that the BMA–ensemble–wavelet–Volterra model were superior to the wavelet–Volterra and ensemble–wavelet–Volterra models, obviously. Therefore, the forecasting accuracy of addressed research follows the previous researches.

As one of the continuous projects for streamflow forecasting, the different forecasting models (e.g., seasonal autoregressive integrated moving average (SARIMA) [66,67] and bootstrap aggregation (bagging) [68]), which demonstrated their superiority for temporal forecasting in previous literature, can be applied to compare and evaluate the performance accuracy of BMA model. In addition, different nature-inspired evolutionary algorithms and data pre-processing approaches can be joined with the BMA model to increase the forecasting accuracy of hydrological processes including streamflow, water stage, and groundwater, etc. Thus, to boost the forecasting accuracy of undergoing project, the continuous researches utilizing the BMA model, evolutionary algorithms, and data pre-processing techniques should be recommended for daily streamflow forecasting.

5. Conclusions

Accurate streamflow forecasting is a major problem of interest related to water resources and hydrology. This research evaluated the efficiency of Bayesian model averaging (BMA) model for daily streamflow forecasting in two different streams including Hongcheon and Jucheon stations, South Korea. Six categories (i.e., M1–M6) of input combination using different antecedent times were employed for streamflow forecasting. Additionally, the forecasting accuracy of the BMA model were compared with those of other models (i.e., MARS, M5Tree, and KELM) with respect to root mean square error (RMSE), Nash-Sutcliff efficiency (NSE), correlation coefficient (R), and mean absolute error (MAE).

The forecasting accuracy confirmed that the best BMA model (i.e., BMA2) increased an achievement by 5.2% (MARS1 model), 5.8% (M5Tree4 model), and 3.4% (KELM3 model) based on RMSE values among the best models during testing phase in Hongcheon station. Additionally, the best BMA model (i.e., BMA1) enhanced an accuracy by 6.7% (MARS1 model), 9.5% (M5Tree1 model), and 3.7% (KELM1 model) based on the best models during testing phase in Jucheon station. In addition, the best BMA models (i.e., BMA2 in Hongcheon station and BMA1 in Jucheon station) permitted the uncertainty estimation, and accomplished higher uncertainty of peak flows compared to that of low flows.

The addressed research outcomes suggested that the BMA model could be successfully employed for streamflow forecasting with different antecedent times for sustainable and efficient water management. For the continuous research, the different hybrid approaches such as coupling BMA model, evolutionary algorithm, and data pre-processing, can be recommended as a potential alternative methodology to enhance the forecasting accuracy based on diverse hydrological processes.

Author Contributions: Conceptualization, S.K. and N.W.K.; methodology, S.K. and M.A.; software, M.A.; validation, S.K. and N.W.K.; formal analysis, S.K. and O.K.; investigation, S.K. and O.K.; data curation, S.K.; writing—original draft preparation, S.K.; writing—review and editing, N.W.K. and O.K.; visualization, S.K. and M.A.; supervision, N.W.K.; funding acquisition, N.W.K. All authors have read and agreed to the published version of the manuscript.

Funding: This research was funded by the Korea Institute of Civil Engineering and Building Technology, grant number 20200027-001.

Acknowledgments: The authors would like to reveal our extreme appreciation and gratitude to the Water Resources Management Information System (http://www.wamis.go.kr/), South Korea. This is for providing the meteorological information. This research was supported by a grant (20200027-001) from a Strategic Research Project (Development of Hydrological Safety Assessment System for Hydraulic Structures) funded by the Korea Institute of Civil Engineering and Building Technology.

Conflicts of Interest: The authors declare no conflict of interest.

References

1. Seo, Y.; Kim, S.; Kisi, O.; Singh, V.P. Daily water level forecasting using wavelet decomposition and artificial intelligence techniques. *J. Hydrol.* **2015**, *520*, 224–243. [CrossRef]
2. Yaseen, Z.M.; Jaafar, O.; Deo, R.C.; Kisi, O.; Adamowski, J.; Quilty, J.; El-Shafie, A. Stream-flow forecasting using extreme learning machines: A case study in a semi-arid region in Iraq. *J. Hydrol.* **2016**, *542*, 603–614. [CrossRef]
3. Zakhrouf, M.; Bouchelkia, H.; Stamboul, M.; Kim, S. Novel hybrid approaches based on evolutionary strategy for streamflow forecasting in the Chellif River, Algeria. *Acta Geophys.* **2020**, *68*, 167–180. [CrossRef]
4. Zakhrouf, M.; Bouchelkia, H.; Stamboul, M.; Kim, S.; Singh, V.P. Implementation on the evolutionary machine learning approaches for streamflow forecasting: Case study in the Seybous River, Algeria. *J. Korea Water Resour. Assoc.* **2020**, *53*, 395–408.
5. Badrzadeh, H.; Sarukkalige, R.; Jayawardena, A.W. Intermittent stream flow forecasting and modelling with hybrid wavelet neuro-fuzzy model. *Hydrol. Res.* **2018**, *49*, 27–40. [CrossRef]
6. Zhou, J.; Peng, T.; Zhang, C.; Sun, N. Data pre-analysis and ensemble of various artificial neural networks for monthly streamflow forecasting. *Water* **2018**, *10*, 628. [CrossRef]
7. Tikhamarine, Y.; Souag-Gamane, D.; Ahmed, A.N.; Kisi, O.; El-Shafie, A. Improving artificial intelligence models accuracy for monthly streamflow forecasting using grey Wolf optimization (GWO) algorithm. *J. Hydrol.* **2020**, *582*, 124435. [CrossRef]
8. Yaseen, Z.M.; Allawi, M.F.; Yousif, A.A.; Jaafar, O.; Hamzah, F.M.; El-Shafie, A. Non-tuned machine learning approach for hydrological time series forecasting. *Neural. Comput. Appl.* **2018**, *30*, 1479–1491. [CrossRef]
9. Luo, X.; Yuan, X.; Zhu, S.; Xu, Z.; Meng, L.; Peng, J. A hybrid support vector regression framework for streamflow forecast. *J. Hydrol.* **2019**, *568*, 184–193. [CrossRef]
10. Cheng, M.; Fang, F.; Kinouchi, T.; Navon, I.M.; Pain, C.C. Long lead-time daily and monthly streamflow forecasting using machine learning methods. *J. Hydrol.* **2020**, *590*, 125376. [CrossRef]
11. Yu, X.; Wang, Y.; Wu, L.; Chen, G.; Wang, L.; Qin, H. Comparison of support vector regression and extreme gradient boosting for decomposition-based data-driven 10-day streamflow forecasting. *J. Hydrol.* **2020**, *582*, 124293. [CrossRef]
12. Papacharalampous, G.A.; Tyralis, H. Evaluation of random forests and prophet for daily streamflow forecasting. *Adv. Geosci.* **2018**, *45*, 201–208. [CrossRef]
13. Rezaie-Balf, M.; Kisi, O. New formulation for forecasting streamflow: Evolutionary polynomial regression vs. extreme learning machine. *Hydrol. Res.* **2018**, *49*, 939–953. [CrossRef]
14. Zakhrouf, M.; Bouchelkia, H.; Stamboul, M.; Kim, S.; Heddam, S. Time series forecasting of river flow using an integrated approach of wavelet multi-resolution analysis and evolutionary data-driven models. A case study: Sebaou River (Algeria). *Phys. Geogr.* **2018**, *39*, 506–522. [CrossRef]
15. Li, F.F.; Wang, Z.Y.; Qiu, J. Long-term streamflow forecasting using artificial neural network based on preprocessing technique. *J. Forecast.* **2019**, *38*, 192–206. [CrossRef]
16. Fu, M.; Fan, T.; Ding, Z.A.; Salih, S.Q.; Al-Ansari, N.; Yaseen, Z.M. Deep learning data-intelligence model based on adjusted forecasting window scale: Application in daily streamflow simulation. *IEEE Access* **2020**, *8*, 32632–32651. [CrossRef]
17. Al-Sudani, Z.A.; Salih, S.Q.; Yaseen, Z.M. Development of multivariate adaptive regression spline integrated with differential evolution model for streamflow simulation. *J. Hydrol.* **2019**, *573*, 1–12. [CrossRef]
18. Adamowski, J.; Chan, H.F.; Prasher, S.O.; Sharda, V.N. Comparison of multivariate adaptive regression splines with coupled wavelet transform artificial neural networks for runoff forecasting in Himalayan micro-watersheds with limited data. *J. Hydroinformatics* **2012**, *14*, 731–744. [CrossRef]

19. Tyralis, H.; Papacharalampous, G.; Langousis, A. Super ensemble learning for daily streamflow forecasting: Large-scale demonstration and comparison with multiple machine learning algorithms. *Neural. Comput. Appl.* **2020**, 1–16. [CrossRef]

20. Solomatine, D.P.; Xue, Y. M5 model trees and neural networks: Application to flood forecasting in the upper reach of the Huai River in China. *J. Hydrol. Eng.* **2004**, *9*, 491–501. [CrossRef]

21. Štravs, L.; Brilly, M. Development of a low-flow forecasting model using the M5 machine learning method. *Hydrol. Sci. J.* **2007**, *52*, 466–477. [CrossRef]

22. Sattari, M.T.; Pal, M.; Apaydin, H.; Ozturk, F. M5 model tree application in daily river flow forecasting in Sohu Stream, Turkey. *Water Resour.* **2013**, *40*, 233–242. [CrossRef]

23. Adnan, R.M.; Yuan, X.; Kisi, O.; Adnan, M.; Mehmood, A. Stream flow forecasting of poorly gauged mountainous watershed by least square support vector machine, fuzzy genetic algorithm and M5 model tree using climatic data from nearby station. *Water Resour. Manag.* **2018**, *32*, 4469–4486. [CrossRef]

24. Yaseen, Z.M.; Kisi, O.; Demir, V. Enhancing long-term streamflow forecasting and predicting using periodicity data component: Application of artificial intelligence. *Water Resour. Manag.* **2016**, *30*, 4125–4151. [CrossRef]

25. Yin, Z.; Feng, Q.; Wen, X.; Deo, R.C.; Yang, L.; Si, J.; He, Z. Design and evaluation of SVR, MARS and M5Tree models for 1, 2 and 3-day lead time forecasting of river flow data in a semiarid mountainous catchment. *Stoch. Environ. Res. Risk. Assess.* **2018**, *32*, 2457–2476. [CrossRef]

26. Kisi, O.; Choubin, B.; Deo, R.C.; Yaseen, Z.M. Incorporating synoptic-scale climate signals for streamflow modelling over the Mediterranean region using machine learning models. *Hydrol. Sci. J.* **2019**, *64*, 1240–1252. [CrossRef]

27. Rezaie-Balf, M.; Kim, S.; Fallah, H.; Alaghmand, S. Daily river flow forecasting using ensemble empirical mode decomposition based heuristic regression models: Application on the perennial rivers in Iran and South Korea. *J. Hydrol.* **2019**, *572*, 470–485. [CrossRef]

28. Rezaie-Balf, M.; Naganna, S.R.; Kisi, O.; El-Shafie, A. Enhancing streamflow forecasting using the augmenting ensemble procedure coupled machine learning models: Case study of Aswan High Dam. *Hydrol. Sci. J.* **2019**, *64*, 1629–1646. [CrossRef]

29. Lima, A.R.; Cannon, A.J.; Hsieh, W.W. Forecasting daily streamflow using online sequential extreme learning machines. *J. Hydrol.* **2016**, *537*, 431–443. [CrossRef]

30. Yadav, B.; Ch, S.; Mathur, S.; Adamowski, J. Discharge forecasting using an online sequential extreme learning machine (OS-ELM) model: A case study in Neckar River, Germany. *Measurement* **2016**, *92*, 433–445. [CrossRef]

31. Niu, W.J.; Feng, Z.K.; Cheng, C.T.; Zhou, J.Z. Forecasting daily runoff by extreme learning machine based on quantum-behaved particle swarm optimization. *J. Hydrol. Eng.* **2018**, *23*, 04018002. [CrossRef]

32. Vrugt, J.A.; Robinson, B.A. Treatment of uncertainty using ensemble methods: Comparison of sequential data assimilation and Bayesian model averaging. *Water Resour. Res.* **2007**, *43*, W01411. [CrossRef]

33. Duan, Q.; Ajami, N.K.; Gao, X.; Sorooshian, S. Multi-model ensemble hydrologic prediction using Bayesian model averaging. *Adv. Water Resour.* **2007**, *30*, 1371–1386. [CrossRef]

34. Jiang, S.; Ren, L.; Hong, Y.; Yong, B.; Yang, X.; Yuan, F.; Ma, M. Comprehensive evaluation of multi-satellite precipitation products with a dense rain gauge network and optimally merging their simulated hydrological flows using the Bayesian model averaging method. *J. Hydrol.* **2012**, *452*, 213–225. [CrossRef]

35. Wang, Q.J.; Schepen, A.; Robertson, D.E. Merging seasonal rainfall forecasts from multiple statistical models through Bayesian model averaging. *J. Clim.* **2012**, *25*, 5524–5537. [CrossRef]

36. Rathinasamy, M.; Adamowski, J.; Khosa, R. Multiscale streamflow forecasting using a new Bayesian Model Average based ensemble multi-wavelet Volterra nonlinear method. *J. Hydrol.* **2013**, *507*, 186–200. [CrossRef]

37. Liu, Z.; Merwade, V. Accounting for model structure, parameter and input forcing uncertainty in flood inundation modeling using Bayesian model averaging. *J. Hydrol.* **2018**, *565*, 138–149. [CrossRef]

38. Friedman, J. Multivariate adaptive regression splines. *Ann. Stat.* **1991**, *19*, 1–67. [CrossRef]

39. Zhang, W.G.; Goh, A.T.C. Multivariate adaptive regression splines for analysis of geotechnical engineering systems. *Comput. Geotech.* **2013**, *48*, 82–95. [CrossRef]

40. Solomatine, D.P.; Dulal, K.N. Model trees as an alternative to neural networks in rainfall—Runoff modelling. *Hydrol. Sci. J.* **2003**, *48*, 399–411. [CrossRef]

41. Loh, W.Y. Classification and regression trees. *Wiley Interdiscip. Rev. Data Min. Knowl. Discov.* **2011**, *1*, 14–23. [CrossRef]

42. Huang, G.B.; Zhu, Q.Y.; Siew, C.K. Extreme learning machine: Theory and applications. *Neurocomputing* **2006**, *70*, 489–501. [CrossRef]

43. Alizamir, M.; Kim, S.; Kisi, O.; Zounemat-Kermani, M. Deep echo state network: A novel machine learning approach to model dew point temperature using meteorological variables. *Hydrol. Sci. J.* **2020**, *65*, 1173–1190. [CrossRef]

44. Alizamir, M.; Kisi, O.; Ahmed, A.N.; Mert, C.; Fai, C.M.; Kim, S.; Kim, N.W.; El-Shafie, A. Advanced machine learning model for better prediction accuracy of soil temperature at different depths. *PLoS ONE* **2020**, *15*, e0231055. [CrossRef] [PubMed]

45. Huang, G.B.; Zhou, H.; Ding, X.; Zhang, R. Extreme learning machine for regression and multiclass classification. *IEEE Trans. Syst. Man Cybern. Syst.* **2012**, *42*, 513–529. [CrossRef] [PubMed]

46. Seo, Y.; Kim, S.; Singh, V.P. Comparison of different heuristic and decomposition techniques for river stage modeling. *Environ. Monit. Assess* **2018**, *190*, 392. [CrossRef] [PubMed]

47. Raftery, A.E.; Madigan, D.; Hoeting, J.A. Bayesian model averaging for linear regression models. *J. Am. Stat. Assoc.* **1997**, *92*, 179–191. [CrossRef]

48. Sloughter, J.M.L.; Raftery, A.E.; Gneiting, T.; Fraley, C. Probabilistic quantitative precipitation forecasting using Bayesian model averaging. *Mon. Weather Rev.* **2007**, *135*, 3209–3220. [CrossRef]

49. Kisi, O.; Alizamir, M.; Gorgij, A.D. Dissolved oxygen prediction using a new ensemble method. *Environ. Sci. Pollut. Res.* **2020**, *27*, 9589–9603. [CrossRef]

50. Kisi, O.; Alizamir, M.; Trajkovic, S.; Shiri, J.; Kim, S. Solar radiation estimation in Mediterranean climate by weather variables using a novel Bayesian model averaging and machine learning methods. *Neural Process. Lett.* **2020**, *52*, 2297–2318. [CrossRef]

51. Baran, S. Probabilistic wind speed forecasting using Bayesian model averaging with truncated normal components. *Comput. Stat. Data Anal.* **2014**, *75*, 227–238. [CrossRef]

52. Raftery, A.E.; Gneiting, T.; Balabdaoui, F.; Polakowski, M. Using Bayesian model averaging to calibrate forecast ensembles. *Mon. Weather Rev.* **2005**, *133*, 1155–1174. [CrossRef]

53. Willmott, C.J.; Matsuura, K. Advantages of the mean absolute error (MAE) over the root mean square error (RMSE) in assessing average model performance. *Clim. Res.* **2005**, *30*, 79–82. [CrossRef]

54. Dawson, C.W.; Abrahart, R.J.; See, L.M. HydroTest: A web-based toolbox of evaluation metrics for the standardised assessment of hydrological forecasts. *Environ. Model. Softw.* **2007**, *22*, 1034–1052. [CrossRef]

55. Deo, R.C.; Şahin, M.; Adamowski, J.F.; Mi, J. Universally deployable extreme learning machines integrated with remotely sensed MODIS satellite predictors over Australia to forecast global solar radiation: A new approach. *Renew. Sust. Energ. Rev.* **2019**, *104*, 235–261. [CrossRef]

56. Nash, J.E.; Sutcliffe, J.V. River flow forecasting through conceptual models, Part 1 – A discussion of principles. *J. Hydrol.* **1970**, *10*, 282–290. [CrossRef]

57. Wilcox, B.P.; Rawls, W.J.; Brakensiek, D.L.; Wight, J.R. Predicting runoff from rangeland catchments: A comparison of two models. *Water Resour. Res.* **1990**, *26*, 2401–2410. [CrossRef]

58. Legates, D.R.; McCabe, G.J. Evaluating the use of "goodness-of-fit" measures in hydrologic and hydroclimatic model validation. *Water Resour. Res.* **1999**, *35*, 233–241. [CrossRef]

59. Chai, T.; Draxler, R.R. Root mean square error (RMSE) or mean absolute error (MAE)?—Arguments against avoiding RMSE in the literature. *Geosci. Model Dev.* **2014**, *7*, 1247–1250. [CrossRef]

60. Salas, J.D.; Kim, H.S.; Eykholt, R.; Burlando, P.; Green, T.R. Aggregation and sampling in deterministic chaos: Implications for chaos identification in hydrological processes. *Nonlinear Process. Geophys.* **2005**, *12*, 557–567. [CrossRef]

61. Yaseen, Z.M.; El-Shafie, A.; Jaafar, O.; Afan, H.A.; Sayl, K.N. Artificial intelligence based models for stream-flow forecasting: 2000–2015. *J. Hydrol.* **2015**, *530*, 829–844. [CrossRef]

62. Rasouli, K.; Hsieh, W.W.; Cannon, A.J. Daily streamflow forecasting by machine learning methods with weather and climate inputs. *J. Hydrol.* **2012**, *414*, 284–293. [CrossRef]

63. Tongal, H.; Booij, M.J. Simulation and forecasting of streamflows using machine learning models coupled with base flow separation. *J. Hydrol.* **2018**, *564*, 266–282. [CrossRef]

64. McCuen, R.H. *Microcomputer Applications in Statistical Hydrology*, 1st ed.; Prentice Hall: Eaglewood Cliffs, NJ, USA, 1993; pp. 20–48.

65. Akaike, H. A new look at the statistical model identification. *IEEE Trans. Automat. Contr.* **1974**, *19*, 716–723. [CrossRef]

66. Thiyagarajan, K.; Kodagoda, S.; Ranasinghe, R.; Vitanage, D.; Iori, G. Robust sensor suite combined with predictive analytics enabled anomaly detection model for smart monitoring of concrete sewer pipe surface moisture conditions. *IEEE Sens. J.* **2020**, *20*, 8232–8243. [CrossRef]
67. Thiyagarajan, K.; Kodagoda, S.; Van Nguyen, L.; Ranasinghe, R. Sensor failure detection and faulty data accommodation approach for instrumented wastewater infrastructures. *IEEE Access* **2018**, *6*, 56562–56574. [CrossRef]
68. Melesse, A.M.; Khosravi, K.; Tiefenbacher, J.P.; Heddam, S.; Kim, S.; Mosavi, A.; Pham, B.T. River water salinity prediction using hybrid machine learning models. *Water* **2020**, *12*, 2951. [CrossRef]

Publisher's Note: MDPI stays neutral with regard to jurisdictional claims in published maps and institutional affiliations.

sustainability

Article

Evaluating Irrigation Performance and Water Productivity Using EEFlux ET and NDVI

Usha Poudel, Haroon Stephen and Sajjad Ahmad *

Department of Civil and Environmental Engineering and Construction, University of Nevada Las Vegas, 4505 S. Maryland Pkwy, Las Vegas, NV 89154, USA; poudeu1@unlv.nevada.edu (U.P.); haroon.stephen@unlv.edu (H.S.)
* Correspondence: sajjad.ahmad@unlv.edu; Tel.: +1-702-895-5456

Abstract: Southern California's Imperial Valley (IV) faces serious water management concerns due to its semi-arid environment, water-intensive crops and limited water supply. Accurate and reliable irrigation system performance and water productivity information is required in order to assess and improve the current water management strategies. This study evaluates the spatially distributed irrigation equity, adequacy and crop water productivity (CWP) for two water-intensive crops, alfalfa and sugar beet, using remotely sensed data and a geographical information system for the 2018/2019 crop growing season. The actual crop evapotranspiration (ETa) was mapped in Google Earth Engine Evapotranspiration Flux, using the linear interpolation method in R version 4.0.2. The approx() function in the base R was used to produce daily ETa maps, and then totaled to compute the ETa for the whole season. The equity and adequacy were determined according to the ETa's coefficient of variation (CV) and relative evapotranspiration (RET), respectively. The crop classification was performed using a machine learning approach (a random forest algorithm). The CWP was computed as a ratio of the crop yield to the crop water use, employing yield disaggregation to map the crop yield, using county-level production statistics data and normalized difference vegetation index (NDVI) images. The relative errors (RE) of the ETa compared to the reported literature values were 7–27% for alfalfa and 0–3% for sugar beet. The average ETa variation was low; however, the spatial variation within the fields showed that 35% had a variability greater than 10%. The RET was high, indicating adequate irrigation; 31.5% of the alfalfa and 12% of the sugar beet fields clustered in the Valley's central corner were consuming more water than their potential visibly. The CWP showed wide variation, with CVs of 32.92% for alfalfa and 25.4% for sugar beet, signifying a substantial scope for CWP enhancement. The correlation between the CWP, ETa and yield showed that reducing the ETa to approximately 1500 mm for alfalfa and 1200 mm for sugar beet would help boost the CWP without decreasing the yield, which is nearly equivalent to 44.52M cu. m (36,000 acre-ft) of water. The study's results could help water managers to identify poorly performing fields where water conservation and management could be focused.

Keywords: EEFlux; irrigation performance; CWP; water conservation; NDVI

Citation: Poudel, U.; Stephen, H.; Ahmad, S. Evaluating Irrigation Performance and Water Productivity Using EEFlux ET and NDVI. *Sustainability* **2021**, *13*, 7967. https://doi.org/10.3390/su13147967

Academic Editor: Ozgur Kisi

Received: 10 June 2021
Accepted: 13 July 2021
Published: 16 July 2021

1. Introduction

Irrigated agriculture is the major consumer of freshwater supplies, being attributed to 65% of total water withdrawals [1,2]. With a rising population, the demand for food production is growing, whereas the share of irrigation water for agriculture is declining. Primarily in the western United States, water management has become a complex issue [3–6] which is further aggravated by a semi-arid climate, periodic drought and low precipitation [2,7,8]. A collective large-scale irrigation scheme is often established to manage the irrigation in such regions; hence, evaluating and improving the performance of the system is a critical step towards establishing better water management practices [9].

The proper evaluation of an existing system is also one of many ways to achieve several of the Sustainable Development Goals (SDGs) developed by the United Nations

(UN) in 2015. Goal no. 2 aims to "end hunger, achieve food security and improved nutrition, and promote sustainable agriculture". Insufficient irrigation water is a major threat to food security in water-scarce areas; hence, the study of crop water consumptions and finding ways to ensure 'more crop per drop' are ways to attain this goal. Moreover, studies of this kind will also assist us in identifying the proportion of agricultural area under productive and sustainable agriculture, which is one of the indicators developed by the UN to track the progress for this goal. In addition to this, goal no. 6 aims to "ensure availability and sustainable management of water and sanitation for all". At the moment, population growth, agricultural intensification and urbanization are beginning to overwhelm the available freshwater resources [10–14]. With irrigation water being the largest consumer of freshwater resources, saving even a fraction of this can significantly ease the strain on other sectors [15,16]. The integrated challenge of maintaining water and food security to achieve SDGs 2 and 6 in such a short period of time requires the extensive study of existing agricultural areas.

The concept of irrigation performance assessment has shifted during the last 25 years, from traditional irrigation efficiency measurements to performance indicators [17–19]. Several performance indicators have been introduced, based on adequacy [20–22], equity [20,22,23] reliability [24], productivity [20,25] and sustainability [26,27]. Indicators based on adequacy and equity are greater in number and have been employed in several studies [18]. Roerink et al. utilized the concept of relative evapotranspiration (RET) to investigate irrigation adequacy and water deficiency severity [23]. A similar concept was employed in other studies [2,28]. Likewise, coefficients of variation for actual evapotranspiration were used as an equity measure by numerous studies [2,17,29]. More recently, the attention to performance indicators based on productivity is also growing, primarily in regions where water is limited. Crop water productivity (CWP) provides information about how effectively water is being expended [30]. CWP, along with water use, has been used to assess water savings measures at different scales in the past, including the basin level [31–33] the irrigation scheme level [29,34–36] and the administrative division level [37].

Important aspects of irrigation performance indicators are the accurate estimation of the crop evapotranspiration (ET) and its spatial distribution. Recent improvements in remote sensing and satellite image products offer effective ways to estimate the spatial variation of ET [11,38]. Over the last few decades, a number of remote sensing techniques have been developed and used to estimate ET in large areas [39], including vegetation index (VI) methods and surface energy balance (SEB) methods. In a VI-based method, a relationship between the crop coefficient (kc) and VI is developed, and the ET is calculated based on the Food and Agriculture Organization (FAO) approach. [40] utilized the normalized difference vegetation index (NDVI) to determine the kc and crop ET, and they derived ET maps on a regional scale. With a slightly different approach, [41] employed additional parameters, including the crop cover fraction and the soil evaporation, to establish a relationship with the kc of a wheat crop. Unlike energy balance methods, these methods avoid complex processes of parameter estimation. However, the relationship developed between VI and crop coefficients varies with the location. Hence, it may require the modification and validation of the relationship in new setting [31].

Commonly used SEB models include the Surface Energy Balance Algorithm for Land (SEBAL) [42], Mapping Evapotranspiration at High Resolution using Internalized Calibration (METRIC) [43], and the Simplified Surface Energy Balance Index (S-SEBI) [44]. SEB models estimate the actual crop ET (ETa) as a residual of the surface energy budget and capture the impacts of poor water management on ETa [31,43]. Singh et al. evaluated different SEB models over the Midwestern US in the calculation of instantaneous ETa for irrigated maize crops [45]. The reported relative errors, when compared with flux tower measurements, were less than 10% for all of the models. A modification of the METRIC version, wet METRIC, was used conjunctively by [46] in the Midwest US to estimate the seasonal ETa. An R^2 value of 0.91 was obtained when comparing the modelled ET with the eddy covariance tower measured ET. Although the residual methods performed well, with promising accuracy [31], the complexity of the calibration, data entry and manipulations, and executions of these models requires a certain level of expertise. A recently introduced automated version of the METRIC algorithm in the EEFlux platform is a promising approach to obtain ETa maps without complex calculations. Costa et al. estimated the maize water consumption for different stages of maize growth development based on the EEFlux platform, with promising results [47]. Likewise, Venancio et al. [48] assessed soybean ET using EEFlux ETa, and the variation of the ETa in field was accurately estimated by the model. While the traditional version of the METRIC algorithm has been widely used and tested, the automated version of METRIC lacks extensive studies. Its applicability for regional water management is scarce and yet to be studied.

Yield mapping is another important aspect for the computation of an indicator based on productivity. Primarily, two different approaches, including crop growth models and empirical models, have been used in the past to estimate the crop yield using remote sensing [49,50]. The first method, though accurate, is limited by data availability [51]. The empirical method, on the other hand, can be utilized to assess the within-field variability in a simple and effective way [52]. Past studies showed that VI can explain up to 80% of the within-field yield variability [53,54]. The NDVI is one of the most widely used vegetation indices for the estimation of crop yield. Crop yield highly correlates with NDVI at specific growth stages [37]. Hence, some studies have directly utilized the production statistics from census data and successfully extrapolated it to the pixel level using NDVI as a medium [31,37]. A gap is often seen to utilize these two data sources conjunctively, because national statistics are often used only at the time of RS data interpretation [31,55]. The statistics on the district levels are collected in an organizational framework, are regularly available, and are widely accepted. Therefore, a simple disaggregating approach of published yield statistics to the pixel level, using remote sensing, can also help fill the gaps between two data sources. This may assist stakeholders to make use of the RS technique effectively.

Functioning as one of the country's largest irrigation projects, the Imperial Valley (IV) provides a significant contribution to the economy, as well as the nation's diet, health, and wellbeing [56]. Although the valley is highly dynamic and productive, the dry environment, water-intensive crops, and limited water supply render challenges in its water management. It is one of the major areas stressed by the Federal Government and State of California for water conservation [57]. Recently, a water transfer agreement was introduced that requires the transfer of about 10% of the IV's total Colorado river allotment to other Southern Californian regions. This new strain has further amplified the challenge water that managers are facing in the Valley. The prospects of the development of a new water supply are very limited in this scenario. However, the hospitable conditions of the Valley's environment support the growth of multiple crops, which would otherwise have been imported from other countries, adding a high economic value to its agriculture. Therefore, the proper assessment of the existing irrigation systems in the valley may help to identify the fields with low performance levels, where water management could be focused.

In this study, we quantified irrigation performance indicators based on adequacy, equity, and productivity, utilizing EEFlux ETa, NDVI and county-level statistics of the crop yield in the IV. The study focuses on fields where alfalfa and sugar beet crops are grown. A linear interpolation was performed in order to generate daily ETa maps for the crops' growing seasons in 2018 and 2019, which was further summated to obtain the total ET for the whole year. The computed ET was verified with literature-reported values and the ET computed from crop coefficient-reference evapotranspiration (kc-ETo) approach. Landsat NDVI images corresponding to the early growth stages for both crops were utilized to disaggregate the county-level yield statistics to the pixel level. A crop map was utilized to generate the crop specific ETa and yield map. We computed the seasonal RET as a ratio of the actual to the potential ET in order to assess the adequacy of the water in the fields. The United States Bureau of Reclamation (USBR) field boundary was utilized to estimate the coefficient of variation (CV) of the actual ET within and among fields, as a measure of equity. A CWP map was produced as a ratio of the crop yield and ETa. High and low zones of the indicators were identified, and possible reasons, along with prospects for improvement and implications were devised. Additionally, a relationship of the CWP with the yield and ETa was also interpreted in order to identify the possible scope of the water management through CWP enhancement. Research questions of interest include: (1) What is the accuracy of EEFlux ETa compared to the literature-reported values and ET from kc-ETo approach? (2) How do high- and low-performance fields differ in proportion and location in the valley? (3) What is the scope of water conservation through water productivity enhancement for high water-use crops?

The remaining document is organized as follows. In the next section, the datasets used for the research are properly outlined, along with the sources and data characteristics. After this section, the research methods are explained progressively. Subsequently, the research results and findings are explained, followed by a discussion. The references are listed at the end.

2. Study Area

The study area for this research is the IV, within the extent of the Imperial Irrigation District (IID) (Figure 1). The IV covers an approximate area of 2.07B sq. m (512,163 acres). It consists of very productive soils, resulting from the periodic flooding of the Colorado River in the past. Because of the extremely hot and dry climate, scant rainfall, and water-intensive crops, the Valley requires an extensive amount of water for its agriculture to thrive. Approximately 3.8B m^3/year (3.1M aft/year) of irrigation water is imported from the Colorado River, and its delivery to the field is managed by the IID. More than 4828 km (3000 miles) of canals and drains have been constructed for this purpose. The availability of irrigation water and rich soil makes it possible to grow hundreds of crops year-round amidst harsh climatic conditions. The major crops include alfalfa, sugar beet, sudan grass hay, winter vegetables, wheat and corn. Alfalfa supports huge industries of cattle and dairy production, and sugar beet is only produced in the IV among western US states. Both crops are grown year round, require intensive irrigation and have high field coverage. Hence, the study of the water use for alfalfa and sugar beets is the focus of this study.

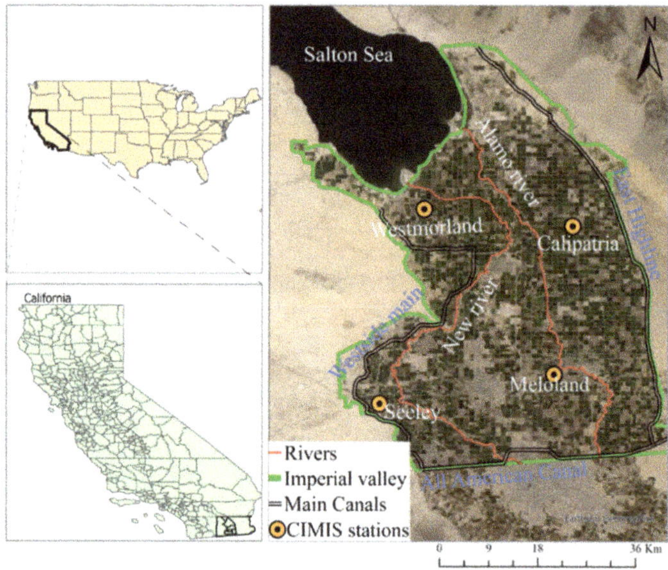

Figure 1. Study area location map that shows the major rivers flowing through the valley, as well as the major canals and meteorological stations.

Alfalfa is usually planted from mid-August to mid-March and is harvested three to four times a year. Sugar beets are planted between early August and October and are harvested between mid-April and mid-August. This study covers a full growing season of sugar beets from October 2018 to August 2019. Because alfalfa has a 3–4-year cropping cycle, the 2019 crop year was considered for the study. Hence, any crop-specific information and mapping presented in this study corresponds to each crop's growing season.

The weather conditions throughout the study period are shown in Figure 2. The weather data were obtained from the California Irrigation Management and Information System (CIMIS) (https://cimis.water.ca.gov/, accessed on 9 November 2020) for four stations located in the IV (Figure 1), and averaged. The average minimum (1.35 m/s) and maximum wind speeds (3.33 m/s) were observed during December and May, respectively. The mean solar radiation was high during most of the spring and summer months, and relatively low during the winter. Likewise, the average monthly temperatures were high most of the year, and ranged between 53 and 93 °F. The warmest month of the year was August (93 °F), whereas January and December (53 °F) were the coldest. The valley experienced the most rainfall during late summer and winter. Precipitation as high as 17.18 mm was observed on 25 September 2019, whereas June, July, August and October did not experience any rainfall in 2019.

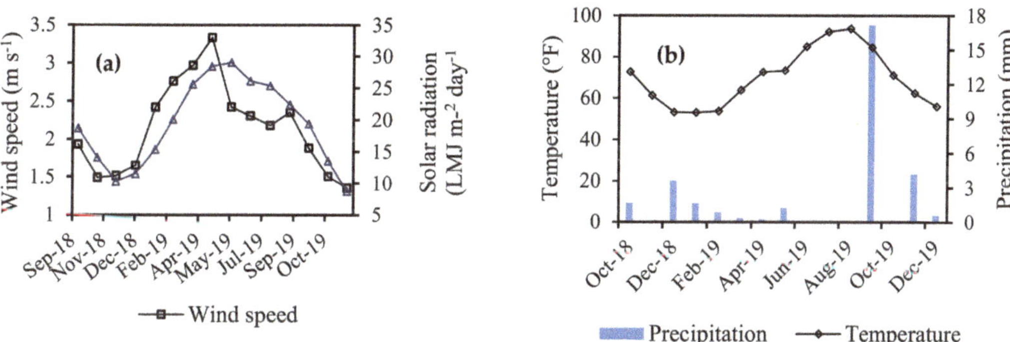

Figure 2. Weather conditions in the IV during the study period. (**a**) Monthly average windspeed and solar radiation, (**b**) monthly average temperature and precipitation. The value denotes the averages from four stations located in the valley.

3. Materials and Methods

The research approach to quantify the irrigation performance indicators based on adequacy, equity and productivity utilizing the EEFlux ET, NDVI and county-level statistics of the crop yield in the IV is presented in this section. Seven sub-sections are designated to explain the process. The Section 3.1 provides information on the key datasets. The image preprocessing steps are explained in the Section 3.2. The crop classification approach is explained in the Section 3.3. The description of the mapping of the seasonal ET is presented in the Section 3.4. The Section 3.5 explains the approach for the yield mapping, followed by the accuracy assessment and validation in the Section 3.6. The Section 3.7 explains the method used to compute the performance indicators. The estimation of each indicator is explained in a separate section.

3.1. Datasets

This section provides the description of the datasets used in the study. Each sub-section presented below summarizes the data types, the sources and their specifications.

3.1.1. Google EEFlux Datasets

The Landsat images for the ET calculation (Table 1) used in this study were processed on EEFlux/METRIC version 2.0.3 on the EEFlux website. EEFlux can be accessed freely at https://EEFlux-level1.appspot.com/ (accessed on 10 November 2020). EEFlux utilizes the thermal and short-wave infrared band of Landsat to estimate the surface energy balance, vegetation amount, albedo and surface roughness. The ETa is computed as a residual of the surface energy balance [43], then calibrated mechanically using the gridded weather data. The ET is expressed in terms of ETrf, which represents the ET as a fraction of the reference ETr (alfalfa reference ET). Additional Details on the EEFlux METRIC processing are provided in Section 3.2. We processed a series of 25 Landsat scenes for the ETrf computation on the EEFlux platform. Both the Landsat 7-Enhanced Thematic Mapper (ETM) and the Landsat 8-Operational Land Imager (OLI), along with the Thermal Infrared Sensor (TIRS) scenes with cloud cover less than 10%, were collected to compensate for large data gaps during the interpolation. The images from December 2018 for both platforms were unusable due to high cloud cover, and hence were not considered. All of the Landsat 7 images obtained after 31 May 2003 had continuous data gaps due to the failure of the Scan Line Corrector (SLC). Therefore, the data gaps in the Landsat 7 scenes must be corrected by the user.

Table 1. Details of the Landsat images acquired in this study for processing in the EEFlux platform.

Satellite	Image Acquisition Dates			
Landsat-7 ETM	17 February 2019			
	8 May 2019			
	24 May 2019			
Landsat-8 OLI	4 October 2018	29 March 2019	19 July 2019	23 October 2019
and TIRS	5 November 2018	16 April 2019	4 August 2019	8 November 2019
	21 November 2018	30 April 2019	20 August 2019	24 November 2019
	24 January 2019	1 June 2019	5 September 2019	10 December 2019
	25 February 2019	17 June 2019	21 September 2019	
	13 March 2019	3 July 2019	7 October 2019	

3.1.2. Satellite Images for NDVI Mapping

Landsat-8 OLI and TIRS images for 13 March 2019 were obtained for the crop-specific NDVI computation. The images were downloaded from the United States Department of Geological Survey (USGS) website. Primarily, the Near Infrared (NIR) and red bands were utilized for the NDVI calculation. In this study, we utilized NDVI images to perform linear regression with the crop yield data for the yield mapping. Previous studies have shown that NDVI correlates well with the alfalfa yield during the second cutting and 10% bloom [58]. Similarly, a good correlation with the sugar beet biomass was observed during the crop development stage (NDVI < 0.85) before leaf senescence in a study by [1]. In semi-arid areas like California, the first alfalfa cutting period is approximately 60 days, and takes place during January and February [59]. After the first cutting, alfalfa is cut every 30 days, starting from March. Therefore, the image of 13 March belongs approximately to the early growth stage of alfalfa. For the sugar beets, we computed the average NDVI in the sugar beet fields during several months, starting from January. An average NDVI of 0.8 was observed for the 13 March image, and hence was considered for the NDVI mapping.

3.1.3. Reference Evapotranspiration

The reference ET was obtained from the CIMIS website. The CIMIS is a database maintained by the California Department of Water Resources (DWR). The ET dataset that CIMIS provides is a standardized ET based on the alfalfa (ETr) or grass surfaces (ETo) where each CIMIS station is located. The estimates of reference ET were carried out by CIMIS based on the CIMIS Penman's method. The CIMIS Penman method is the modified version of Penman equation created by [60]. The detailed steps used to compute the CIMIS

Penman's ET are described here: https://cimis.water.ca.gov/Content/PDF/CIMIS%20Equation.pdf (accessed on 5 May 2021).

In this study, we retrieved the daily ETr and ETo values over the study period (4 October 2018 to 10 December 2019) from four stations located in the IV. The EEFlux utilizes ETr during the calibration of the algorithm [48]. Because CIMIS has only one station based on the alfalfa's surface, the daily ETos from the remaining stations were converted to ETr by multiplying them by a factor of 1.23 [59] in order for the further processing to be consistent with EEFlux.

3.1.4. Data for the Crop Mapping and Validation

The mapping of the crop areas is required in order to quantify and map the crop-specific water consumption, yield and performance indicators. In this study, Sentinel-2 Level-2A images were acquired from the European Space Agency (ESA) (Paris, France) on the Sentinel Scientific Data Hub website for crop mapping. The satellites systematically acquire optical imagery including 13 spectral bands. The four visible bands have spatial resolutions of 10-m, the six infrared bands are at 20-m resolutions, and the remaining three bands are at 60-m resolutions. Level 2A products do not require atmospheric correction, as they are corrected for atmospheric effects before their delivery to users.

In addition, as a reference dataset to collect the training samples for crop-specific and non-crop categories, the Cropland Data Layer (CDL) products were acquired from the National Agricultural Statistics Service (NASS), United States Department of Agriculture (USDA) (Washington, U.S). The field-level crop data were obtained from Lower Colorado Water Accounting System (LCRAS), United States Bureau of Reclamation (USBR) (Washington, U.S), in order to assess the classification accuracy. The dataset contains the Geographical Information System (GIS) layer, which shows the crops grown throughout the year within the extent of the IID. The crop acreage estimates for each year were obtained from crop reports published by the IID's water department for the validation of the extracted area.

3.1.5. Other Datasets

No other spatially explicit ETa data were available to assess the accuracy of the ETa map prepared in this study. Hence, the ET reported in Table 1 of [57] were extracted for comparison with the EEFlux-computed Eta (Table 3). It includes the ET computed for the IV, as well as other western states for the two crops studied. Although the reported values represented point measurements, and were associated with methodological differences, the datasets were the best available reference to access the accuracy. In addition, the ETa was also compared with the ET computed using the kc-based procedure (ETc), in which the ETo is multiplied by kc to produce an estimate of the ETc. In order to do so, the kc values of alfalfa and sugar beet at different growth stages were obtained from various reference studies in the Western US and worldwide, as listed in Table 2. Crop-specific yield data for 2019 was obtained from Agriculture Commissioner reports. The shape files of the field boundaries were provided by USBR.

Table 2. Kc values from the literature used in this study.

Crop	Kc			Location	References
	Ini	Mid	Late		
Alfalfa	0.87	0.91	0.86	Argentina, semi-arid area	[61]
	0.6	1.1	1.1	California	[62]
	0.3	1	0.95	Idaho	[63]
	0.4	1.04	0.98	-	[59]
Sugar beet	0.2	1.17	1.12	California	[64]
	0.35	1.24	0.78	-	[59]

3.2. Image Preprocessing

The Google EEFlux platform was used to process the Landsat scenes for the ETrf calculations. EEFlux utilizes the METRIC algorithm, which computes the energy expended during the evapotranspiration process as a residual of the surface energy balance according to Equation (1):

$$LE = R_n - G - H \tag{1}$$

where LE is the latent heat flux or energy consumed by ET (W m^{-2}), R_n is the net radiation, G is the soil heat flux (W m^{-2}), and H is the sensible heat flux (W m^{-2}).

In order to extrapolate the LE for each pixel from the exact moment of the passage of the satellite to the instantaneous value, the LE is divided by the latent heat of vaporization using Equation (2):

$$ET_{inst} = 3600\frac{LE}{\lambda\rho_w} \tag{2}$$

where ET_{inst} is the instantaneous ET (mm h.$^{-1}$), λ is the latent heat of vaporization (J kg^{-1}), and ρ_w is the density of water.

The resulting ET_{inst} is expressed as ETrf, which represents the fraction of the reference evapotranspiration (ETr) (Equation (3)). ETr is the reference evapotranspiration based on alfalfa, as defined by the ASCE Standardized Penman-Monteith equation [65]. EEFlux computes ETr using gridded hourly and daily weather data stored in the Earth Engine.

$$ETrf = \frac{ET_{inst}}{ETr} \tag{3}$$

ETrf can be used to estimate the actual ET for any period by multiplying it with the ETr for nearby stations. In the EEFlux platform, the image size that can currently be downloaded is limited and does not cover the entire study area. Hence, the resulting ETrf maps were mosaicked for each image date in ArcGIS Pro. In addition, the data gaps in the images from the Landsat-7 platform were filled using the nibble tool in ArcGIS Pro. The nibble tool replaces the cells of the raster with the values from the nearest neighbors.

3.3. Crop Classification

We classified a Level-2A S2 image of 6 April 2019 using a machine learning approach, i.e., a random forest (RF) algorithm. The image corresponds to the pre-harvest period for most of the crops in the field. Before classification, the 10 bands of S2 (aerosol bands not included) were resampled to 10-m resolutions using a bilinear resampling method. Five crop classes grown in the Valley were considered, including alfalfa, mixed grasses, wheat, corn, mixed crops and sugar beets. Mixed grasses include hays, excluding alfalfa (turf grass, bermuda grass, klein grass, etc.). All of the vegetables and remaining crops were included in the class of mixed crops. The non-crop classes included built-up, water bodies, and fallow. Though the non-crop classes were included during the classification process, the results were only interpreted in terms of the alfalfa and sugar beets, the main focus of this study. The CDL layer from the USDA was used as the reference dataset to collect the training samples for crop-specific and non-crop categories. Polygons, representing various crop types, were sampled from the Sentinel image, in which the identification of the crop type was performed based on a cropland map. We utilized the 'randomForest' function in the 'randomForest' package of R version 4.0.2 for the crop classification. Independent reflectance values from each band of S2, along with NDVI, were used as features during the crop categorization. A confusion matrix was prepared between the ground truth datasets provided by the USBR and Sentinel-identified crop types in order to evaluate the classification accuracy. The predicted crop acreages were validated with acreage statistics provided by the IID. The crop map was resampled to a 30-m resolution after accessing accuracy to use with 30-m resolution Landsat-derived products. Then, alfalfa and sugar beet polygons were extracted from the resampled classified map, and two distinct crop layers representing the crops were prepared to be used for further analysis.

3.4. Mapping the Seasonal ET

The series of ETrf maps of the study area, obtained from the EEFlux platform, were used as a vehicle to extrapolate the ETa for the whole season. At the beginning, daily ETrf maps were generated for each day between the image dates by means of linear interpolation. The interpolation was performed in R statistical software version 4.2. The Approx() function in the base R package was utilized. Two sets of computations were performed in order to obtain daily ETrf maps, with each corresponding to the growing seasons of the crops studied. We interpolated images from 24 January 2019 to 10 December 2019, which were utilized later for the ETa computation of alfalfa, whereas images from 4 October 2018 to 20 August 2019 were interpolated in order to compute the ETa for sugar beets. The daily ETrf images obtained after interpolation were multiplied by the ETr of each day, then totaled to obtain the cumulative ET for the whole season (Equation (4)). The ETrs from four weather stations were averaged for each day.

$$ET_{period} = \sum_{i=m}^{n} [(ETrf_i) \times (ETr_{24i})] \text{ (Allen et al., 2007)} \tag{4}$$

where ET_{period} is the cumulative ET starting from day m to n, $ETrf_i$ is the interpolated ETrf for day i, and ETr_{24i} is the 24 h ETr for day i.

An overlay of the seasonal ET map with the classified crop map of the study area was made in order to obtain a seasonal ET map for individual crop classes.

3.5. NDVI and Yield Mapping

The Landsat images utilized to compute the NDVI for 13 March 2019 were atmospherically corrected before further processing. The Digital Number (DN) was converted to the at-surface reflectance using the formula provided by the USGS. The required information, including the multiplicative rescaling factor, additive rescaling factor and local sun elevation angle, were obtained from the metadata files. Then, the NDVI was computed using NIR and red band.

In order to obtain the crop yield map at the pixel level, the yield information of the crops obtained from the crop report was disaggregated using the Landsat NDVI data as a bridge [31]. An assumption was made that the NDVI of the crops during the growing season is directly related to the yield. The higher the NDVI of the crops, the higher the yield would be. Hence, a weighting factor (WF) was defined as the ratio of the pixel-wise NDVI for the crop of interest to the average NDVI for the Valley (Equation (5)). WF was related to the observed yield statistics using Equation (6). The equation generated a crop yield map with a resolution of 30 m.

$$WF = \frac{NDVI_{pixel}}{NDVI_{avg}} \tag{5}$$

$$Yield_{pixel} = WF \times Yield_{obs} \times \text{Area of one pixel} \tag{6}$$

where $NDVI_{pixel}$ and $NDVI_{avg}$ are the NDVI of an individual pixel and the average NDVI for the crop of interest, respectively. A crop-specific map of $NDVI_{pixel}$ was masked out utilizing the crop map. $Yield_{obs}$ is the observed yield from the report, and $Yield_{pixel}$ is the yield of any given pixel for the crop of interest in kg/m^2.

3.6. Validation of ETa and Yield

The Mean Absolute Deviation (MAD) and Relative Error (RE) were computed in order to assess the associated differences between the computed Eta and the values reported in the literature for the IV. In addition, we computed the ETc for each crop using the kc values from the other reference studies, as well as those from [59], and compared this with the EEFlux ETa. The grass reference ETs (ETo) from three CIMIS stations were obtained for days corresponding to initial, mid- and late growth stage of the crops, then multiplied with the kc from Table 2 to obtain the ETc values. The FAO-56 specified kc values represented the standard climate, with the mean daily minimum relative humidity (RHmin) equal to 45% and the mean daily wind speed (WS) equal to 2 ms^{-1}. When the mean weather differs from the standard, kcmid and kclate has to be adjusted as described in Allen et al. (1998) [66]. This procedure was followed in this study, and Table 2 represents the adjusted values. The mean ETa from EEFlux were extracted for similar days. The Root Mean Square Error (RMSE) and Mean Absolute Error (MAE) were computed between the two, and the results were analyzed. The pixel-wise yield assessment was also restricted due to limited data. Therefore, for validation, the pixel-wise yield values were summed and compared with the reported total crop production. Because the model is the linear extrapolation of the field data reported by the county itself, the quality of the yield map produced can be considered acceptable.

3.7. Computation of the Performance Indicators

Several performance indicators are available to evaluate the existing practices in the field and identify room for efficient water management improvement. Hence, the selection of the appropriate indicators is needed, which should be based on the purpose of the assessment and the availability of data [67]. Surface energy balance models are direct indicators of equity and adequacy [18]. Productivity indicators based on yield and ETa can provide valuable insights for the identification of the scope of water management, where water is the limiting factor [34]. Therefore, considering the aforementioned factors, three indicators were chosen in this study, and their computation methods are explained in several subsections below.

3.7.1. Water Consumption Uniformity (WCU)

WCU is the indicator of the irrigation equity or the uniformity of water consumption [18]. Measurements of equity based on water consumption, rather than on the supply side, are considered more relevant in water-scarce regions [17]. The WCU was evaluated by computing the CV of the ETa at two levels in this study. The availability of field boundaries from the USBR allowed for the calculation of the CV within the fields (CVw). The zonal statistics tool in ArcGIS Pro 2.5 (ESRI, 2020) was utilized for this purpose. The crop map prepared in this study contained several scattered pixels, which might have resulted from misclassification. Therefore, in order to avoid the inclusion of such pixels during the CVw computation, the fields with less than 40 pixels (~36,000 m^2) were masked out from the crop-specific ETa map beforehand. The threshold was set after visually analyzing the crop map prepared. Hence, the mean and standard deviation of the ETa were obtained for fields with areas larger than 36,000 m^2. The number of alfalfa fields studied was reduced from 4183 to 2481, and the number of sugar beet fields from 817 to 478 after masking out the redundant pixels. In addition to CVw, we also computed the CV of ETa among the fields (Cva).

3.7.2. Relative Evapotranspiration (RET)

RET is the indicator of the water adequacy in the field, and it provides essential information on crop stress and water shortages. In the present study, RET is computed as the ratio of seasonal ETa to ETp. ETp refers to the maximum crop evapotranspiration under the optimal crop growing conditions, with no limitation based on plant growth. It is similar to the theoretically computed ET. Therefore, ETp was expressed as a product of seasonal ETo and kc [31]. The daily ETo measurements from the CIMIS stations were summed to obtain the seasonal value at each station. Then, the station location information was utilized, and seasonal ETp raster maps were generated from the point measurements using Inverse Distance Weighting (IDW) interpolation. The algorithm was implemented in ArcGIS Pro. A seasonal kc value of 0.9 was used for alfalfa [7,59]. The seasonal kc for sugar beets was computed using kc values from [59]. The weighted average of kc computed according to the growing season length resulted in a seasonal kc of 0.9 for sugar beets.

3.7.3. CWP

The CWP was estimated for the study area using the crop specific ETa and yield maps from Equation (7). Any differences in units from the yield map and ETa were adjusted to calculate the CWP in kg/m^3.

$$CWP\ (kg/m^3) = \frac{Crop\ yield\ (ton/acre)}{ETa\ (mm)} \tag{7}$$

In order to identify the field or areas with significant scope of improvement, the CV of the CWP in the study area was computed. A low CV of CWP indicates homogeneity and limited scope for improvement, whereas a high CV for an area indicates the opportunity of water management [34]. The relationships of CWP with ETa and yield were also looked at in order to evaluate the scope of CWP enhancement in the field under the current cropping conditions.

4. Results

The results and discussion are organized into five sections. Section 1, the results of the crop classification are presented. The ETa computation and spatial distribution are presented and discussed in the Section 2. The accuracy assessment of the computed ET is discussed in the Section 3. The yield map prepared using NDVI is presented in the Section 4, along with the validation results. Following this section, the outputs from the performance indicators are explained. The Section 5 primarily discusses the scope of the water management improvements in the Valley, as depicted by the performance indicator results.

4.1. Crop Classification

The overall accuracy of the crop mapping obtained was 85%. The individual assessment of the crop prediction accuracy showed that sugar beets were the most accurately mapped, with a producer accuracy of 99.5%. Alfalfa was also mapped reasonably well, given its wide distribution in the field. The accuracy of alfalfa was 85.2%. Several mixed grasses and mixed crops in the field were incorrectly labeled as alfalfa, which resulted in an alfalfa prediction commission error. The researchers mostly identified the confusion between alfalfa and mixed grasses. A similarity in the spectral signatures between alfalfa and mixed grasses could be the reason for it. The comparisons between the predicted and observed crop acreages showed that, for S2, there was underestimation of the alfalfa crop area by 16.65%, and of the sugar beets by 4.38%. The slightly high difference in the alfalfa acreage could be attributed to the predominant presence of alfalfa in the field. Hence, the accuracies obtained for alfalfa and sugar beets were considered reasonable for further study.

4.2. Spatial Distribution of ET and Yield

Figure 3 shows the spatial distribution of the seasonal ETa for both crops in the Valley. Alfalfa showed a large spatial variation, with a CV of 23.86 %. The mean ETa for alfalfa was observed to be 1388.26 mm. Though alfalfa's ETa ranged from 39.09 mm to 2064.93 mm (Figure 3a), less than 7.2% of the pixels were in the range of 39.09 mm to 849.43 mm. A small proportion of the alfalfa field clustered on the north corner exhibited a low ETa (849.43 mm–1659.77 mm). About 45% of the alfalfa fields had an ETa in the range of 1254.60 mm to 1659.77 mm, and this was scattered around the valley. However, fields with a high ETa, up to 2064.93 mm, were found to be visibly clustered largely on the eastern corner of the Valley. The high spatial variation for alfalfa may be attributed to the periodic cutting of alfalfa during the growing season. Similar to the alfalfa, the proportion of sugar beets with a low ETa, ranging from 124.87 mm to 763.25 mm (Figure 3b), occupied less than 8.5%. For sugar beets, the mean ETa was observed to be 1126.95 mm, and the CV was 21.36%. Approximately 78% of the sugar beet fields exhibited ETas from 763.25 mm to

1401.64 mm, clustered around northeast corner of the Valley. Less than 13% had high ETas, up to 1720.83 mm.

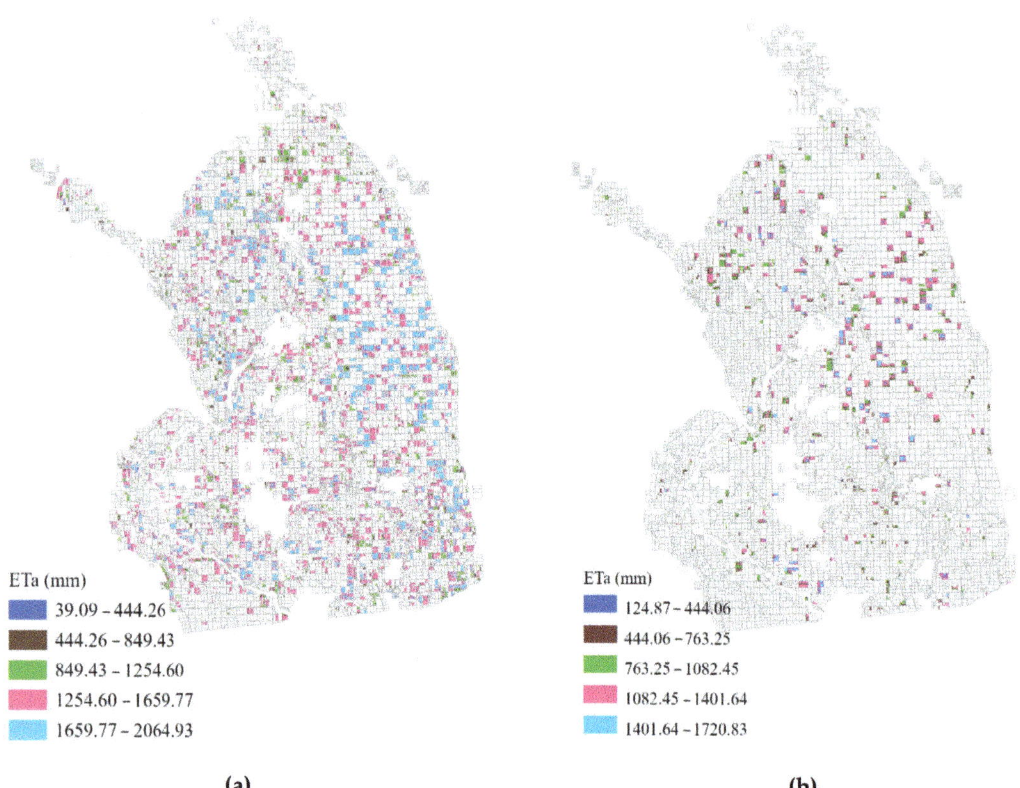

Figure 3. Spatial distribution of the growing season ETa for (**a**) alfalfa and (**b**) sugar beet. The ETa maps are overlaid with the USBR field boundary.

The alfalfa and sugar beet productivity maps resulting from disaggregation are shown in Figure 4. The average alfalfa yield for the study period was observed to be 0.395 kg/m^2. Less than 2% of the areas exhibited yields in the range of 0.004–0.198 kg/m^2; hence, they are not visible in the spatial map presented (Figure 4a). The estimated range of the alfalfa yield in this study was close to the range predicted by [58] in Saudi Arabia, which was 0.179–0.628 kg/m^2. A few regions with visibly low yields (the brown patch in Figure 4a) were observed in the northern corner of the Valley. It should be noted that the ETa was also low in this area (Figure 3a). Besides this, it has also been identified that areas without a significantly high ETa (pink patch in Figure 3a) also exhibited high yields for alfalfa. In regard to sugar beets, the average yield was 2.351 kg/m^2. Though 88% of the areas exhibited yields greater than 2.5 kg/m^2, noticeable spatial variation among the fields can be seen after this range (Figure 4b). Regions with low yields were clustered at the northeast of the Valley (green patch in Figure 4b). Continuous fields of high yield were perceived in the northwest of the Valley, near the Salton Sea.

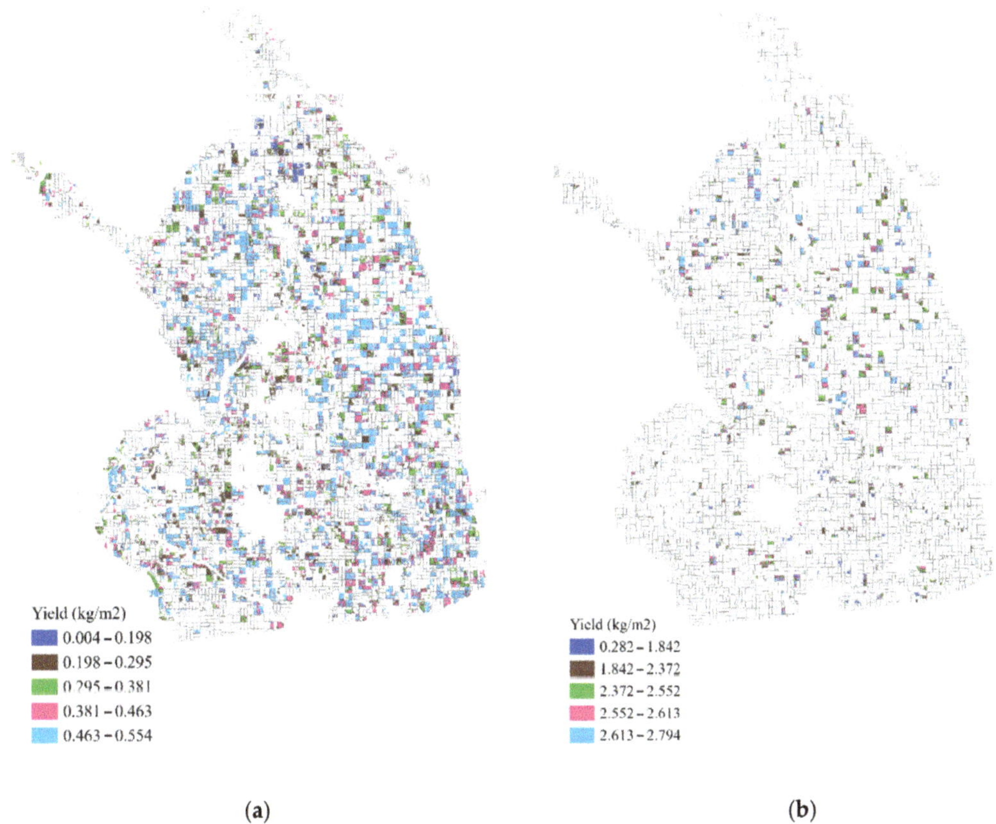

(**a**) (**b**)

Figure 4. Pixel-based distribution of the growing season yields for (**a**) alfalfa and (**b**) sugar beet, in kg/m^2.

In order to analyze the extent of the differences between the observed and disaggregated production values, the distributed yield was aggregated. The absolute differences with the observed production were 13.55% and 2.9%, respectively. The resulting differences were because of the underestimation of the planted area during the crop classification. The pixel-level validity was limited due to a lack of ground truth data. However, because the modelled yield is based on the linear extrapolation of district data, it mostly relies on the input data consistency [31], which is officially accepted in the current study.

4.3. ET Validation

Section 4.3.1 describes the results of the validation of the computed Eta with the literature reported values. Additionally, the results of the comparison with the ET from the kc-ETo approach are explained in Section 4.3.2.

4.3.1. Comparison with the ET from the Literature

The alfalfa ET values obtained from the literature were mostly point measurements ranging from 1295.4 mm to 1889.8 mm. Table 3 shows the comparison between the predicted and observed ET from the literature. The MAD and RE of the alfalfa ETa were found to be as low as 46.43 mm/year and 7%, respectively, with the USDA. The computed values also showed reasonable agreement with [7], with an MAD of 134.37 mm/year and an RE of 16%. With the IID and USBR, the deviation ranged from 220.27 mm to 250.77 mm, and the RE ranged from 24% to 27%, respectively. For sugar beets, a very good

ETa agreement was obtained when compared with the point representative methods. The MAD and RE were found to be nearly null with the IID.

Table 3. Comparison of the EEFlux-derived ETa with ET values reported in the literature. The ET values for columns three to five were extracted from Table 1 of [57]. IID*, Imperial Irrigation District; USBR*, United States Bureau of Reclamation; USDA*, United States Department of Agriculture.

Crops	EEFlux Eta (mm)	ET (mm) from Literatures			
		IID*	USBR*	USDA*	[7]
Alfalfa	Mean = 1388.26 SD = 331.34 Max = 2064.9	1828.8	1889.8	1295.4	1657
MAD		220.27	250.77	46.43	134.37
RE		0.24	0.27	0.07	0.16
Sugar beet	Mean = 1126.95 SD = 240.82 Max = 1720.83	1127	1097.3	660.4	N/A
MAD		0.02	14.83	233.28	-
RE		0.00	0.03	0.71	-

Similarly, with the USBR values, an MAD of 14.83 mm/year and an RE of 3% was attained. Slightly high differences were observed with the USDA-measured values. The USDA-derived values were based on the Salt River Valley and Arizona; hence, climatological differences with the IV may have resulted in greater variances. The literature shows that the ETa computed from remote sensing can vary, with ground measurements in the range of 1 to 20% [68]. With METRIC, an RE of up to 13.7% was observed by [45], and up to 16% in the study by [69]. This shows that the difference exhibited in this study is within the plausible range. It should be noted that the values identified in the literature are not only associated with the methodological differences, scale of measurements, weather conditions, and year when the measurements were taken but also varied, which results in a complexity of comparison. Based on this fact, a relatively good agreement of the ETa with the point measurements provides strong evidence that EEFlux may be a valuable tool for the mapping of seasonal ETa.

4.3.2. Comparison with the FAO-56-Computed ETc

Figure 5 presents the mean ETa values during the initial, mid- and late season stages obtained from the present work, along with the ETc computed using the kc values from other reference works around California and other western states. The initial, mid- and late growth stages for alfalfa correspond to 1, 13 and 29 March 2019, respectively. Likewise, for sugar beets, they correspond to 5 November 2018, 1 February 2019, and 12 August 2019 for the respective growth stages. The values of the ET obtained from EEFlux for alfalfa have an initial phase value of 3.58 mm, an intermediate phase value of 3.25 mm and a final phase of 3.91 mm. Similarly, the values of 2.14 mm, 2.41 mm and 4.32 mm were obtained for sugar beets for the corresponding growth phases.

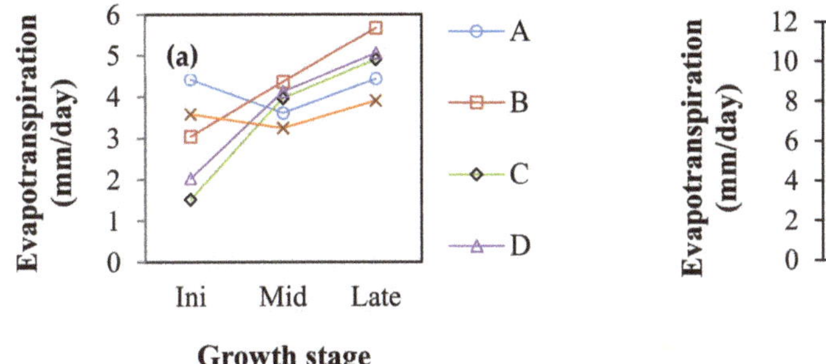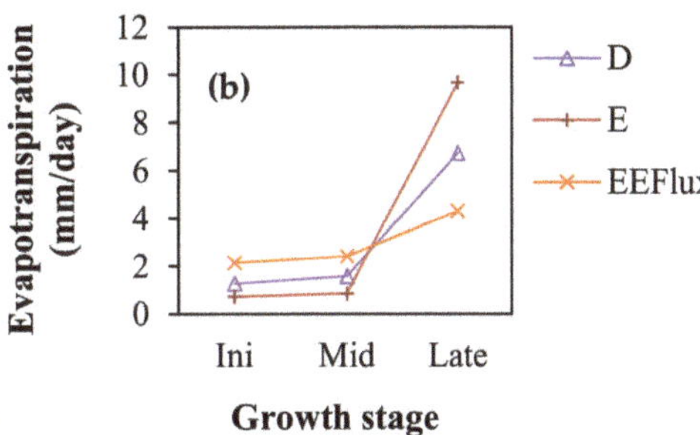

Figure 5. Comparison of the ET calculated from EEFlux METRIC with the ET computed using the kc-ETo approach for (**a**) alfalfa and (**b**) sugar beet. The kc values were obtained from A: [61]; B: [62]; C: [63]; D: [59] and E: [64].

It was observed that ETa showed a wide variation with the observed data. The RMSE and MAE variations for alfalfa from those in the literature ranged from 0.6 to 1.3 mm/day and 0.5 to 1.25 mm/day, respectively. When compared with the computed ETc using kc values from [59], the resulting RMSE was 1.22 mm/day, and the MAE was 1.19 mm/day. The ETa was overestimated during the initial growth stage, whereas it was underestimated in later stages. The sugar beet ETas computed in this study were like those obtained from the literature during the initial and mid-growth stages. The resulting RMSE and MAE were 1.5–3.3 mm/day and 1.3–2.7 mm/day, respectively, with the lower value range computed with the kc from [59]. The overall results suggest that EEFlux tends to underestimate the ETa during the mid and late stages of alfalfa growth, as well as the late growth stage for sugar beets. A similar result was obtained in the study by [47] for maize crops, and by [48] for soybean crops, using the EEFlux tool. In addition, [70,71] also reported low accuracy for cotton and coffee crops, respectively, using the SEBAL algorithm.

The low level of accuracy in general for EB algorithms, when compared to the computation based on the kc from [59] can be attributed to several reasons. First, EB algorithms consider the spatial variability of ET and kc, unlike other methods. Differences in the ET values may also be attributed to the variability of kc according to local growing conditions and land use management, as well as the rainfall and atmospheric conditions, including the air temperature, wind speed and vapor deficit [72]. In addition, FAO- kc values are derived from multi-day data (average values), while EEFlux METRIC obtains its kc, i.e., ETrf values, at the satellite's passing time [47,48]. Because the evaporation from soil is usually higher immediately after rain or irrigation events, higher overall ET concentrations can be anticipated during this time [61]. The average kc values from the literature therefore consider both higher and lower evaporation rates from wet soil surfaces, whereas EEFlux ETrf represents the actual conditions of the satellite overpass time [47].

4.4. Performance Indicators

In this study, three performance indicators based on equity, adequacy and productivity were studied for alfalfa and sugar beet fields. The results for each are explained in the separate subsections below.

4.4.1. WCU

ETa maps and field boundaries were used to calculate the coefficient of variation of the water consumption, also known as WCU, for the alfalfa and sugar beet fields. A problem associated with irrigation uniformity is suggested by the high variation of the water use within the fields [73]. Figure 6 shows the spatial variation of the CVw for both crops. The CVw for alfalfa ranged from 0.5% to 36.1% (Figure 6a), with the average being 9.7%. Although the map shows a CVw as high as 36.1%, only 0.8% of the regions had variations greater than 21%. Hence, this may have resulted from the inclusion of some partial crop coverage fields (Santos et al., 2008). Similarly, for sugar beets, the CVw of the seasonal ETa ranged from 0.2% to 22.2% (Figure 6b), with an average of 3.2%. Less than 0.23% of the regions had CVws greater than 18%. This could be explained by the same reason attributed to alfalfa. We also computed the field variability of the water use for both crops. A high variation of water use among the fields may indicate differences in farmers' irrigation practices [73]. The resulting variations were 19.36% and 19.9% for the alfalfa and sugar beet fields, respectively.

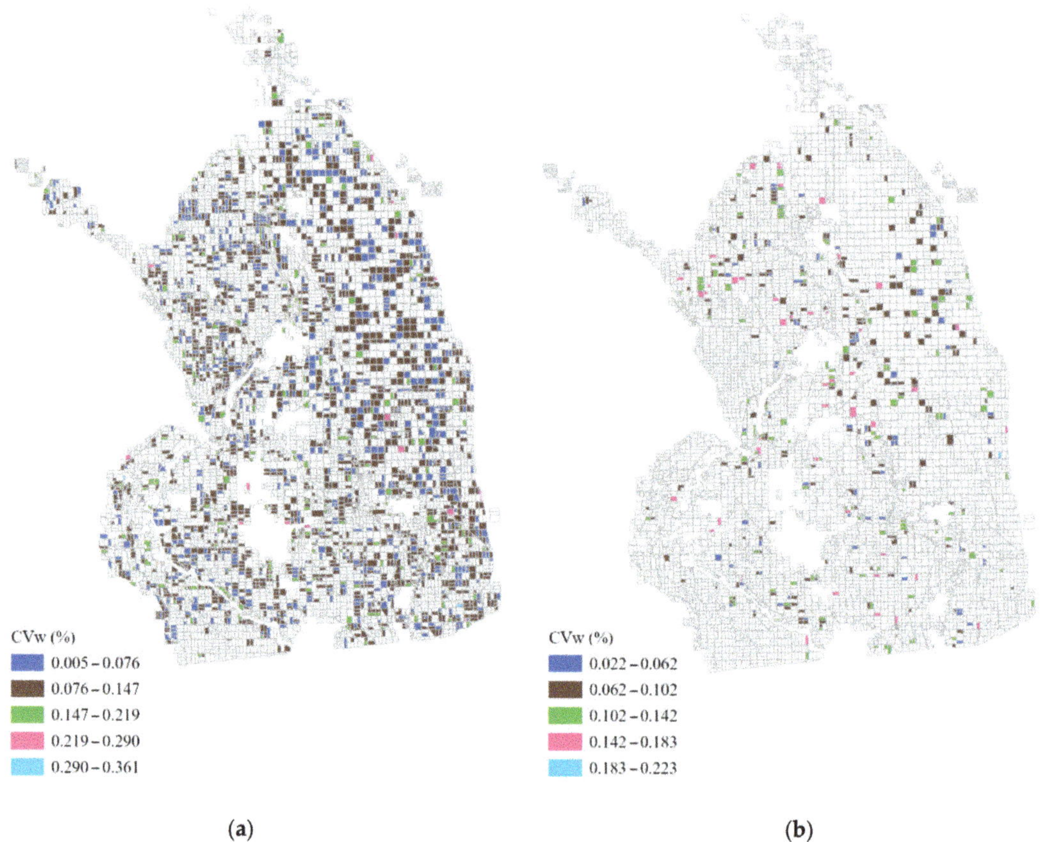

Figure 6. Within-field coefficient of variation for (**a**) alfalfa and (**b**) sugar beet fields, expressed in percentages.

Molden and Gates suggests a CV less than 10% to be good uniformity [74]. Approximately 36.14% of the sugar beets and 34.17% of the alfalfa exhibited CVws greater than 10% in our study. The focus on water management could be placed on those flagged fields, rather than the whole district. The overall greater uniformity in the Valley can be attributed to the application of various tilling methods, such as levelling and sod busting systems [75] for precise field grading. Similarly, greater Cvas for both crop fields suggest that the irrigation equity is slightly poor among the fields. The high Cva among the alfalfa fields could also be attributed to their continuous planting and harvesting. Overall, even though the average performance was satisfactory, the large variation in performance among farmers suggests that there is significant room for improvement.

4.4.2. RET

The ratio of the actual to the theoretical ET was used to calculate the adequacy for both crops, which is also called the RET. The spatial distribution of the RET is presented in Figure 7. The average RETs for the alfalfa and sugar beet fields were 0.844 and 0.797, respectively. The difference between ETa and Etp should be less under ideal growth conditions, and the ratio should be nearly equal to 1. Roerink et al. suggests that, for irrigated agriculture, values of 0.75 and higher are satisfactory [23]. In the current study, more than half of the planted area for both crops exhibited RETs greater than 0.75, suggesting satisfactory adequacy. Focus could be placed on the fields with RETs lower than the optimal value, where crops are experiencing water shortages that could result in poorly developed crops, affecting the yield [68].

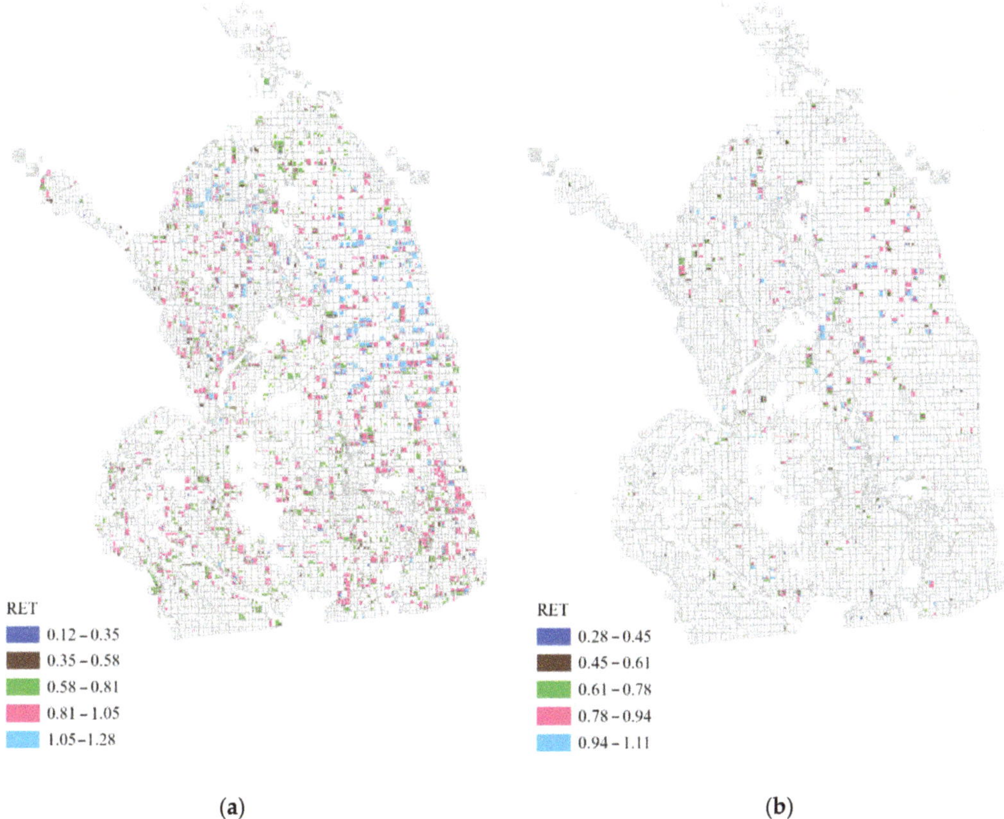

Figure 7. Spatial distribution of RET for (**a**) Alfalfa (**b**) Sugar beet fields.

The RET maps not only allowed us to identify the fields experiencing water shortages but also the fields consuming more water than their potential. About 31.5% of the alfalfa and 12% of the sugar beet fields had RETs of more than one. For alfalfa, these fields are visibly clustered on the east side of the valley (Figure 7a), whereas for sugar beets, fields in eastern regions and a few in the central regions exhibited RETs greater than one (Figure 7b). A slightly high average RET (0.97) was found for the Palo Verde Irrigation District (PVID), which is also located in Imperial County, in the study by [2]. The average was computed for the whole irrigation district, rather than the crop-specific fields, and using the Priestley–Taylor approach, which may have resulted in the differences. Bastiaanssen et al. [76]

reported a comparable average RET value of 0.77 in the Nilo Coelho irrigation system in Brazil, and 0.7 in the Gediz Basin in Turkey in a study by [28].

4.4.3. CWP

The water productivity was computed as a ratio of the crop yield to ETa in this study. Figure 8 shows the spatial map of the CWP for both crops. The average alfalfa CWP for the Valley is 0.328 kg/m^3, with a CV of 32.92%. Although the alfalfa CWP ranged from 0.004 to 4.062 kg/m^3, approximately 99% of the fields had CWPs of less than 0.8 kg/m^3. The remaining fields were likely to be associated with mixed pixels with other crop types during the classification. The sugar beet CWP resulted in an average of 2.387 kg/m^3, with a slightly lower CV (25.4%) than alfalfa. Like alfalfa, though the CWP exhibited a higher range for sugar beets, about 99% of the fields with CWPs of less than 4.6 kg/m^3 were observed, with the attributed reason being similar to that of the alfalfa. The average CWP of alfalfa obtained in the study was close to the range, 0.38–0.43 kg/m^3, recommended by [77] for Saudi Arabia; however, it was slightly lower than that reported in [78], which was 0.55 kg/m^3. The variation in the climate and alfalfa productivity may have caused the differences.

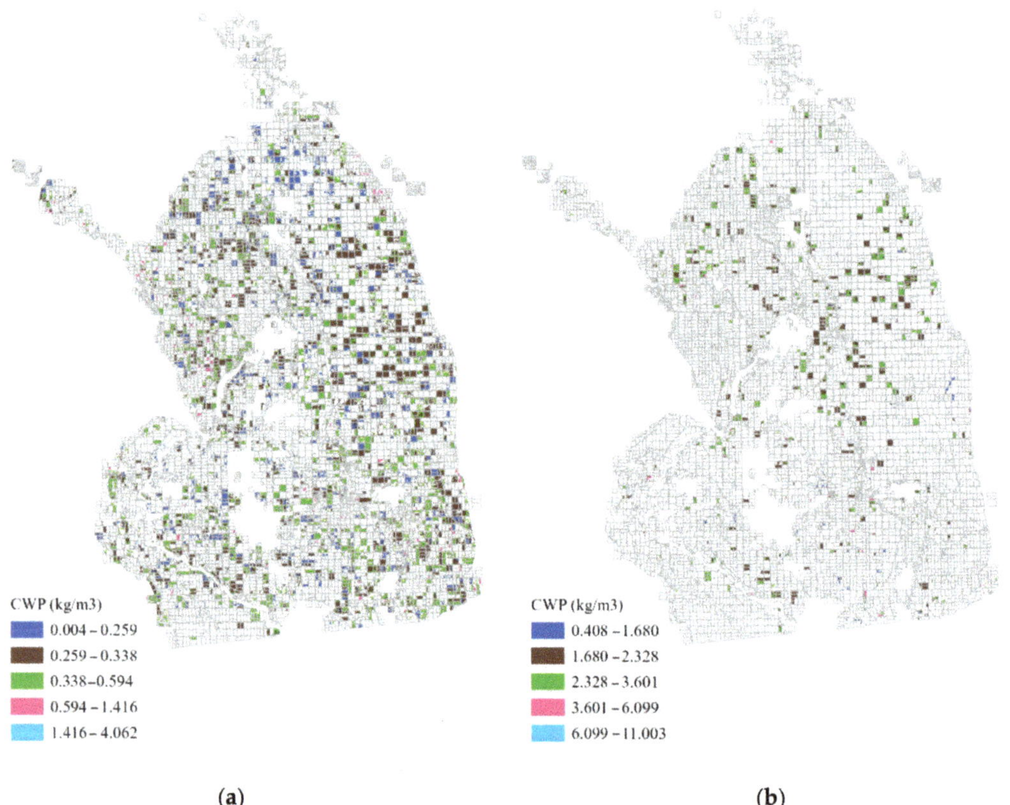

Figure 8. Spatial distribution of CWP for (**a**) alfalfa and (**b**) sugar beet fields.

The crop fields showed a mixed trend of CWP with ETa and yield. The northwest corner of the sugar beet fields (the green patch in Figure 8b) exhibited a high CWP because of its low ETa and high yield. However, the northern corner of the alfalfa fields (the blue patch in Figure 8a), in spite of having a low ETa, exhibited a low CWP as a result of its low yield and inadequate water, as displayed by the RET map. Similarly, the high-water use alfalfa fields (RET > 1) in the eastern regions, with moderate-to-high CWPs, also had moderate-to-high levels of ETa and yield. This suggests that these fields are compensated by high yields resulting in high CWP, despite having more than the required water use. The overall results imply that the factors by which CWP is influenced vary widely in the study area, and there should be careful consideration before making any management decisions on enhancing CWP.

4.5. Analysis on the Scope of Water Conservation

The high CV of CWP for both alfalfa and sugar beet fields implies that by narrowing the variability, there is a wide scope of CWP improvement. CWP enhancement can be achieved either by increasing the yield or maintaining the same yield while reducing the water use [34]. Because yield increment takes time [31], the second option could be more viable in areas where water availability is limited.

In order to better understand the scope of water conservation through CWP improvement, random points were generated from a corresponding spatial map, and scatter plots were prepared among the CWP, yield and ETa in order to observe the association between them (Figure 9). Figure 9a indicates that the distribution range of the alfalfa ETa with yield was high for ETa in the range 500 mm to 1500 mm. However, it is observed that, for ETas above 1500 mm, the variation decreases, and the yield remains constantly high (Figure 9a; rectangular box). This implies that above this range, the reduction of the irrigation water amount would not affect the yield significantly, and its effect would not be adverse. Similar results were observed for sugar beets in the range of Etas from 1200 mm to 1600 mm (Figure 9c; rectangular box), where the yield distribution with ETa was low. The relationship between the CWP and Eta for both crops, alfalfa (Figure 9b) and sugar beet (Figure 9d), showed that the CWP decreases as the ET increases primarily for ETas above 1500 mm for alfalfa and 1200 mm for sugar beet. Therefore, by decreasing the water use above the aforementioned ranges to around 1500 mm and 1200 mm for alfalfa and sugar beet, respectively, we identified a scope of CWP enhancement that keeps the yield constant. Table 4 shows the possible volume of water that could be conserved through CWP enhancement. The results imply that it is possible to save approximately 44.52M cu. m (36,000 acre-ft) of irrigation water volume by reducing the ETa to a range where the yield is not adversely affected and the CWP is enhanced.

Table 4. Possible water saving opportunities by reducing the ETa without adversely affecting the yield.

Alfalfa	
No. of pixels > 1500 mm	214,020
Volume for ETa > 1500 mm	327.73M cu.m
Volume after reducing ETa = 1500 mm	288.92M cu.m
Saved water volume =	38.37M cu.m
Sugar beet	
No. of pixels > 1200 mm	45,357
Volume for ETa > 1200 mm	55.12M cu.m
Volume after reducing ETa = 1200 mm	48.98M cu.m
Saved water volume =	6.14M cu.m
Total volume that can be saved =	44.52M cu.m

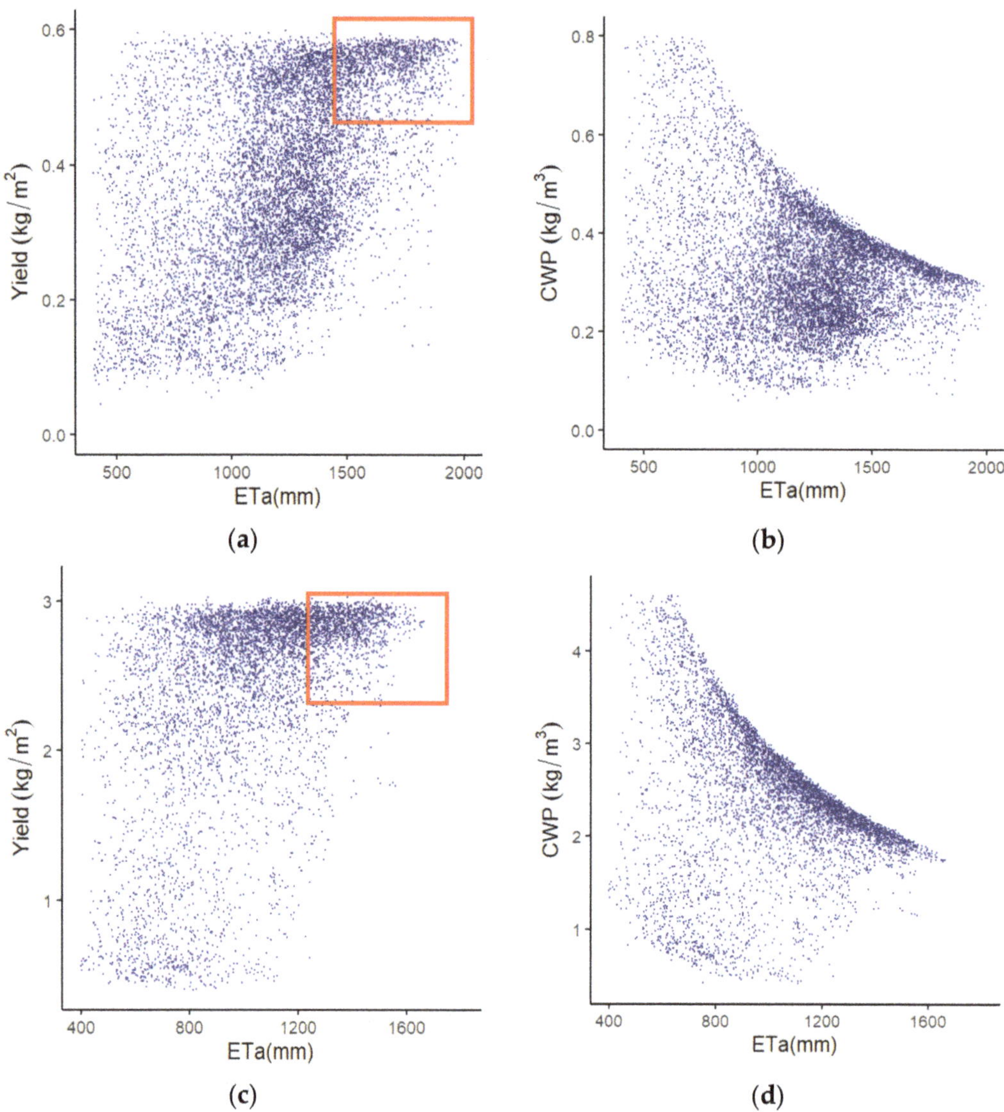

Figure 9. Relationship between (**a**,**c**) yield (kg/m^2) and ETa (mm), and (**b**,**d**) CWP (kg/m^3) and ETa (mm) for alfalfa (top row) and sugar beet (bottom row), respectively. The rectangular red box denotes the area where the yield is nearly constant.

Other that water conservation through CWP enhancement, we also identified high-water-use fields through RET analysis. For nearly 32% of alfalfa and 12% of sugar beet pixels with RET > 1, the water overuse was equivalent to approximately 11M cu. m (8940 acre-ft) collectively for the two crops. Hence, possible water saving opportunities were also identified for high RET fields, and the spatial map presented in this study helps identify those fields.

5. Discussions

The high yield fluctuation with ETa found in this analysis is consistent with the previous research findings. [1] found a high variation of irrigated crop yields, including sugar beets, even when the ETa is constant. Similar high fluxes of alfalfa yield with ETc were observed by [7] for irrigated land in the Californian desert. The inconsistent relationship of the yield with ETa below 1500 mm for alfalfa and 1200 mm for sugar beet infers that the yield is only marginally limited by water consumption for both crops beneath this range. This suggests that there are a number of other important factors that influence crop production and, as a result, yield variability. Those factors may include the irrigation and management practices, as well as the soil, nutrients, water table depth, land preparation and fertilizer application, along with the tolerance capacity of the crop itself for drought and salinity [34,35]. In addition, Smeal et al. discussed the dependence of the alfalfa yield on the cutting number and the accumulated growing day temperatures [79]. Therefore, the challenge associated with enhancing CWP in IV by decreasing the ETa for these ranges would be to maintain a constant yield. Another option would be to reduce the ET in low-yielding fields. This could be a suitable strategy specifically for the IV, where water costs only account 10–20% of the overall alfalfa production costs [7]. Hence, the implementation of water conservation approaches that will result in a greater loss of the hay yield may be undesirable to growers. However, the research may necessitate data from multiple years in order to validate the findings and establish accurate CWP benchmarks.

Although the average spatial variability of the water use within the fields was reasonable for both crops, the use of the average performance value may not inform us about the actual system performance level [76]. Having about 35% of fields with CVws greater than optimal implies spatially heterogeneous water consumption in these fields. Moreover, the high Cva exhibited among the fields implies significant room for improvement in performance among the farmers. Because of the variable growing season, this heterogeneity might represent the somewhat-upper limit for alfalfa. The high variance or non-uniformity does not necessarily imply poor management in all of the fields, as it may also relate to sub-optimal crop husbandry or deficit irrigation practices [73]. Remote sensing alone may not explain the variation. However, the assessment of the variability offered by the distributed nature of the remote sensing aided in locating the non-uniform fields, where further investigation can be performed. Besides fields with water overuse (RET >1), equal attention should be placed on the fields where adequacy is less than optimal (RET < 1). A less salt-tolerant alfalfa crop may be unfavorably affected by salinity accumulations due to the reduced water [80] in comparison to moderately salt-tolerant sugar beets.

Using EEFlux to estimate the water usage in this study provided a valuable method for the quantification of the irrigation system performance at both the crop and field levels. The overall accuracy of the ETa, when compared to the values in the literature, was found to be reasonable. Though the comparison with the kc-ETo-derived ETc for several growth stages resulted in differences, the discrepancies may be associated with methodological differences in computing the kc [47,48]. EEFlux considers the current conditions of the satellite overpass time, whereas the reported kc values are the average kc during the growing phases. In addition, although an attempt was made to reduce the large temporal gaps between the image dates by including Landsat 7/8 images, images were unusable for December due to high cloud cover. The gap between the images might have impacted the interpolation results, and ultimately the seasonal ETa values [61]. The advantage of an eight-day satellite overpass frequency, in comparison to 16 days, to predict the seasonal ET of a cotton crop was observed by [81]. Similarly, we employed a linear interpolation method to generate daily ETa maps. Previous studies have shown that the use of spline interpolation may improve the seasonal ETa estimation, although the results were not statistically significant in either case [46].

This study utilized a yield disaggregation method [31] to map the large-scale yield, using county production statistics and NDVI at specific growth stages. The method links the publicly available district-level statistics to remotely sensed data and helps fill the gap between the two. Because this method avoids complex land surface processes and biophysical parameter estimation [31,37], agricultural mangers can map the yields of large areas efficiently. Although pixel-level validation was not performed in this study, the reasonable accuracy from this method in other studies for wheat [31] and rice [37] provides a confidence in others crops as well. Very few studies have explored the use of NDVI in the prediction of the yields of alfalfa and sugar beet in the past. However, in other crops of medium-to-high canopy sizes, such as corn, soybeans and winter wheat, NDVI showed a good correlation with the cropped biomass per area [82,83]. This may also imply that NDVI is a good metric for the estimation of the biomass in the crops analyzed in this study. The uncertainty in the exact growing cycles of the crops may have resulted some bias in the predicted yield using a single NDVI image, which could be improved in future studies by incorporating accurate field data. Nevertheless, the yield disaggregation method using NDVI is a promising approach that brings simplicity to the crop-yield mapping of large areas.

6. Conclusions

In this study, we computed irrigation performance indicators based on equity, adequacy and water productivity, using remote sensing, NDVI and crop production statistics. All of the available Landsat 7/8 images with low cloud cover were processed in the EEFlux platform to obtain ETrf images of the IV. Linear interpolation was performed in order to obtain daily ETrf images. Seasonal ETa images were produced as a sum of the product of the daily ETrf and ETr. Crop classification was performed using S2 images and the RF algorithm, and crop-specific (alfalfa and sugar beet) ETa images were produced for the growing season by a crop map overlay. We computed the WCU as a measure of irrigation equity, RET as a measure of adequacy, and crop water productivity to reflect on productivity. The relationship between the crop ETa, yield and CWP was also studied in brief, in order to identify the scope of the CWP enhancement and water conservation.

The average WCUs for both crops were found to be uniform; however, spatial variation within the fields showed that 36.14% of the sugar beet and 34.17% of the alfalfa fields had variabilities greater than 10%. Similarly, the among-field variability was approximately 19% for both. The variability within and among the fields implies the variation in irrigation and management practices among the farmers and indicates a wide scope for improvement. Another indicator, RET, showed that more than half of the fields were provided with adequate water (RET > 0.75). However, about 31.5% of the alfalfa and 12% of the sugar beets were consuming more water than necessary (RET > 1), and are therefore where the water conservation should be focused. The results showed that nearly 11M cu. m (8940 acre-ft) of water can be saved by reducing the water overuse in these fields. The CWP showed a wide variation, with a CV of 32.92% for alfalfa and 25.4% for sugar beets, indicating a significant scope for CWP enhancement. Nearly 44.52M cu. m (36,000 acre-ft) of water saving opportunities were identified by reducing the ETa to approximately 1500 mm for alfalfa and 1200 mm for sugar beet, which will enhance the CWP without reducing the yield.

The EEFlux served as a valuable source to compute the ET in a simple and inefficient manner. The MAD and RE, when the mean EEFlux ETa was compared with point representative values from the literature, were as low as 46.43 mm and 0.07, respectively, for alfalfa. Similarly, for sugar beets, the lowest MAD was 0.02. However, it is understood that the EEFlux-computed ETa shows a significant difference with the ETc computed using kc-ETo methods. This may be partly attributed to the methodological difference associated with kc calculation between the two methods. The large gap between the image dates for some months may also have affected the results of the seasonal ETa. The accuracy of ETa mapping is imperative for the accurate estimation of the performance indicators. Further investigation of other interpolation methods, the use of all of the available Landsat images for the growing season, and validation with the ground truth data is recommended for future studies.

Overall, the results of this study confirm the wide scope of water conservation in the valley. Fields with non-uniform irrigation distribution and high RET were visually identified. Similarly, fields with a wide variation in CWP were also predicted, in which—by narrowing the variability—significant CWP enhancement can be achieved. Although the procedure did not provide thorough insight into the reasons for the high variation or high–low values, the bigger picture of irrigation performance across the irrigation district was shown. Policymakers and water authorities may use this information to increase the effectiveness of the water conservation in the IV, which is of primary concern these days.

Author Contributions: Conceptualization, H.S. and S.A.; methodology, U.P.; software, U.P.; validation, U.P., S.A. and H.S.; formal analysis, U.P.; investigation, U.P.; resources, S.A. and H.S.; data curation, U.P.; writing—original draft preparation, U.P.; writing—review and editing, S.A. and H.S.; visualization, U.P.; supervision, H.S. and S.A.; project administration, H.S. and S.A.; funding acquisition, H.S. and S.A. All authors have read and agreed to the published version of the manuscript.

Funding: This research received no external funding.

Institutional Review Board Statement: Not applicable.

Informed Consent Statement: The study did not involve humans.

Data Availability Statement: The data is available upon request.

Conflicts of Interest: The authors declare no conflict of interest.

References

1. González-Dugo, M.P.; Mateos, L. Spectral vegetation indices for benchmarking water productivity of irrigated cotton and sugarbeet crops. *Agric. Water Manag.* **2008**, *95*, 48–58. [CrossRef]
2. Taghvaeian, S.; Neale, C.M.; Osterberg, J.C.; Sritharan, S.I.; Watts, D.R. Remote sensing and GIS techniques for assessing irrigation performance: Case study in Southern California. *J. Irrig. Drain. Eng.* **2018**, *144*, 05018002. [CrossRef]
3. Dawadi, S.; Ahmad, S. Changing climatic conditions in the Colorado River Basin: Implications for water resources management. *J. Hydrol.* **2012**, *430*, 127–141. [CrossRef]
4. Kalra, A.; Sagarika, S.; Pathak, P.; Ahmad, S. Hydro-climatological changes in the Colorado River Basin over a century. *Hydrol. Sci. J.* **2017**. [CrossRef]
5. Rahaman, M.M.; Thakur, B.; Kalra, A.; Ahmad, S. Modeling of GRACE-Derived Groundwater Information in the Colorado River Basin. *Hydrology* **2019**, *6*, 19. [CrossRef]
6. Tamaddun, K.; Kalra, A.; Kumar, S.; Ahmad, S. CMIP5 Models' Ability to Capture Observed Trends under the Influence of Shifts and Persistence: An In-depth Study on the Colorado River Basin. *J. Appl. Meteorol. Climatol.* **2019**. [CrossRef]
7. Bali, K.M.; Grismer, M.E.; Tod, I.C. Reduced-runoff irrigation of alfalfa in Imperial Valley, California. *J. Irrig. Drain. Eng.* **2001**, *127*, 123–130. [CrossRef]
8. Sagarika, S.; Kalra, A.; Ahmad, S. Evaluating the effect of persistence on long-term trends and analyzing step changes in streamflows of the continental United States. *J. Hydrol.* **2014**, *517*, 36–53. [CrossRef]
9. Ghumman, A.R.; Ahmad, S.; Khan, R.A.; Hashmi, H.N. Comparative Evaluation of Implementing Participatory Irrigation Management in Punjab Pakistan. *Irrig. Drain.* **2014**, *63*, 315–327. [CrossRef]
10. Dawadi, S.; Ahmad, S. Evaluating the Impact of Demand-Side Management on Water Resources under Changing Climatic Conditions and Increasing Population. *J. Environ. Manag.* **2013**, *114*, 261–275. [CrossRef]
11. Saher, R.; Stephen, H.; Ahmad, S. Urban evapotranspiration of Green Spaces in Arid Regions through Two Established Approaches: A Review of Key Drivers, Advancements, Limitations, and Potential Opportunities. *Urban Water J.* **2020**. [CrossRef]
12. Bukhary, S.; Batista, J.; Ahmad, S. Analyzing Land and Water Requirements for Solar Deployment in the Southwestern United States. *Renew. Sustain. Energy Rev.* **2018**, *82*, 3288–3305. [CrossRef]
13. Qaiser, K.; Ahmad, S.; Johnson, W.; Batista, J.R. Evaluating the impact of water conservation on fate of outdoor water use: A study in an arid region. *J. Environ. Manag.* **2011**, *92*, 2061–2068. [CrossRef]
14. Qaiser, K.; Ahmad, S.; Johnson, W.; Batista, J.R. Evaluating Water Conservation and Reuse Policies using a Dynamic Water Balance Model. *Environ. Manag.* **2013**, *51*, 449–458. [CrossRef] [PubMed]
15. Ghumman, A.R.; Iqbal, M.; Ahmad, S.; Hashmi, H.N. Experimental and Numerical Investigations for Optimal Emitter Spacing in Drip Irrigation. *Irrig. Drain.* **2018**, *67*, 724–737. [CrossRef]
16. Tamaddun, K.; Kalra, A.; Ahmad, S. Potential of rooftop rainwater harvesting to meet outdoor water demand in arid regions. *J. Arid Land.* **2018**, *10*, 68–83. [CrossRef]
17. Ahmad, M.U.D.; Turral, H.; Nazeer, A. Diagnosing irrigation performance and water productivity through satellite remote sensing and secondary data in a large irrigation system of Pakistan. *Agric. Water Manag.* **2009**, *96*, 551–564. [CrossRef]

18. Bastiaanssen, W.G.; Bos, M.G. Irrigation performance indicators based on remotely sensed data: A review of literature. *Irrig. Drain. Syst.* **1999**, *13*, 291–311. [CrossRef]
19. Murray-Rust, H.; Snellen, W.B. *Irrigation System Performance Assessment and Diagnosis*; IWMI: Anand, India, 1993.
20. Menenti, M.; Visser, T.; Morabito, J.A.; Drovandi, A. *Appraisal of Irrigation Performance with Satellite Data and Georeferenced Information: The Rio Tunuyan Irrigation Scheme*; Institute of Irrigation Studies, Southampton University: Southampton, UK, 1989.
21. Moran, M.S. Irrigation management in Arizona using satellites and airplanes. *Irrig. Sci.* **1994**, *15*, 35–44. [CrossRef]
22. Bastiaanssen, W.G.; Van der Wal, T.; Visser, T.N.M. Diagnosis of regional evaporation by remote sensing to support irrigation performance assessment. *Irrig. Drain. Syst.* **1996**, *10*, 1–23. [CrossRef]
23. Roerink, G.J.; Bastiaanssen, W.G.; Chambouleyron, J.; Menenti, M. Relating crop water consumption to irrigation water supply by remote sensing. *Water Resour. Manag.* **1997**, *11*, 445–465. [CrossRef]
24. Alexandridis, T.; Asif, S.; Ali, S. *Water Performance Indicators Using Satellite Imagery for the Fordwah Eastern Sadiqia (South) Irrigation and Drainage Project*; No. H024895; International Water Management Institute: Colombo, Sri Lanka, 1999.
25. Bastiaanssen, W.G.M.; Thiruvengadachari, S.; Sakthivadivel, R.; Molden, D.J. Satellite remote sensing for estimating producitivities of land and water. *Int. J. Water Resour. Dev.* **1999**, *15*, 181–186. [CrossRef]
26. Thiruvengadachari, S.; Sakthivadivel, R. *Satellite Remote Sensing Techniques to Aid Irrigation System Performance Assessment: A Case study in India*; Research Report 09; International Water Management Institute: Colombo, Sri Lanka, 1997; p. 31.
27. Ambast, S.K.; Singh, O.P.; Tyagi, N.K.; Menenti, M.; Roerink, G.J.; Bastiaanssen, W.G.M. Appraisal of irrigation system performance in saline irrigated command using SRS and GIS. In *Operational remote sensing for sustainable development*; Balkema: Rotterdam, The Netherlands, 1999; pp. 457–461.
28. Karatas, B.S.; Akkuzu, E.; Unal, H.B.; Asik, S.; Avci, M. Using satellite remote sensing to assess irrigation performance in Water User Associations in the Lower Gediz Basin, Turkey. *Agric. Water Manag.* **2009**, *96*, 982–990. [CrossRef]
29. Kharrou, M.H.; Le Page, M.; Chehbouni, A.; Simonneaux, V.; Er-Raki, S.; Jarlan, L.; Chehbouni, G. Assessment of equity and adequacy of water delivery in irrigation systems using remote sensing-based indicators in semi-arid region, Morocco. *Water Resour. Manag.* **2013**, *27*, 4697–4714. [CrossRef]
30. Geerts, S.; Raes, D. Deficit irrigation as an on-farm strategy to maximize crop water productivity in dry areas. *Agric. Water Mana.* **2009**, *96*, 1275–1284. [CrossRef]
31. Cai, X.L.; Sharma, B.R. Integrating remote sensing, census, and weather data for an assessment of rice yield, water consumption and water productivity in the Indo-Gangetic River basin. *Agric. Water Manag.* **2010**, *97*, 309–316. [CrossRef]
32. Immerzeel, W.W.; Gaur, A.; Zwart, S.J. Integrating remote sensing and a process-based hydrological model to evaluate water use and productivity in a south Indian catchment. *Agric. Water Manag.* **2008**, *95*, 11–24. [CrossRef]
33. Yan, N.; Wu, B. Integrated spatial–temporal analysis of crop water productivity of winter wheat in Hai Basin. *Agric. Water Manag.* **2014**, *133*, 24–33. [CrossRef]
34. Zwart, S.J.; Bastiaanssen, W.G. SEBAL for detecting spatial variation of water productivity and scope for improvement in eight irrigated wheat systems. *Agric. Water Manag.* **2007**, *89*, 287–296. [CrossRef]
35. Ahmed, B.M.; Tanakamaru, H.; Tada, A. Application of remote sensing for estimating crop water requirements, yield, and water productivity of wheat in the Gezira Scheme. *Int. J. Remote Sens.* **2010**, *31*, 4281–4294. [CrossRef]
36. Gorantiwar, S.D.; Smout, I.K. Performance assessment of irrigation water management of heterogeneous irrigation schemes: 1. A framework for evaluation. *Irrig. Drain. Syst.* **2005**, *19*, 1–36. [CrossRef]
37. Usman, M.; Liedl, R.; Shahid, M.A. Managing irrigation water by yield and water productivity assessment of a rice-wheat system using remote sensing. *J. Irrig. Drain. Eng.* **2014**, *140*, 04014022. [CrossRef]
38. Zhang, K.; Kimball, J.S.; Running, S.W. A review of remote sensing based actual evapotranspiration estimation. *Wiley Interdiscip. Rev. Water* **2016**, *3*, 834–853. [CrossRef]
39. Kustas, W.P.; Norman, J.M. Use of remote sensing for evapotranspiration monitoring over land surfaces. *Hydrol. Sci. J.* **1996**, *41*, 495–516. [CrossRef]
40. Reyes-González, A.; Kjaersgaard, J.; Trooien, T.; Hay, C.; Ahiablame, L. Estimation of crop evapotranspiration using satellite remote sensing-based vegetation index. *Adv. Meteorol.* **2018**, 4525021. [CrossRef]
41. Er-Raki, S.; Chehbouni, A.; Duchemin, B. Combining satellite remote sensing data with the FAO-56 dual approach for water use mapping in irrigated wheat fields of a semi-arid region. *Remote Sens.* **2010**, *2*, 375–387. [CrossRef]
42. Bastiaanssen, W.G.; Menenti, M.; Feddes, R.A.; Holtslag, A.A.M. A remote sensing surface energy balance algorithm for land (SEBAL). 1. Formulation. *J. Hydrol.* **1998**, *212*, 198–212. [CrossRef]
43. Allen, R.G.; Tasumi, M.; Trezza, R. Satellite-based energy balance for mapping evapotranspiration with internalized calibration (METRIC)—Model. *J. Irrig. Drain. Eng.* **2007**, *133*, 380–394. [CrossRef]
44. Roerink, G.J.; Su, Z.; Menenti, M. S-SEBI: A simple remote sensing algorithm to estimate the surface energy balance. *Phys. Chem. Earth B* **2000**, *25*, 147–157. [CrossRef]
45. Singh, R.K.; Senay, G.B. Comparison of four different energy balance models for estimating evapotranspiration in the Midwestern United States. *Water* **2016**, *8*, 9. [CrossRef]
46. Singh, R.K.; Liu, S.; Tieszen, L.L.; Suyker, A.E.; Verma, S.B. Estimating seasonal evapotranspiration from temporal satellite images. *Irrig. Sci.* **2012**, *30*, 303–313. [CrossRef]

47. De Oliveira Costa, J.; José, J.V.; Wolff, W.; de Oliveira, N.P.R.; Oliveira, R.C.; Ribeiro, N.L.; Coelho, R.D.; da Silva, T.J.A.; Silva, E.M.B.; Schlichting, A.F. Spatial variability quantification of maize water consumption based on Google EEflux tool. *Agric. Water Manag.* **2020**, *232*, 106037. [CrossRef]

48. Venancio, L.P.; Eugenio, F.C.; Filgueiras, R.; França da Cunha, F.; Argolo dos Santos, R.; Ribeiro, W.R.; Mantovani, E.C. Mapping within-field variability of soybean evapotranspiration and crop coefficient using the Earth Engine Evaporation Flux (EEFlux) application. *PLoS ONE* **2020**, *15*, e0235620. [CrossRef] [PubMed]

49. Lambert, M.J.; Traoré, P.C.S.; Blaes, X.; Baret, P.; Defourny, P. Estimating smallholder crops production at village level from Sentinel-2 time series in Mali's cotton belt. *Remote Sens. Environ.* **2018**, *216*, 647–657. [CrossRef]

50. Kayad, A.; Sozzi, M.; Gatto, S.; Marinello, F.; Pirotti, F. Monitoring within-field variability of corn yield using Sentinel-2 and machine learning techniques. *Remote Sens.* **2019**, *11*, 2873. [CrossRef]

51. Morel, J.; Todoroff, P.; Bégué, A.; Bury, A.; Martiné, J.F.; Petit, M. Toward a satellite-based system of sugarcane yield estimation and forecasting in smallholder farming conditions: A case study on Reunion Island. *Remote Sens.* **2014**, *6*, 6620–6635. [CrossRef]

52. Sibley, A.M.; Grassini, P.; Thomas, N.E.; Cassman, K.G.; Lobell, D.B. Testing remote sensing approaches for assessing yield variability among maize fields. *Agron. J.* **2014**, *106*, 24–32. [CrossRef]

53. Shanahan, J.F.; Schepers, J.S.; Francis, D.D.; Varvel, G.E.; Wilhelm, W.W.; Tringe, J.M.; Major, D.J. Use of remote-sensing imagery to estimate corn grain yield. *Agron. J.* **2001**, *93*, 583–589. [CrossRef]

54. Tucker, C.J.; Holben, B.N.; Elgin, J.H., Jr.; McMurtrey, J.E., III. Relationship of spectral data to grain yield variation [within a winter wheat field]. *Photogramm. Eng. Remote Sens.* **1980**, *46*, 657–666.

55. Shirsath, P.B.; Sehgal, V.K.; Aggarwal, P.K. Downscaling regional crop yields to local scale using remote sensing. *Agriculture* **2020**, *10*, 58. [CrossRef]

56. Imperial County Planning and Development Services. Agriculture Element. 2015. Available online: https://www.icpds.com/assets/planning/agricultural-element-2015.pdf (accessed on 12 July 2020).

57. Inouye, D. Crop Water Requirements Imperial Valley. 1981. Available online: https://nrm.dfg.ca.gov/FileHandler.ashx?DocumentID=7226 (accessed on 12 July 2020).

58. Kayad, A.G.; Al-Gaadi, K.A.; Tola, E.; Madugundu, R.; Zeyada, A.M.; Kalaitzidis, C. Assessing the spatial variability of alfalfa yield using satellite imagery and ground-based data. *PLoS ONE* **2016**, *11*, e0157166. [CrossRef]

59. Allen, R.G.; Pereira, L.S.; Raes, D.; Smith, M. Crop Evapotranspiration-Guidelines for computing crop water requirements-FAO Irrigation and drainage paper 56. *FAO Rome* **1998**, *300*, D05109.

60. Pruitt, W.O.; Doorenbos, J. *Empirical Calibration: A Requisite for Evapotranspiration Formulae Based on Daily or Longer Mean Climate Data?* Hungarian National Committee: Budapest, Hungary, 1977.

61. Salgado, R.; Mateos, L. Evaluation of different methods of estimating ET for the performance assessment of irrigation schemes. *Agric. Water Manag.* **2021**, *243*, 106450. [CrossRef]

62. Hanson, B.; Putnam, D.; Snyder, R. Deficit irrigation of alfalfa as a strategy for providing water for water-short areas. *Agric. Water Manag.* **2007**, *93*, 73–80. [CrossRef]

63. Wright, J.L. New evapotranspiration crop coefficients. *J. Irrig. Drain. Div.* **1982**, *108*, 57–74. [CrossRef]

64. Pruitt, W.O.; Lourence, F.; Von Oettingen, S. Water use by crops as affected by climate and plant factors. *Calif. Agric.* **1972**, *26*, 10–14.

65. Walter, I.A.; Allen, R.G.; Elliott, R.; Jensen, M.E.; Itenfisu, D.; Mecham, B.; Martin, D. ASCE's standardized reference evapotranspiration equation. In *Watershed Management and Operations Management*; American Society of Civil Engineers: Reston, VA, USA, 2000; pp. 1–11.

66. Allen, R.G.; Clemmens, A.J.; Burt, C.M.; Solomon, K.; O'Halloran, T. Prediction accuracy for projectwide evapotranspiration using crop coefficients and reference evapotranspiration. *J. Irrig. Drain. Eng.* **2005**, *131*, 24–36. [CrossRef]

67. Bos, M.G.; Burton, M.A.; Molden, D.J. *Irrigation and Drainage Performance Assessment: Practical Guidelines*; CABI Publishing: Oxford, UK, 2005.

68. Blatchford, M.L.; Mannaerts, C.M.; Zeng, Y.; Nouri, H.; Karimi, P. Status of accuracy in remotely sensed and in-situ agricultural water productivity estimates: A review. *Remote Sens. Environ.* **2019**, *234*, 111413. [CrossRef]

69. Bhattarai, N.; Shaw, S.B.; Quackenbush, L.J.; Im, J.; Niraula, R. Evaluating five remote sensing based single-source surface energy balance models for estimating daily evapotranspiration in a humid subtropical climate. *Int. J. Appl. Earth Obs. Geoinf.* **2016**, *49*, 75–86. [CrossRef]

70. José, J.V.; Oliveira, N.P.R.D.; Silva, T.J.D.A.D.; Bonfim-Silva, E.M.; Costa, J.D.O.; Fenner, W.; Coelho, R.D. Quantification of cotton water consumption by remote sensing. *Geocarto Int.* **2020**, *35*, 1800–1813. [CrossRef]

71. Costa, J.D.O.; Coelho, R.D.; Wolff, W.; José, J.V.; Folegatti, M.V.; Ferraz, S.F.D.B. Spatial variability of coffee plant water consumption based on the SEBAL algorithm. *Sci. Agric.* **2019**, *76*, 93–101. [CrossRef]

72. Kamble, B.; Kilic, A.; Hubbard, K. Estimating crop coefficients using remote sensing-based vegetation index. *Remote Sens.* **2013**, *5*, 1588–1602. [CrossRef]

73. Santos, C.; Lorite, I.J.; Tasumi, M.; Allen, R.G.; Fereres, E. Integrating satellite-based evapotranspiration with simulation models for irrigation management at the scheme level. *Irrig. Sci.* **2008**, *26*, 277–288. [CrossRef]

74. Molden, D.J.; Gates, T.K. Performance measures for evaluation of irrigation-water-delivery systems. *J. Irrig. Drain. Eng.* **1990**, *116*, 804–823. [CrossRef]

75. Inouye, D.; Yoha, R.E. Preliminary Evaluation of Soils and Irrigation' Practices in the Imperial Valley. 1981. Available online: https://nrm.dfg.ca.gov/FileHandler.ashx?DocumentID=9024 (accessed on 19 November 2020).
76. Bastiaanssen, W.G.M.; Brito, R.A.L.; Bos, M.G.; Souza, R.A.; Cavalcanti, E.B.; Bakker, M.M. Low-cost satellite data for monthly irrigation performance monitoring: Benchmarks from Nilo Coelho, Brazil. *Irrig. Drain. Syst.* **2001**, *15*, 53–79. [CrossRef]
77. Patil, V.C.; Al-Gaadi, K.A.; Madugundu, R.; Tola, E.H.; Marey, S.; Aldosari, A.; Gowda, P.H. Assessing agricultural water productivity in desert farming system of Saudi Arabia. *IEEE J. Sel. Top. Appl. Earth Obs. Remote Sens.* **2014**, *8*, 284–297. [CrossRef]
78. Madugundu, R.; Al-Gaadi, K.A.; Tola, E.; Patil, V.C.; Biradar, C.M. Quantification of agricultural water productivity at field scale and its implication in on-farm water management. *J. Indian Soc. Remote Sens.* **2017**, *45*, 643–656. [CrossRef]
79. Smeal, D.; Kallsen, C.E.; Sammis, T.W. Alfalfa yields as related to transpiration, growth stage and environment. *Irrig. Sci.* **1991**, *12*, 79–86. [CrossRef]
80. Maas, E.V.; Hoffman, G.J. Crop salt tolerance—Current assessment. *J. Irrig. Drain. Div.* **1977**, *103*, 115–134. [CrossRef]
81. French, A.N.; Hunsaker, D.J.; Thorp, K.R. Remote sensing of evapotranspiration over cotton using the TSEB and METRIC energy balance models. *Remote Sens. Environ.* **2015**, *158*, 281–294. [CrossRef]
82. Lokupitiya, E.; Lefsky, M.; Paustian, K. Use of AVHRR NDVI time series and ground-based surveys for estimating county-level crop biomass. *Int. J. Remote Sens.* **2010**, *31*, 141–158. [CrossRef]
83. Meng, J.; Du, X.; Wu, B. Generation of high spatial and temporal resolution NDVI and its application in crop biomass estimation. *Int. J. Digit. Earth* **2013**, *6*, 203–218. [CrossRef]

Article

Prognostication of Shortwave Radiation Using an Improved No-Tuned Fast Machine Learning

Isa Ebtehaj [1], Keyvan Soltani [2], Afshin Amiri [3], Marzban Faramarzi [4], Chandra A. Madramootoo [5] and Hossein Bonakdari [1,*]

[1] Department of Soils and Agri-Food Engineering, Université Laval, Québec, QC G1V 0A6, Canada; isa.ebtehaj.1@ulaval.ca

[2] Department of Civil Engineering, Razi University, Kermanshah 6714967346, Iran; keyvansoltanii@gmail.com

[3] Department of Remote Sensing and GIS, University of Tehran, Tehran 1417935840, Iran; afshinamirii@yahoo.com

[4] Rangeland and Watershed Management Group, Faculty of Agriculture, Ilam University, Ilam 69315516, Iran; m.faramarzi@ilam.ac.ir

[5] Department of Bioresource Engineering, McGill University, Quebec, QC H9X 3V9, Canada; chandra.madramootoo@mcgill.ca

* Correspondence: hossein.bonakdari@fsaa.ulaval.ca; Tel.: +1-418-656-2131

Abstract: Shortwave radiation density flux (SRDF) modeling can be key in estimating actual evapotranspiration in plants. SRDF is the result of the specific and scattered reflection of shortwave radiation by the underlying surface. SRDF can have profound effects on some plant biophysical processes such as photosynthesis and land surface energy budgets. Since it is the main energy source for most atmospheric phenomena, SRDF is also widely used in numerical weather forecasting. In the current study, an improved version of the extreme learning machine was developed for SRDF forecasting using the historical value of this variable. To do that, the SRDF through 1981–2019 was extracted by developing JavaScript-based coding in the Google Earth Engine. The most important lags were found using the auto-correlation function and defined fifteen input combinations to model SRDF using the improved extreme learning machine (IELM). The performance of the developed model is evaluated based on the correlation coefficient (R), root mean square error (RMSE), mean absolute percentage error (MAPE), and Nash–Sutcliffe efficiency (NSE). The shortwave radiation was developed for two time ahead forecasting (R = 0.986, RMSE = 21.11, MAPE = 8.68%, NSE = 0.97). Additionally, the estimation uncertainty of the developed improved extreme learning machine is quantified and compared with classical ELM and found to be the least with a value of ±3.64 compared to ±6.9 for the classical extreme learning machine. IELM not only overcomes the limitation of the classical extreme learning machine in random adjusting of bias of hidden neurons and input weights but also provides a simple matrix-based method for practical tasks so that there is no need to have any knowledge of the improved extreme learning machine to use it.

Keywords: water resources; Daymet V3; Google Earth Engine; improved extreme learning machine (IELM); sensitivity analysis; shortwave radiation flux density; sustainable development

Citation: Ebtehaj, I.; Soltani, K.; Amiri, A.; Faramarzi, M.; Madramootoo, C.A.; Bonakdari, H. Prognostication of Shortwave Radiation Using an Improved No-Tuned Fast Machine Learning. *Sustainability* **2021**, *13*, 8009. https://doi.org/10.3390/su13148009

Academic Editor: Roger Jones

Received: 12 June 2021
Accepted: 16 July 2021
Published: 17 July 2021

Publisher's Note: MDPI stays neutral with regard to jurisdictional claims in published maps and institutional affiliations.

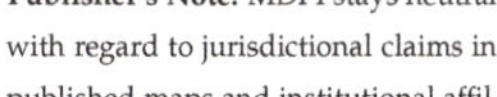

1. Introduction

Shortwave radiation is of essential significance in climate research since it controls the complete energy exchange between the land/ocean surface and atmosphere [1]. It plays a crucial role in biogeochemical, physical, ecological, and hydrological processes [2]. Shortwave radiation is the energy source that causes photosynthesis, transpiration, evaporation, and other significant process connected to agriculture systems. It is incredibly variable (both temporally and spatially) on the earth's surface. Precise shortwave radiation is essential for evapotranspiration models, which are employed to construct irrigation plans to improve crop yield while saving water and minimizing herbicide, fertilizer, and pesticide applications [3].

Reliable calculation of shortwave radiation is essential as a significant component of the energy budget to understand global change [4–7]. In the clear sky, most of the shortwave radiation is caused by direct sunlight. Shortwave radiation is divided into direct and diffuse components. The direct component is the energy that emanates directly from the sun's rays, and particles and molecules disperse the diffuse component in the air [8]. The amount of diffused shortwave radiation is affected by the height of the sun. Shortwave solar input stands as an essential component of the surface energy balance and is considered to be the principal source of energy on Earth [9–12]. The available energy for hydrological processes such as evaporation and transpiration is strongly affected by solar radiation. Additionally, biological phenomena such as photosynthesis and the carbon cycle are also dependent on solar radiation (direct and diffused radiation) [13,14].

Re-analysis datasets are constantly being improved with increasing access to observational data and advances in modeling and data assimilation systems. The use of re-analysis data has considerable potential for studying areas with a shortage of terrestrial data. Re-analysis data are a combination of ground observations, field surveys, and various models [15]. Different re-analysis datasets are available on a continental and global scale today, e.g., Twentieth-Century Reanalysis, Daily Surface Weather and Climatological Summaries (Daymet), and the Global Land Data Assimilation System. These datasets can provide continuous data and compensate for the gap between terrestrial data.

The Daymet model includes a set of tools designed to calculate estimates of daily weather parameters in Canada, the United States, and Mexico. The main advantages of Daymet are (1) covering the vast majority of North America, (2) providing data on a daily scale for a long period starting from 1980, (3) low spatial resolution (i.e., one kilometer), which is higher than other Reanalysis data. According to the mentioned reasons and increasing the modeling accuracy, Daymet was used to provide shortwave radiation flux density (SRFD) information.

In the last decade, applying the machine learning-based approach to modeling nonlinear complex problems in hydrology and environmental science has attracted many scholars [16–19]. The main advantages of these approaches are high accuracy, low human intervention, and continuous improvement [20–22]. One of the commonly used machine-learning-based techniques is the feedforward neural network, which has been successfully applied in different fields of science. The main training algorithm in this approach is backpropagation. The key advantage of this method is the nonlinear mapping of the independent input variables and dependent output variable(s), which overcomes the limitation of the classical regression-based approaches. It should be noted that although non-linear mapping has several advantages, it preserves some of the limitations of the parent independent input variables. For example, it can preserve the expected value of the autocorrelation function but not the higher-order joint moments and time-asymmetry [23]. The feedforward neural network trained with the backpropagation algorithm is a well-known machine learning method. According to easy implementation, suitable performance, and inherent simplicity [24], it has been successfully applied in different fields of science [25–28]. To overcome the drawbacks of this approach, including slow convergence, time-consuming training [29], and trapping in local minima that leads to low generalizability [30], the extreme learning machine (ELM) [31] was introduced. The ELM is a single-layer feedforward neural network with a fast training process. The main pros of this algorithm are high accuracy, robustness, least user intervention, rapid learning rate, a learning process that requires only a single iteration, and high generalization [32]. However, the main drawback of the ELM is the random generation of the two main matrices, including bias of hidden neurons and input weights. To remove the main limitation of the ELM, an improved version of the ELM is developed based on the orthogonal of the random generation matrices and the definition of an iteration parameter.

According to current knowledge about the influence of non-renewable energies on the environment, renewable energy sources have attracted scholars for their research interests. It is principally because renewable energies are stabilized by natural procedures, which do not contribute to climate change, global warming, and greenhouse gases. With the advancement of science and the discovery of solar radiation as a sustainable source of renewable energy, it has been considered to overcome the problems caused by fossil fuels. Moreover, the Sustainable Development Goals, which include various economic, social, and environmental goals, provide a global framework that the United Nations member states are committed to achieving [33]. The seventh goal of sustainable development is to improve access to clean and affordable energy, which aims to ensure that sustainable, reliable, modern, and affordable energy is accessible to everyone. To achieve this goal, energy consumption and efficiency must be controlled and monitored. Solar radiation as a renewable energy source can contribute to the seventh goal of sustainable development [34]. In this regard, it is necessary to develop practical and appropriate models for each region to achieve sustainable development goals. This study tries to establish a suitable model for SRFD estimation in Nunavik, which is one of the coldest regions of the world, so that it can be used to play an effective role in environmental processes such as vegetation management, water resources management, lake management, control of changes in land use, and so on.

The primary purpose of the current study is to build an improved version of the ELM known as IELM for monthly short-term prediction of the Shortwave Radiation Flux Density (SRFD) in the Nunavik region, Quebec, Canada. It has been shown that key hydrological-cycle processes (such as evapotranspiration, temperature, and precipitation) exhibit the so-called long-range dependence that also affects the fractal short-range dependence [35], and thus, prediction. The novelty of this study is four-fold. (1) Introducing the IELM to overcome the main constraint of the ELM in random generation of the bias of hidden neurons and input weights matrices: In this model, iteration parameters and calculation of orthogonal random generation matrices are considered to find the most reliable results in terms of simplicity and accuracy simultaneously. The main advantages of this model are high generalization capability and fast training samples for a large number of iterations. (2) Coding a JavaScript-based code in the Google Earth Engine environment to extract SRFD data from the gridded Daymet product: Daymet supplies long-term and continuous estimates of daily weather and climatology factors generated using ground-based observations through statistical modeling techniques. (3) Time series-based modeling of the SRFD is used without requiring other independent variables to develop a simple model. (4) Apractical matrix-based equation for practical applications is provided.

2. Materials and Methods

2.1. Case Study

SRFD is the main component of energy exchanges between the atmosphere, the Earth's surface, and the ocean. Therefore, it affects the context of temperature, atmospheric and oceanic circulation, and the hydrological cycle [1]. The selected case study is located in the northern part of Quebec in Canada. The Eastern Hudson Bay Basin is situated in the east of Hudson Bay (longitudes 76°40′ W–71°30′ W and latitudes 54°50′ N–57°10′ N), which is a part of the Nunavik area (Figure 1). Nunavik lies in both the subarctic and Arctic climate zones. Due to the climatic conditions of this area, access to the site is complex, and measuring hydrological variables in this area is not only easy but also costly. The basin's elevation is between zero and 594 m relative to the mean sea level, and its average elevation is 300 m. The most important natural phenomena of this basin are clearwater lakes composed of two separate lakes that occupy the middle part of the basin. The deepest part of these lakes is 178 m. Annual precipitation in this basin is between 600 to 852 mm with an average of 726 mm.

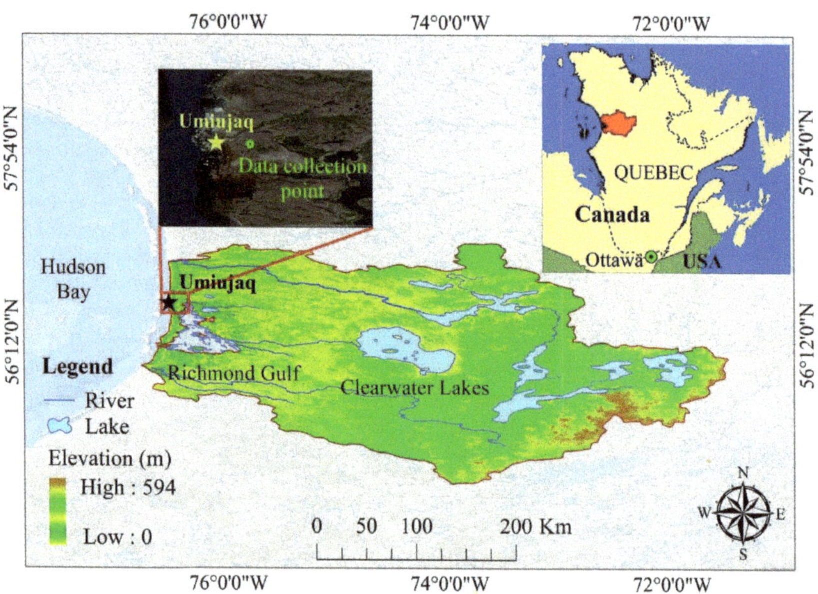

Figure 1. The geographic location of the study area.

Based on land cover data (MOD12Q1, 500 m × 500 m) obtained from NASA Land Processes Distributed Active Archive Center, open shrubland mostly covers the northern parts of the area with a height of one to two meters, the southern parts mostly include savannas areas with 10% to 30% tree cover, and most of the cover is grasslands in coastal regions (Figure 2).

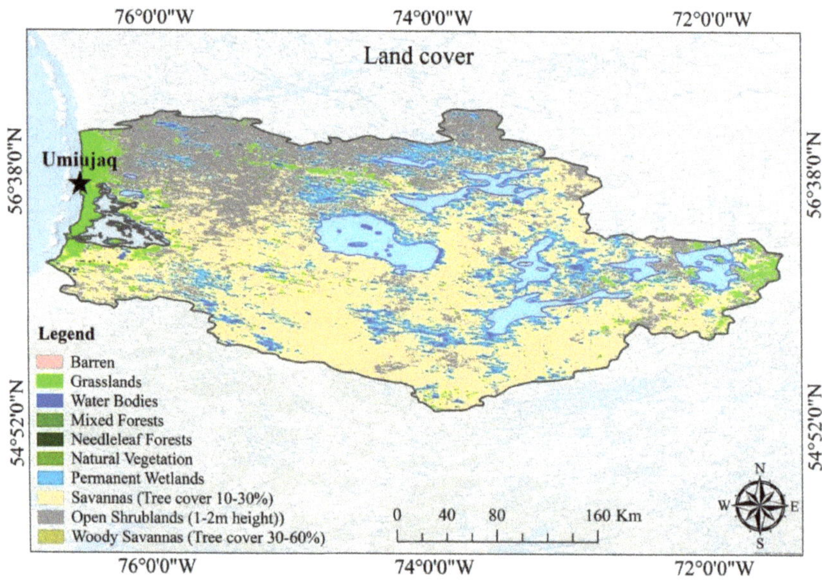

Figure 2. Land cover of the study area.

Solar energy is a clean source that does not introduce any pollutants to the environment, and it may even overcome the problems caused by fossil fuels. It should be noted that this is mostly possible if the solar energy is combined with a water-energy nexus battery (e.g., hydroelectric dam) to be effective and sustainable as an energy source. In this regard, its efficient use requires practical and effective knowledge [36]. SRFD data provide information about the impact of solar energy on the Earth in a given area over a period of time. It should be noted that due to the high cost of observing and measuring, this parameter is not easily accessible. Therefore, there is a need to create an alternative for estimating this data as well as predicting it [37,38]. Over the past decade, SRFD measurements as well as sunshine have declined in western Canada and elsewhere. Since the amount of SRFD is critical for calculating the evapotranspiration, soil melt, snowmelt radiation, and other hydrological cycle components, it is necessary to provide practical and accurate methods to estimate it [39]. Because a large volume of the study area is composed of different vegetation and water, solar radiation has an influential role in the area's environment. Changes in solar radiation in each region may affect different parameters such as plant growth period, photosynthesis, and melting outside the natural time of ice and snow (especially important due to the cold region) [40,41].

Since the study area is one of the cold regions and its period of frost and snowfall is high, the management of melting snow and ice is necessary for this region. Therefore, comprehension of snow and ice melting levels is essential for the proper management of available water resources, including short-term or seasonal flow forecasts, for hydrological studies and ice–snow mass balance. SRFD is an effective parameter in this regard [42,43].

2.2. Daymet V3 (Daily Surface Weather and Climatological Summaries)

Daymet V3 provides daily ground-level weather parameters in North America (Canada, Mexico, USA, Hawaii, and Puerto Rico) (https://daymet.ornl.gov/, accessed on 31 March 2020). This data is available with a spatial resolution of 1000 m and a time interval of one day. Daymet includes data on daily minimum 2-meter air temperature (°C), daily maximum 2-meter air temperature (°C), precipitation, the partial pressure of water vapor, duration of the daylight period, snow water equivalent, and shortwave radiation flux [44].

Shortwave radiation flux density (SRFD) was extracted from the gridded Daymet product for the study area. These data were processed in the Google Earth Engine environment and provide a monthly time series to apply as input in subsequent analysis. Google Earth is free, it has up-to-date maps and data, it is available on a wide array of devices, it is incredibly detailed, and it is very user-friendly. This dataset covers the period from January 1980 to December 2020. To access the SRFD data provided in the Google Earth Engine by NASA, a JavaScript-based code was offered to extract data for the desired location in Nunavik.

Google Earth Engine is a cloud platform for global spatial data analysis that allows the processing of large amounts of data in various fields, including drought, natural disaster water management, deforestation, vegetation, agriculture, soil studies, climate monitoring, and conservation of the environment [45]. The Google Earth Engine provides convenient conditions for developing algorithms and receiving results quickly with easy access and a user-friendly environment. It improves accessibility and usability by offering Earth observation data to a wide range of research fields. Additionally, the Google Earth Engine cloud platform also provides access to relevant data and scripts for users who do not have the necessary data or computing tools [46,47].

For the current study, 468 SRFD monthly data between 1981–2019 were used to model SRFD, so that 336 samples (from January 1982 to December 2008) were considered for the training phase, while the other 132 samples (from January 2009 to December 2019) were applied to check the performance of the calibrated model. The different values of the SRFD for both phases are provided in Figure 3. Additionally, the statistical characteristics of the total, training, and testing data are presented in Table 1.

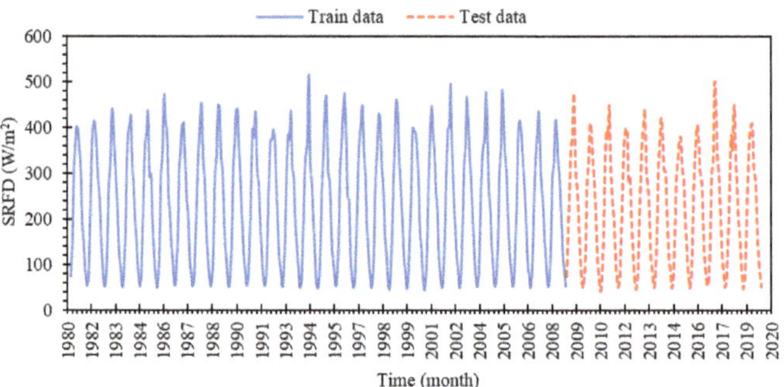

Figure 3. SRFD time series plot for test and training data.

Table 1. Statistical attributes of SRFD data.

	Nbr.	Min.	Max.	1st Q.	Median	3rd Q.	Mean	$\sigma(n)$	γ_1	γ_2
Total	468	41.89	516.18	106.77	243.38	350.39	231.67	131.50	0.11	−1.31
Train	336	43.41	516.18	106.78	245.02	356.37	232.74	131.06	0.08	−1.38
Test	132	41.89	497.03	105.92	232.70	329.67	227.24	128.67	0.08	−1.28

Nbr.: Number of data, Min. and Max.: minimum and maximum of data, 1st Q. and 3rd Q.: first and third quartiles, $\sigma(n)$: standard deviation, $\gamma 1$: skewness, $\gamma 2$: kurtosis.

2.3. Improved Extreme Learning Machine (IELM)

The backpropagation is a well-established training algorithm for the feed-forward neural network to solve many nonlinear complex problems leading to acceptable results [48]. However, similar to any machine learning-based approach, the BP has some drawbacks with its generalization and implementation, including local minima, learning rate, overfitting, low generalization ability [49], and time-consuming training [50].

To overcome the abovementioned limitations, Huang et al. [30] presented the extreme learning machine (ELM) as a training algorithm for a single-layer feedforward neural network. Using ELM is so simple that no parameters need to be set other than defining the network architecture. Therefore, many of the complexities of tuning parameters in gradient algorithms are not present in this algorithm. Additionally, as the modeling speed in ELM is so high, most of the training takes a short time with a large amount of data, which is not easy to model with the classical neural network; it takes about a few minutes. Therefore, the most important advantages of this method are the least user intervention, the learning process needing a single iteration, robustness, the fast learning rate, avoiding local minimizations, high generalization, and high accuracy [51].

The modeling process in ELM consists of three main stages: (1) random determination of input weights and bias matrices of hidden neurons, (2) calculation of the hidden layer outputs matrix using randomly generated matrices and activation function, and (3) calculation of output weights through a linear process. Accordingly, among three different matrices that need to be quantified, only the output weights are calculated analytically, and the other two matrices are randomly set. The least-square solution of a linear system is considered to define the output weights matrix.

To model a problem, it is first necessary to specify the inputs and outputs of the problem. For the input matrix with d independent inputs and the output of the problem with m, the outputs are $InV_i \in R^{DM}$ and $Ta_i \in R^z$, respectively. The number of S training samples are defined as $\{(InV_i, Ta_i)\}_{i=1}^{S}$. By considering the $f(\cdot)$ as activation function and NHN as number of hidden neurons, the output of the feedforward neural network is considered as follows:

$$O_i = \sum_{j=1}^{NHN} \beta_j f(a_j \times InV_i + b_j,), \qquad i = 1, 2 \ldots, S \qquad (1)$$

where NHN is the number of hidden neurons, S is the number of training samples, $\beta_j \in R^z$ is the output weight matrix that connects the jth hidden node to the corresponding hidden node, z is the number of the output variable, and O_i and InV_i are the output and input variables, respectively. $b_j \in R$ is the bias of the jth hidden neuron, and $a_j \in R^{DM}$ is the input weights that link the input variables to the jth hidden node and $a_j \cdot InV_i$ denotes the inner product of the a_j and InV_i. The mathematical form of the different activations applied in the current study are as follows:

Sine

$$f(OutW, b, InV) = \sin(OutW \times InV + b) \tag{2}$$

Radial Basis Function

$$f(OutW, b, InV) = \exp(-(OutW \times InV + b)^2) \tag{3}$$

Triangular Basis Function

$$f(OutW, b, InV) = \begin{cases} 1 - |OutW \times InV + b| & OutW \times InV + b^3 0 \\ 0 & \text{otherwise} \end{cases} \tag{4}$$

Hardlimit

$$f(OutW, b, InW) = \begin{cases} 1 & \text{If } OutW \times InV + b^3 0 \\ 0 & \text{otherwise} \end{cases} \tag{5}$$

Sigmoid

$$f(OutW, b, InV) = \frac{1}{1 + \exp(-(OutW \times InV + b))} \tag{6}$$

Tangent hyperbolic

$$f(OutW, b, InV) = \tanh(OutW, b, InV) = \frac{\exp(2(OutW \times InV + b)) - 1}{\exp(2(OutW \times InV + b)) + 1} \tag{7}$$

The matrix-based form of Equation (1) with N separated equations is defined as follows:

$$T\beta = O \tag{8}$$

where T, β, and O ($O = [O_1, \ldots, O_N]^T$) denote hidden neurons' output, output weight, and output, respectively. The hidden neurons output matrix (T) is defined as follows:

$$T(OuTW_1, \ldots, OuTW_L, INV_1, \ldots, INV_N, b_1, \ldots, b_L,)$$
$$= \begin{bmatrix} f(OuTW_1 \times INV_1 + b_1) & L & f(OuTW_{NHN} \times INV_1 + b_{NHN}) \\ M & O & M \\ f(OuTW_1 \times INV_{DM} + b_1) & L & f(OuTW_{NHN} \times INV_{DM} + b_{NHN}) \end{bmatrix}_{DM \times NHN} \tag{9}$$

The two matrices T and β in Equation (8) are unknown. The H is calculated using input weights and the bias of hidden neuron matrices that both of them are randomly assigned. Therefore, T is calculated without the experience of training samples. Consequently, the output weights matrix (β) is the only unknown matrix that should be calculated through the training phase. To find this matrix, Equation (8) as a linear system should be solved.

The dimension of the T is DM × NHN. The NHN is generally higher than the input variables (DM), and therefore, the H is not a square matrix. Thus, finding the output weights matrix using Equation (8) is not simple [30]. To find the output weights matrix, the optimal least square solution of the output weights matrix by loss function minimization is calculated as follows:

$$E_{ELM} = MIN\|O - T\beta\| \tag{10}$$

where the output weights matrix calculated through the least square solution of the above equation is calculated as:

$$\hat{\beta} = T^+ y \tag{11}$$

where T^+ denotes the Moore–Penrose generalized inverse of the T [52]. The solution of Equation (10) for NHN < S is as follows:

$$\hat{\beta} = \left(T^T T\right)^{-1} T^T O \tag{12}$$

The dimensions of the bias of hidden neurons, input weights, and output weights matrices are "1 × NHN", "DM × NHN", and "1 × NHN", respectively. Therefore, the number of all tuned parameters through the training phase of the ELM is as follows:

$$k = NHN + DM \times NHN + NHN = NHN(DM + 2) \tag{13}$$

The ratio of the randomly generated parameters that are related to the input weights (i.e., DM × NHN) and bias of hidden neurons (i.e., 1 × NHN) to the output weight (i.e., 1 × NHN) is as follows:

$$R = \frac{NHN + DM \times NHN}{NHN(DM + 2)} = \frac{NHN(d + 1)}{NHN(d + 2)} = 1 - \frac{1}{DM + 2} \tag{14}$$

If the number of input variables is one, the R is more than 0.66. Therefore, it is observed that two-thirds of the parameters tuned in the modeling using ELM are randomly determined, which has a significant impact on the modeling results. An inaccurate value of these parameters may reduce the generalizability of the developed model. Therefore, in this study, two competencies are performed on the original ELM: (1) considering iteration parameter for ELM and (2) applying the orthonormal basis for the range of input weights and bias of hidden neuron matrices. Using these two competencies, the new version of the ELM is named improved ELM (IELM). The flowchart of the developed IELM is presented in Figure 4.

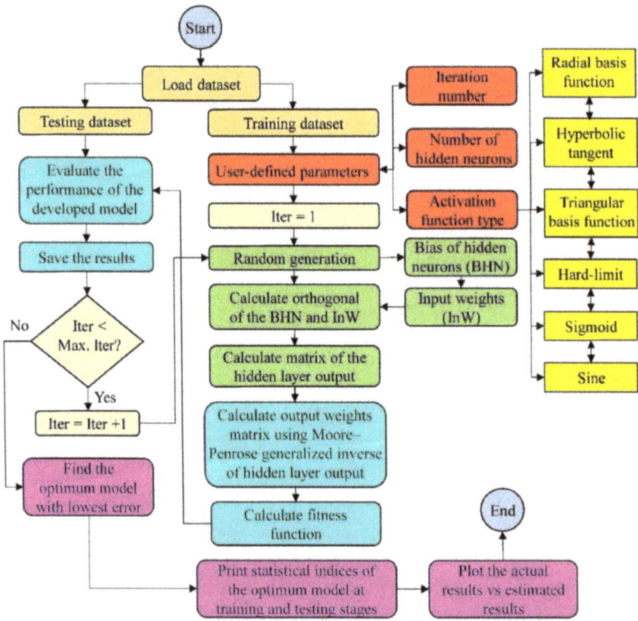

Figure 4. The flowchart of the developed IELM.

2.4. Uncertainty Analysis

The quantitative appraisal of the uncertainty analysis (UA) in the estimation of the shortwave radiation flux density (SRFD) has employed the developed improved extreme learning machine (IELM) instead of the original ELM. For a fair comparison of the different IELM-based models, the UA is utilized to the test data [53,54]. The use of test data in calculating and checking the UA of the developed IELM model has the advantage that its performance is examined for testing data without any role in model training. The generalizability of this model can be confirmed. The first step in UA is the calculation of the individual estimation error (IEE) as follows:

$$E_i = SRFD_{E,i} - SRFD_{O,i} \tag{15}$$

where $SRFD_{E,i}$, and $SRFD_{O,i}$ are ith estimated and observed SRFD, respectively, and E_i is the IEE of the ith sample. The IEE for all samples are applied to calculate mean estimation error (MEE) as follows:

$$\overline{E} = \sum_{i=1}^{S} E_i \tag{16}$$

where $\overline{E}$ is the MEE and S is the number of samples. Using the calculated $\overline{E}$ and IEE for all samples (E_i), the standard deviation of the estimation error (SDEE) is calculated as follows:

$$SDEE = \sqrt{\frac{1}{N-1}\sum_{i=1}^{N}(E_i - \overline{E})} \tag{17}$$

The negative (or positive) value of the MEE demonstrates that the estimator model underestimated (or overestimated) the observed values of the SRFD. To approximate a confidence band around the estimated values of an error, the ± 1.96 SDEE is calculated.

2.5. Workflow Approach

In this section, the workflow approach of the current study is presented. This workflow comprises four main steps: data collecting, model definition, tuning IELM parameters, and performance evaluation. Performing all four steps will lead to achieving the optimal model in SRFD estimation.

The first step is collecting data. To collect data, SRFD is extracted from the Daymet dataset by developing a JavaScript-based code in Google Engine Cloud. Using the developed code, the daily SRFD dataset from January 1981 to December 2019 is selected. The second step is the definition of the inputs to apply developed IELM-based MATLAB code to SRFD estimation. To do that, the auto-correlation function is employed [55]. Using this function, the most effective lags of the SRFD are found and defined as different combinations of these lags to find the best model. The third step is the definition of IELM parameters for all input combinations, defined in the previous step. To define IELM parameters, the type of activation function, the number of hidden neurons, and iteration number must be pre-defined by the user. It should be noted that the maximum number of hidden neurons should be considered. As the maximum allowable value of the hidden neurons should be considered, the number of optimal tuning parameters through the training phase will be less than the training samples. According to that, the number of columns and rows in the input weighs is identical to the number of input variables (InV) and the number of hidden neurons (NHN), respectively. The bias of hidden neurons (BHN) and output weights are two matrices with one column in which the number of rows is equal to NHN. The maximum allowable NHN is calculated as follows:

$$Max. NHN < \frac{TrSa}{InV + 2} \tag{18}$$

where TrSa is the number of training samples.

Besides the maximum allowable NHN, the activation function type and iteration numbers also must be pre-defined. The iteration number is considered in the range of 100 to 100,000. At the same time, the six different activation functions, including hyperbolic tangent, Sigmoid, hard limit, Sin, radial basis function, and triangular basis function, are investigated to find the optimum one. The final step of the IELM modeling is the performance evaluation of the developed models to find the optimum one. In this step, different statistical indices and uncertainty analysis are employed to find the optimum one. The schematic workflow for shortwave radiation flux density modeling is provided in Figure 5.

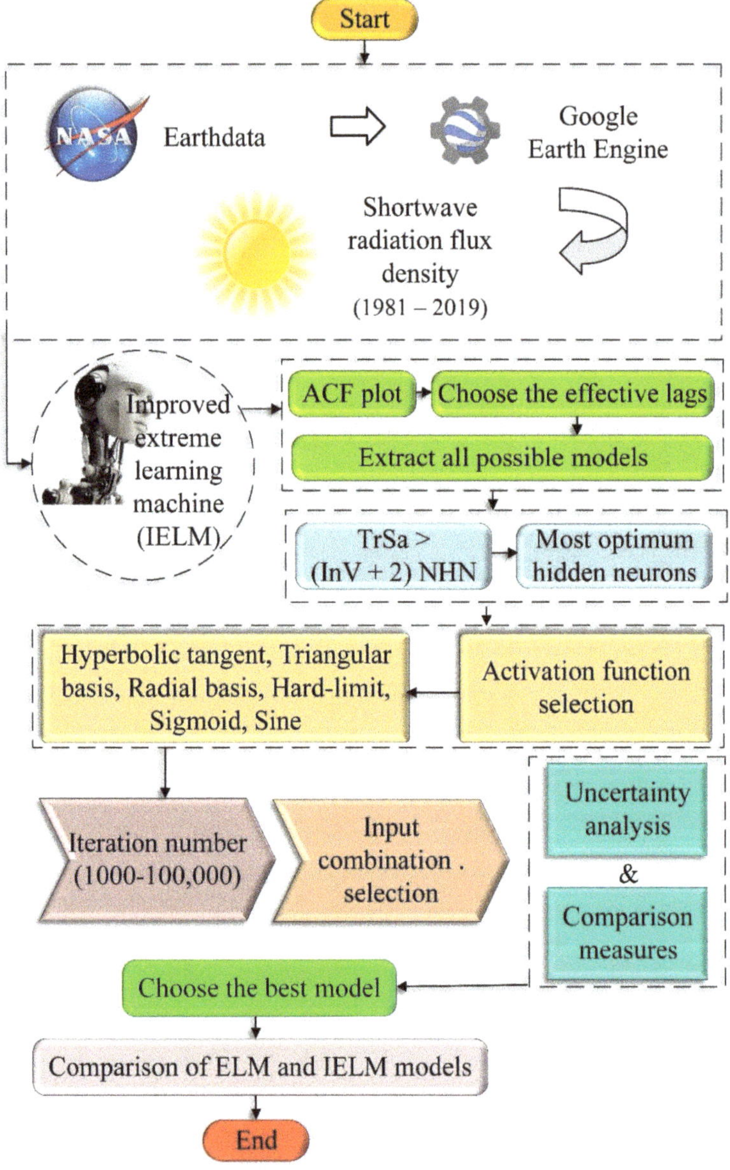

Figure 5. Schematic workflow for shortwave radiation flux density modeling.

2.6. Comparison Measures

In this section, four different statistical indices, including correlation coefficient (R), root mean squared error (RMSE), mean absolute percentage error (MAPE), and Nash–Sutcliffe model efficiency coefficient (NSE), are used to assess the performance of the developed IELM model in SRFD estimation. The RMSE index is of considerable importance in studying environmental and climatic parameters and is widely used in these fields [56]. However, this index is not practical enough to check the average performance of a model, which may be a misleading index of the average error [57]. Scale dependency is one of the main drawbacks of RSME, allowing outliers to have a significant impact on the obtained results of this index, which will make it difficult to evaluate the model, and any fraction of the data will cause fundamental changes in the results [58]. Therefore, it is necessary to use other indices along with it. Scholars often employ MAPE because of its intuitive understanding in terms of relative error [59]. In addition to these two relative (MAPE) and absolute (RMSE) indices, two correlation-based indices (i.e., R and NSE) are also used. According to the characteristics of each indicator, their simultaneous application can be an excellent approach to assess the efficiency of the developed IELM-based models in SRFD estimation. The mathematical definition of the R, RMSE, MAPE, and NSE is defined as follows:

$$R = \left(\frac{\sum_{i=1}^{S} \left(SRFD_{O,i} - \overline{SRFD}_O\right)\left(SRFD_{E,i} - \overline{SRFD}_E\right)}{\sqrt{\sum_{i=1}^{S} \left(SRFD_{O,i} - \overline{SRfD}_O\right)^2 \sum_{i=1}^{S} \left(SRFD_{E,i} - \overline{SRFD}_E\right)}} \right) \tag{19}$$

$$RMSE = \frac{1}{S} \sqrt{\sum_{i=1}^{S} \left(SRFD_{O,i} - SRFD_{E,i}\right)^2} \tag{20}$$

$$MAPE = \frac{100}{S} \sum_{i=1}^{S} \left(\left| \frac{SRFD_{O,i} - SRFD_{E,i}}{SRFD_{O,i}} \right| \right) \tag{21}$$

$$NSE = 1 - \frac{\sum_{i=1}^{S} \left(SRFD_{O,i} - SRFD_{E,i}\right)^2}{\sum_{i=1}^{S} \left(SRFD_{O,i} - \overline{SRFD}_O\right)} \tag{22}$$

where $SRFD_{O,i}$ and $SRFD_{E,i}$ are the observed and estimated values of the ith samples of the SRFD, respectively, N is the number of samples, and $\overline{SRFD}_O$ and $\overline{SRFD}_E$ are the mean of the observed and estimated SRFD, respectively.

3. Results and Discussion

3.1. SRFD Modeling

Before starting the modeling, the input parameters need to be determined. In the current study, historical values of the SRFD are used to estimate this parameter at future times. Indeed, the time series concept is used to solve the problem. Hence, effective lags are determined using the auto-correlation function. It indicates the correlations between the past and future values of the desired parameters (i.e., SRFD in the current study). The auto-correlation function of the SRFD time series was shown the most critical lags are 1 and 2. Additionally, Lag 12 is also evident as the periodic term. In addition to these three lags and to find a more reliable model, Lag 3 is also considered as one of the input parameters in the IELM-based modeling:

$$SRFD(t) = f(SRFD(t-1),\ SRFD(t-2),\ SRFD(t-3),\ SRFD(t-12)) \tag{23}$$

According to the above equation, the number of fifteen input combinations with one to four input variables is defined as provided in Figure 6.

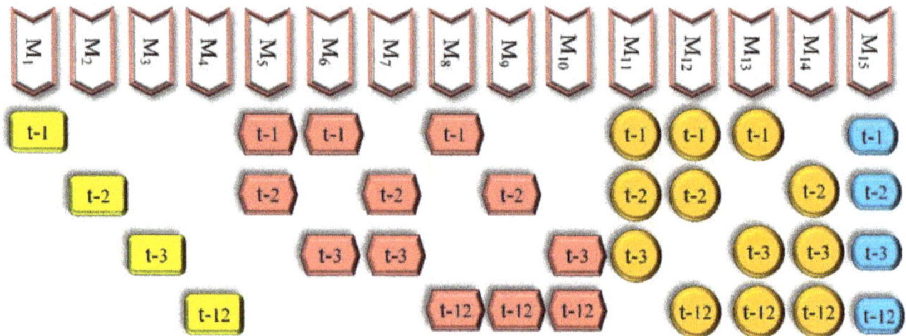

Figure 6. Input combinations of historical data defined for SRDF modeling.

3.2. Most Optimum Hidden Neurons

After defining different input combinations, finding the optimal number of neurons for SRFD estimation is required. For this purpose, Model 15 with SRFD (t − 1), SRFD (t − 2), SRFD (t − 3), and SRFD (t − 12) as input parameters (Figure 6) is employed. Additionally, by keeping other tuning parameters constant, including the activation function (i.e., Sigmoid) and iteration number (10,000 iterations), the optimal number of neurons is found. It should be noted that this number can be changed in the range of 1 to Max. NHN (Equation (18)). The reason for limiting the number of hidden layer neurons to Max is that increasing NHN could result in higher generalizability of the model so that if this limitation is not taken into account, the generalizability of this model may be doubtful.

The total number of data is 468, 70% of which have been selected as training samples, and 30% of the samples were considered testing samples. Taking into account the delays created in the modeling process, the value of TrSa is 336 (this number can be varied by changing the inputs). Given that the modeling process was initially performed using all lags (Model 15), the value of InV is equal to four. Consequently, the Max. NHN is calculated as 55 (Equation (18)).

The statistical indices of the developed IELM with different hidden neurons are provided in Figure 7. The minimum values of NSE and R are less than 0.6 and 0.8, respectively. As the number of the hidden layer neurons increases, the value of these two indices increases significantly. In NHN > 10, the value of both indices is more than 0.9, which is an acceptable value. Although the growth of the value of these two indices is also observed in most models with NHN > 10, the growth rate compared to NHN < 10 has a significant decrease. The upward trend presented in the correlation-based indices (i.e., R and NSE) is also observed as a similar downward trend in RMSE and MAPE (%). A significant point is a sharp decrease in the value of R and NSE as well as the increase in the RMSE and MAPE (%) at NHN = 31, which is significantly different from the values of its neighbors. One of the reasons for the performance of Model 31 may be the number of iterations, so for this model, the number of more iterations was examined, but there was no effective change in the performance of the model in the testing stage. Another reason can be the lack of accurate assignment of the input weights and bias of hidden neuron parameters, which has led to a significant reduction in the generalizability of this model. The results of this figure show that the best performance is obtained at NHN = 27 (R = 0.98; NSE = 0.96, RMSE = 25.02; MAPE (%) = 10.64). Therefore, the number of hidden neurons in the following modeling process is considered to be 27.

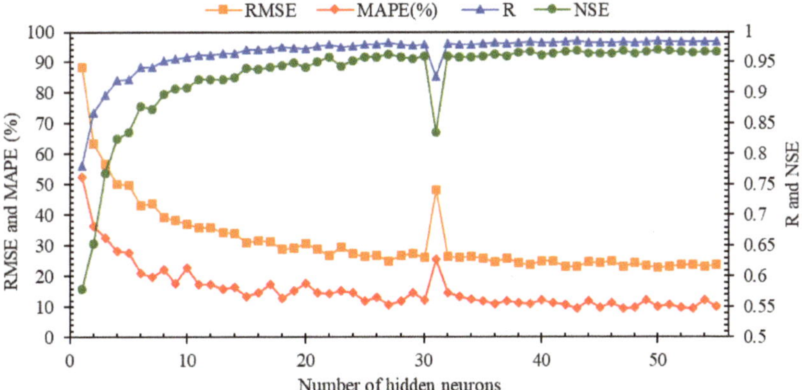

Figure 7. Statistical indices of the developed IELM with different hidden neurons.

3.3. Activation Function Selection

After determining the optimum number of hidden neurons, six activation functions, including hyperbolic tangent, sigmoid, radial basis function, triangular basis, hard limit, and sine, are evaluated in this section. The statistical indices of the IELM with defined input combinations in Model 15 and different activation functions are provided in Figure 8. As shown in this figure, although hyperbolic tangent, hard-limit, and sigmoid activation functions have close results, the RMSE value for sigmoid is improved by 12.2% and 4.2% compared to hard-limit and hyperbolic tangent, respectively, which indicates sigmoid function performs better than the other two functions. The RMSE for the other three functions (i.e., sine, radial basis, and triangular basis) is about ten times the value of this index for sigmoid. Moreover, the value of MAPE error index for sigmoid is 10.64% and for hard limit and hyperbolic tangent functions are equal to 12.44% and 11%, respectively. The NSE index shows the same results for the three mentioned activation functions. Additionally, the correlation coefficient value indicates the equality of this index in the hyperbolic tangent and sigmoid functions with a value of 0.98, while this value for the hard limit is 0.97. Therefore, it can be concluded that sigmoid has shown better performance compared to the other activation functions.

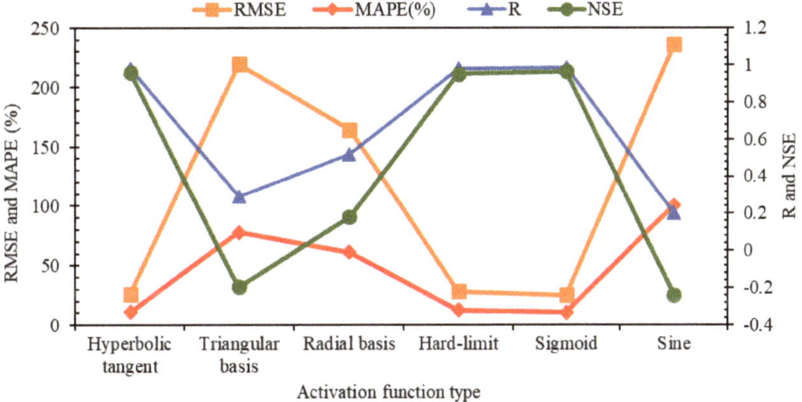

Figure 8. Statistical indices of the developed IELM with different activation functions.

3.4. Iteration Number

The third parameter is used to determine the iteration number. Many iterations in IELM modeling are used to overcome the problems caused by randomly determining the two matrices of bias of hidden neurons and input weights, which include at least 66% of the total parameters optimized during the training process (Equation (14)).

In the next step, the effect of iteration number in SRFD modeling was investigated. To investigate the impact of changes in the iteration number, a boxplot diagram was used to obtain the best number of iterations in the modeling process. In this study, 15 different values were used for the iteration number in the range of 10,000–100,000, similar to Table 2. The distribution of RMSE values for various iterations is shown in Figure 9. According to the high value of RMSE for some models, the maximum value of RMSE in this figure was plotted in two ranges, $[0, 2\times10^7]$ and $[0, 160]$. Additionally, the boxplot parameters, including minimum, maximum, median, first quartile (Q1), and third quartile (Q3), are provided in Table 2. The difference between Q1 and Q3 is known as IQR. Using IQR, the minimum and maximum values are calculated as $Q1 - 1.5 \times IQR$ and $Q3 + 1.5 \times IQR$, respectively. Values smaller than the minimum and larger than the maximum are known as outliers. Indeed, if a number outside this range is recorded, it indicates that the amount of model error that is considered as RMSE in some cases might be very small or very large. Figure 9 shows that at iteration number = 30,000, the RMSE value also reaches 2×10^{-7}, a high value for this index. By limiting the error range to $[0, 160]$, it can be seen that in all the values defined for the iteration number, a very large number of iterations record an error value greater than the maximum value (=Q3 + 1.5 Q IQR). Except for iteration numbers = 10,000 and 30,000, in other cases, the changes' range minimum, maximum, Q1, and Q3 are almost constant.

Table 2. Boxplot parameters for 15 iteration numbers.

No.	It. Number	Max	Min	Median	Q1	Q3
1	1000	50,000	29.93	47.46	42.15	54.57
2	2000	51,000	30.57	47.90	42.71	54.75
3	3000	1,201,000	28.47	47.83	42.69	55.16
4	5000	619,000	29.16	47.68	42.37	54.86
5	10,000	1000	21.11	62.40	47.64	81.13
6	15,000	619,000	29.16	48.02	42.66	55.16
7	20,000	498,000	24.83	47.84	42.60	54.94
8	30,000	20,386,000	44.18	99.93	82.81	120.11
9	40,000	435,000	24.47	47.71	42.57	54.79
10	50,000	1,019,000	27.10	47.72	42.53	54.83
11	60,000	997,000	26.75	47.79	42.59	55.00
12	70,000	1,714,000	25.68	47.66	42.48	54.81
13	80,000	2,108,000	25.91	47.79	42.59	54.99
14	90,000	1,327,000	26.21	47.72	42.48	54.80
15	100,000	2,506,000	26.44	47.76	42.57	54.92

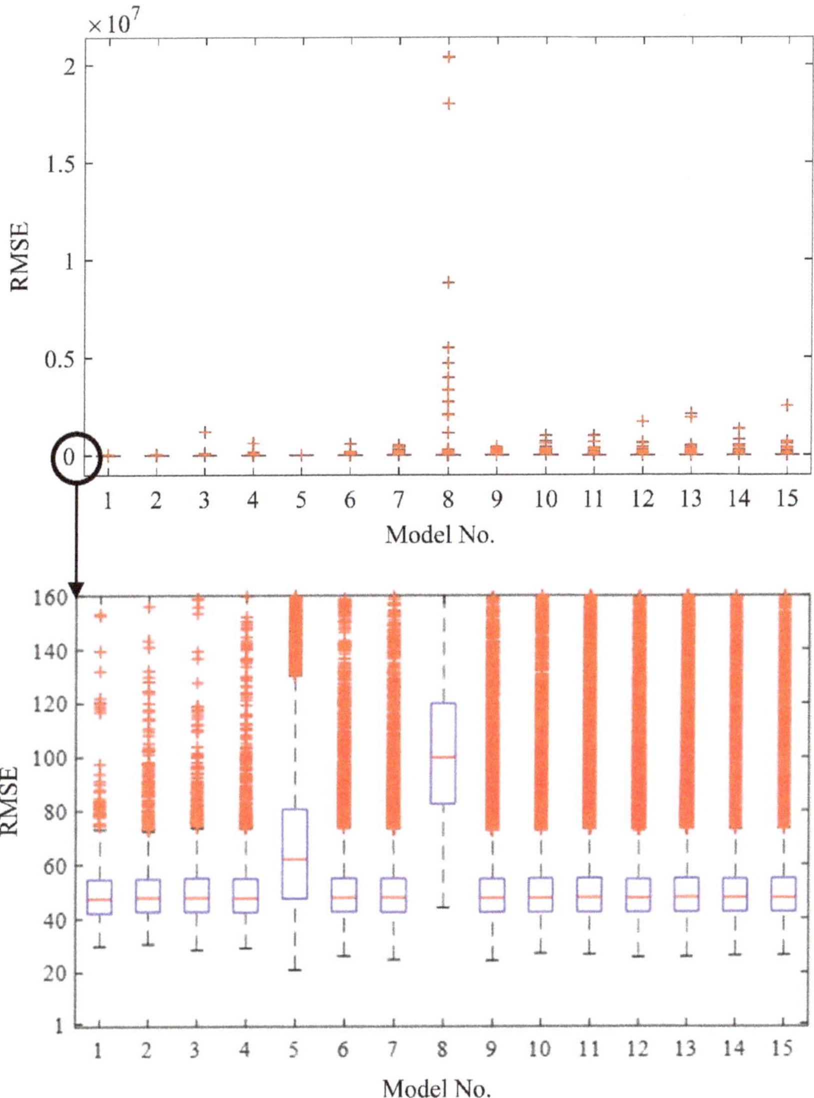

Figure 9. The distribution of the RMSE for different iteration numbers in SRFD modeling by IELM.

In Model 5 (iteration number = 10,000), the median is 62.4, which indicates that 50% of the data model from 10,000 runs is in the range of 47.64 to 81.13, while the lowest error recorded in this case is equal to 21.11. In all other models with different iteration numbers, the minimum error value is greater than 21.11, which indicates the better performance of Model 5 (Figure 9).

Figure 10 signifies the statistical indices of the best IELM-based model with the different iteration numbers. According to this figure, the lowest MAPE (%) was recorded for the iteration number = 10,000 (MAPE (%) = 10.64). The correlation-based indices (i.e., R and NSE) for all iteration numbers except 30,000 are very close together. The lowest and highest values of R are 0.973 and 0.982, respectively, and the lowest and highest of the NSE are 0.945 and 0.964, respectively. The lowest RMSE is related to the 20,000 iteration

(RMSE = 24.83). Considering the values of different indices, the iteration number = 10,000 is selected as the optimum value of the iteration number (R = 0.98, NSE = 0.96, RMSE = 25.02, and MAPE (%) = 10.64).

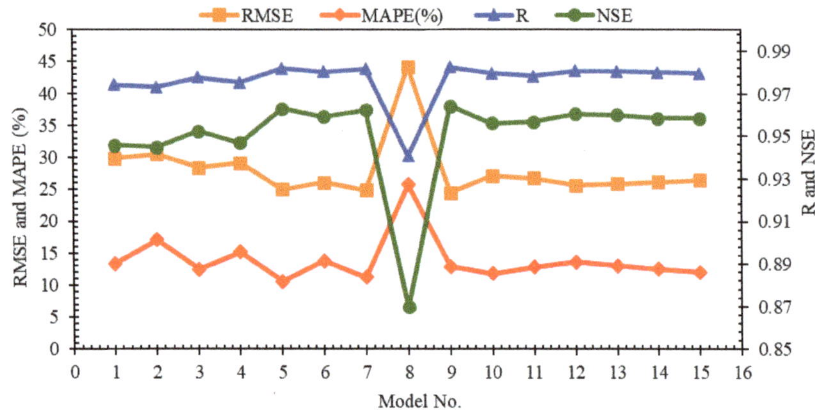

Figure 10. Statistical indices of the best IELM-based model with different iteration numbers.

3.5. Input Combination Selection

After determining the modeling parameters in IELM, the effect of each input is examined by considering 15 input combinations, defined in Figure 6. The presented models in this figure are divided into four general categories: models with one, two, three, and four inputs, which include four, five, four, and one input combinations, respectively.

Models 1 to 4 are single-input models that include Lag 1, 2, 3, and 12. Among the models presented with a single input, it is observed that the highest value of correlation-based indices (i.e., R and NSE) is related to Model 4 (R = 0.983, NSE = 0.966) so that the closest value of these two indices to Model 4 corresponds to Model 1 (R = 0.892; NSE = 0.756). The results of these two indices in Lag 2 and Lag 3 are significantly different from the other two lags because in Lag 3, the NSE index has a negative value (NSE = -0.691). The value of this index in Lag 2 (NSE = 0.106) is approximately 10% of the value of this index in Lag 12. In addition to these two indices, Model 4 (i.e., Lag 12) also offers the best performance in RMSE and MAPE indices, so that the modeling relative error is less than 9% (MAPE (%) = 8.88), while the value of this index for Models 2 and 3, whose inputs are Lag 2 (MAPE (%) = 88.23) and Lag 3 (MAPE (%) = 90.28), is about ten times the value of this index in Model 4. According to the explanations provided, Lag 12 has the most effectiveness, and Lag 1 is in the second place for one-step-ahead modeling of SRFD by considering only one lag.

Although Lag 2 had the weakest performance among all lags, its combination with Lag 1 (Model 5) (R = 0.956; RMSE = 37.645; MAPE (%) = 20.038; 0.911) increased the performance of Model 1, which uses only Lag 1 so that the values of R, RMSE, MAPE, and NSE indices have increased by 7.24%, 34.42%, 55.88%, and 20.47%, respectively, compared to Model 1.

The combination of Lag 3 with Lag 1 has also led to Model 6 (R = 0.975; RMSE = 28.833; MAPE (%) = 13.352; 0.948), which is more accurate than Model 1. According to the results presented for single-input models, the weakest performance was observed for Lag 3. Therefore, it was expected that using Lag 2 in combination with Lag 1 would provide better performance than combining Lag 3 with Lag 1, but Model 6, whose inputs are Lag 1 and Lag 3, compared to Model 5, whose inputs are Lag 1 and Lag 2, provided better performance. Therefore, it is concluded that to achieve the optimal model, it is necessary to examine the synergy of different parameters with each other.

Considering that in single-input models, the best performance was obtained for Lag 12, which was significantly different from other lags, it is expected that the combination of this lag with one of the other lags (i.e., Lags 1, 2, and 3 as Models 8 to 10, respectively) leads to a more accurate result compared to Models 5 to 7. The results presented in Figure 11 show that the performance of Model 4, whose only input was Lag 12, has been slightly improved by Models 8 to 10 (two-input models). In Models 8 to 12, correlation based-indices (i.e., R and NSE) and RMSE experienced an increase in accuracy compared to Model 4, which has only one input (i.e., Lag 12). Still, for MAPE, this trend is not incremental in all three models. In Models 8 and 10, which use Lag 1 and Lag 3 (respectively) as input in addition to Lag 12, the MAPE is increased by about 2.5%, while in Model 9, which uses Lags 2 and 12 to estimate SRFD, the value of this index has decreased. Therefore, it is concluded that the simultaneous combination of Lag 12 and Lag 2 offers the best performance between models with 1 to 2 inputs. As the number of model inputs increases to 3 and 4 (Models 11 to 15), it is observed that in all models, correlation-based indices decrease, and both RMSE and MAPE indices are increased. Therefore, it is concluded that Model 9 is the best input combination for SRFD estimation.

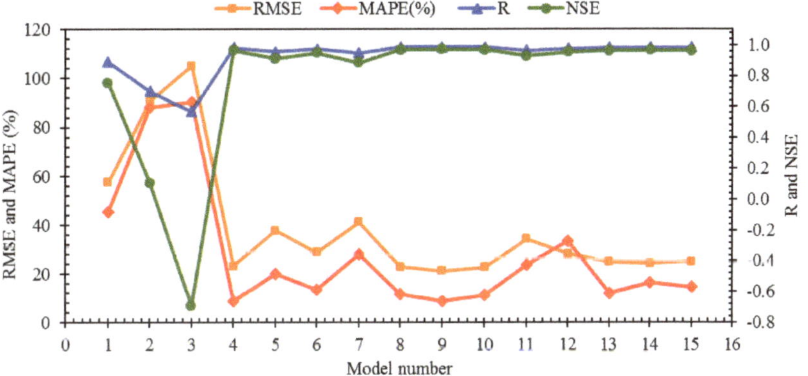

Figure 11. Error and correlation coefficient values for different input combinations.

3.6. Comparison of the IELM with ELM

Figure 12 illustrates the scatter plot of the observed and predicted SRFD by ELM and IELM. According to this figure, most of the estimated samples by IELM are in the range of ±10%, while the results provided by ELM in different SRFD ranges have many errors. Most of the estimated samples by the ELM in the range of 180–350 are fixed so that in this range, the estimated points do not follow the 45-degree line and are presented almost in a horizontal line.

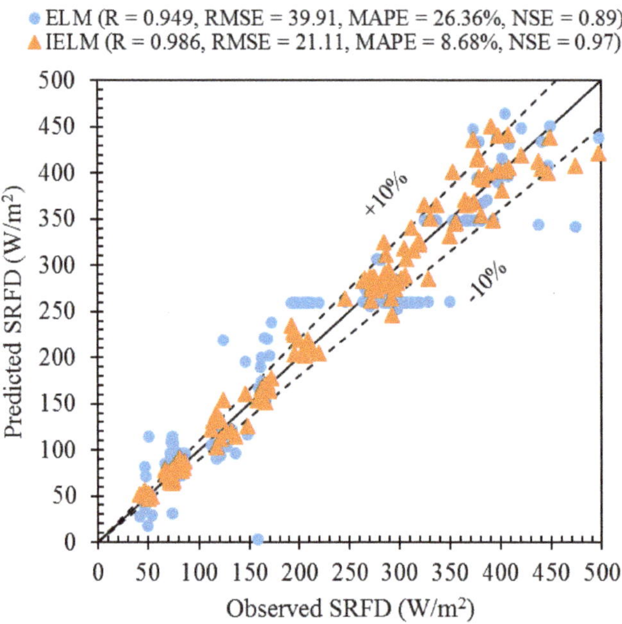

Figure 12. Scatter plot of the observed and predicted SRFD by ELM and IELM.

In both ELM and IELM methods, the errors are presented in both under and overestimate forms, and the only significant difference is the modeling error values. The indices presented in the figure show that correlation-based indices (i.e., R and NSE) in the IELM method have increased by about 3.9% and 9% compared to ELM. An increase of about 47% and more than 17% was observed in RMSE and MAPE, respectively. Therefore, it is determined that the developed method in the current study (i.e., IELM) has well overcome the limitations of the ELM method.

3.7. Uncertainty Analysis of the IELM Models Versus ELM

The uncertainty analysis (UA) results are summarized in Table 3. The mean estimation error (MEE), standard deviation of estimation error (SDEE), 95% estimation error interval (EEI), and width of uncertainty band (WUB) are provided in this table. According to this table, the MEE for IELM$_4$ and ELM are positive, while the value of this index for others is negative. Therefore, it is concluded that these two models over-estimated the SRFD, while the other models underestimated the SRFD. Given that the positive and negative values of the error are added together by considering their sign, the use of this model cannot be used as an index to check the accuracy of the model. Therefore, other indices need to be evaluated. The SDEE for all IELM models indicates that this index's lowest and highest values are associated with 21.12 and 105.47, related to the IELM$_9$ and IELM$_3$, respectively. The widths of uncertainty bands of the IELM are in the ranges of [$\pm$17.56, $\pm$3.64] so that the lowest and highest ones are related to the IELM$_3$ and IELM$_9$, respectively. The widths of uncertainty bands for ELM is $\pm$6.9, which is 52% higher than the best of the IELM model (WUB (IELM$_9$) = $\pm$3.64).

Table 3. Uncertainty analysis of IELM versus ELM.

Model	MEE	SDEE	95% EEI	WUB
$IELM_1$	-2.40	57.56	$(-11.91\cdots 7.11)$	± 9.51
$IELM_2$	-3.55	91.00	$(-18.65\cdots 11.54)$	± 15.10
$IELM_3$	-4.86	105.47	$(-22.42\cdots 12.7)$	± 17.56
$IELM_4$	0.07	23.24	$(-3.93\cdots 4.07)$	± 4
$IELM_5$	-4.71	37.48	$(-10.93\cdots 1.51)$	± 6.22
$IELM_6$	-4.58	28.57	$(-9.34\cdots 0.17)$	± 4.76
$IELM_7$	-5.83	40.91	$(-12.64\cdots 0.99)$	± 6.81
$IELM_8$	-2.73	22.81	$(-6.66\cdots 1.19)$	± 3.93
$IELM_9$	-1.73	21.12	$(-5.36\cdots 1.91)$	± 3.64
$IELM_{10}$	-3.01	22.43	$(-6.87\cdots 0.84)$	± 3.86
$IELM_{11}$	-4.70	34.05	$(-10.37\cdots 0.97)$	± 5.67
$IELM_{12}$	-1.63	28.37	$(-6.52\cdots 3.25)$	± 4.89
$IELM_{13}$	-3.57	24.68	$(-7.82\cdots 0.68)$	± 4.25
$IELM_{14}$	-4.36	23.97	$(-8.49\cdots -0.23)$	± 4.13
$IELM_{15}$	-4.27	24.75	$(-8.53\cdots -0.01)$	± 4.26
ELM	0.093	40.07	$(-6.81\cdots 6.99)$	± 6.9

MEE = mean prediction error, SDEE = standard deviation of estimation error, 95% EEI = 95% estimation error interval, WUB = width of uncertainty band.

The universal form of the IELM is as follows:

$$SRFD(t) = \left[\frac{1}{(1 + \exp(InW \times InV + BHN))}\right]^T \times OuTW \qquad (24)$$

where the InW, InV, BHN, and OutW are the input weights, input variables, bias of hidden neurons, and output weights, respectively. The InW, InV, BHN, and OutW are defined as follows:

$$InV = \begin{bmatrix} SRFD(t-2) \\ SRFD(t-12) \end{bmatrix}$$

$$BHN = \begin{bmatrix} 0.0028 \\ 0.0798 \\ 0.0282 \\ -0.2781 \\ 0.0971 \\ 0.0618 \\ -0.2191 \\ -0.298 \\ -0.0738 \\ 0.2939 \\ -0.1102 \\ 0.061 \\ 0.1924 \\ -0.1088 \\ -0.1521 \\ -0.3477 \\ -0.2962 \\ -0.3353 \\ -0.2547 \\ -0.0322 \\ -0.2454 \\ -0.0749 \\ -0.1349 \\ -0.1087 \\ -0.2937 \\ -0.1236 \\ 0.0312 \end{bmatrix} \quad InW = \begin{bmatrix} -0.3155 & 0.0897 \\ 0.1958 & -0.0233 \\ -0.0134 & 0.006 \\ 0.0698 & 0.183 \\ 0.1113 & 0.2568 \\ -0.1232 & -0.3535 \\ 0.0043 & 0.1263 \\ 0.2896 & 0.0352 \\ -0.2924 & -0.1488 \\ 0.2656 & -0.1874 \\ -0.2286 & 0.2698 \\ -0.113 & -0.2295 \\ 0.1747 & -0.0049 \\ 0.1458 & -0.1498 \\ 0.0484 & -0.202 \\ 0.3529 & -0.0543 \\ -0.3473 & -0.0036 \\ 0.1246 & -0.0774 \\ -0.2028 & -0.3509 \\ -0.0016 & 0.0059 \\ -0.019 & -0.1279 \\ 0.0384 & 0.313 \\ -0.0116 & -0.1706 \\ -0.2961 & 0.1584 \\ 0.0461 & -0.1289 \\ -0.1318 & -0.3874 \\ 0.2436 & -0.1908 \end{bmatrix} \quad OutW = \begin{bmatrix} 10.6 \\ 34962.01 \\ -228.72 \\ 31039.42 \\ 32876.78 \\ 0.0013 \\ 30746.25 \\ 32791.86 \\ 0.0427 \\ 8.6694 \\ -12.476 \\ 8.1247 \\ -133843.68 \\ -35.032 \\ 573.72 \\ -61940.9 \\ -38.53 \\ 36.8 \\ 1.68 \times 10^{-5} \\ 944.98 \\ 417363.84 \\ 32980.31 \\ -399887.66 \\ 63.81 \\ -2.006 \\ 2.87 \times 10^{-5} \\ -13.614 \end{bmatrix}$$

4. Conclusions

Shortwave radiation flux density (SRFD) modeling can be a key factor in the better control of environmental parameters such as evaporation, transpiration, use in solar structures, etc. In the current study, an improved version of the ELM (IELM) is proposed to overcome the limitation of this method due to random generation of the two matrices, including input weights and bias of hidden neurons. Additionally, the Google Earth Engine environment is employed to extract the monthly satellite-based data of the SRFD from the gridded Daymet product. The datasets were from January 1981 to December 2019. Because the time series concept was used in the SRFD modeling, using the SRFD historical data, the most important lags were found to determine the input combinations. The auto-correlation function was applied to find the most effective lags. This function indicated that Lags 1, 2, 3, and 12 are the most effective ones. Using these lags, fifteen different input combinations were obtained. The best input combination was found through the modeling phase, with Lags 2 and 12 as input variables. It should be noted that the sigmoid was found as the best activation function, and the optimum number of hidden neurons was 27. By considering different iteration numbers from 1000 to 100,000, the best results were obtained at 10,000. Comparison of the developed IELM (R = 0.986, RMSE = 21.11, MAPE = 8.68%, NSE = 0.97) with original ELM (R = 0.949, RMSE = 39.91, MAPE = 26.36%, NSE = 0.89) proved the higher performance of the developed ones. Additionally, the uncertainty analysis results indicated that the widths of uncertainty bands for IELM (WUB = ±3.64) are almost half of this index's calculated value for the original ELM (WUB = ±6.9). A matrix-based equation for the optimum IELM-based model is provided to calculate SRFD for one month ahead using Lags 2 and 12. Using this model in predicting the amount of SRFD can be a practical solution in energy and water resources management. In the current situation, where environmental factors and climate change are occurring, using this type of model can be an effective way to advance the management of goals and achieve a roadmap in the future with a sustainable development approach. In the current study, the developed model was checked for only one station. The developed IELM could be applied to other stations and other real-world problems. The developed IELM applied an iterative process to overcome the limitation of the classical ELM in random generation of the two main matrices (i.e., input weights and bias of hidden neurons). However, an implementation of the developed ELM to overcome the mentioned drawback could also be made in a follow-up study by using the new developed evolutionary algorithms such as sperm swarm optimization (SSO), conscious neighborhood-based crow search algorithm (CCSA), and other evolutionary-based algorithms to optimize the randomly generated parameters for the input weights and bias of hidden neurons matrices. Moreover, the ACF was applied to track the most effective lags dependence structure of the SRFD. As the ACF has a large statistical bias that tends to underestimate the second-order dependence structure, it is recommended to apply alternative methods.

Author Contributions: Conceptualization, I.E., K.S., and H.B.; methodology, I.E., A.A., and H.B.; software, I.E. and A.A.; validation, I.E., K.S., and H.B.; formal analysis, I.E., K.S., A.A., and H.B.; resources, K.S., A.A., and H.B.; data curation, K.S. and A.A.; writing—original draft preparation, I.E., M.F., C.A.M., and H.B.; writing—review and editing, I.E., M.F., C.A.M., and H.B.; visualization, I.E. and K.S.; supervision, H.B.; project administration, H.B.; funding acquisition, H.B. All authors have read and agreed to the published version of the manuscript. Please turn to the CRediT taxonomy for the term explanation. Authorship must be limited to those who have contributed substantially to the work reported.

Funding: This research was funded by the Natural Science and Engineering Research Council of Canada (NSERC) Discovery Grant (#RGPIN-2020-04583).

Institutional Review Board Statement: Not applicable.

Informed Consent Statement: Not applicable.

Data Availability Statement: The data that support the findings of this study are available from the corresponding author upon reasonable request.

Conflicts of Interest: The authors declare no conflict of interest.

References

1. Tang, B.; Li, Z.L.; Zhang, R. A direct method for estimating net surface shortwave radiation from MODIS data. *Remote Sens. Environ.* **2006**, *103*, 115–126. [CrossRef]
2. Wu, H.; Ying, W. Benchmarking machine learning algorithms for instantaneous net surface shortwave radiation retrieval using remote sensing data. *Remote Sens.* **2019**, *11*, 2520. [CrossRef]
3. Klassen, S.; Bugbee, B. Shortwave radiation. *Micrometeorol. AES Syst.* **2005**, *47*, 43–57.
4. Hatzianastassiou, N.; Matsoukas, C.; Fotiadi, A.; Pavlakis, K.G.; Drakakis, E.; Hatzidimitriou, D.; Vardavas, I. Global distribution of Earth's surface shortwave radiation budget. *Atmos. Chem. Phys.* **2005**, *5*, 2847–2867. [CrossRef]
5. Ceppi, P.; Zelinka, M.D.; Hartmann, D.L. The response of the Southern Hemispheric eddy-driven jet to future changes in shortwave radiation in CMIP5. *Geophys. Res. Lett.* **2014**, *41*, 3244–3250. [CrossRef]
6. Wang, B.; Zheng, L.; Liu, D.L.; Ji, F.; Clark, A.; Yu, Q. Using multi-model ensembles of CMIP5 global climate models to reproduce observed monthly rainfall and temperature with machine learning methods in Australia. *Int. J. Climatol.* **2018**, *38*, 4891–4902. [CrossRef]
7. Rabehi, A.; Guermoui, M.; Lalmi, D. Hybrid models for global solar radiation prediction: A case study. *Int. J. Ambient Energy* **2020**, *41*, 31–40. [CrossRef]
8. Wallenberg, N.; Lindberg, F.; Holmer, B.; Thorsson, S. The influence of anisotropic diffuse shortwave radiation on mean radiant temperature in outdoor urban environments. *Urban Clim.* **2020**, *31*, 100589. [CrossRef]
9. Slater, A.G. Surface solar radiation in North America: A comparison of observations, reanalyses, satellite, and derived products. *J. Hydrometeorol.* **2016**, *17*, 401–420. [CrossRef]
10. Soares, J.; Alves, M.; Ribeiro, F.N.D.; Codato, G. Surface radiation balance and weather conditions on a non-glaciated coastal area in the Antarctic region. *Polar Sci.* **2019**, *20*, 117–128. [CrossRef]
11. Wei, Y.; Zhang, X.; Hou, N.; Zhang, W.; Jia, K.; Yao, Y. Estimation of surface downward shortwave radiation over China from AVHRR data based on four machine learning methods. *Solar Energy* **2019**, *177*, 32–46. [CrossRef]
12. Schwarz, M.; Folini, D.; Yang, S.; Allan, R.P.; Wild, M. Changes in atmospheric shortwave absorption as important driver of dimming and brightening. *Nat. Geosci.* **2020**, *13*, 110–115. [CrossRef]
13. Zhang, X.; Liang, S.; Zhou, G.; Wu, H.; Zhao, X. Generating Global Land Surface Satellite incident shortwave radiation and photosynthetically active radiation products from multiple satellite data. *Remote Sen. Environ.* **2014**, *152*, 318–332. [CrossRef]
14. Ryu, Y.; Jiang, C.; Kobayashi, H.; Detto, M. MODIS-derived global land products of shortwave radiation and diffuse and total photosynthetically active radiation at 5 km resolution from 2000. *Remote Sens. Environ.* **2018**, *204*, 812–825. [CrossRef]
15. Schellekens, J.; Dutra, E.; Martínez-de la Torre, A.; Balsamo, G.; Dijk, A.V.; Sperna Weiland, F.; Minvielle, M.; Calvet, J.C.; Decharme, B.; Eisner, S.; et al. A global water resources ensemble of hydrological models: The eartH2Observe Tier-1 dataset. *Earth Syst. Sci. Data* **2017**, *9*, 389–413. [CrossRef]
16. Zeng, L.; Xia, T.; Elsayed, S.K.; Ahmed, M.; Rezaei, M.; Jermsittiparsert, K.; Dampage, U.; Mohamed, M.A. A Novel Machine Learning-Based Framework for Optimal and Secure Operation of Static VAR Compensators in EAFs. *Sustainability* **2021**, *13*, 5777. [CrossRef]
17. Kamolov, A.A.; Park, S. Prediction of Depth of Seawater Using Fuzzy C-Means Clustering Algorithm of Crowdsourced SONAR Data. *Sustainability* **2021**, *13*, 5823. [CrossRef]
18. Haq, I.U.; Khan, Z.Y.; Ahmad, A.; Hayat, B.; Lee, Y.E.; Kim, K.I. Evaluating and Enhancing the Robustness of Sustainable Neural Relationship Classifiers Using Query-Efficient Black-Box Adversarial Attacks. *Sustainability* **2021**, *13*, 5892. [CrossRef]
19. Soltani, K.; Amiri, A.; Zeynoddin, M.; Ebtehaj, I.; Gharabaghi, B.; Bonakdari, H. Forecasting monthly fluctuations of lake surface areas using remote sensing techniques and novel machine learning methods. *Theor. Appl. Climatol.* **2021**, *143*, 713–735. [CrossRef]
20. Aissani, N.; Beldjilali, B.; Trentesaux, D. Use of machine learning for continuous improvement of the real-time heterarchical manufacturing control system performances. *Int. J. Ind. Syst. Eng.* **2008**, *3*, 474–497. [CrossRef]
21. Guyon, I.; Chaabane, I.; Escalante, H.J.; Escalera, S.; Jajetic, D.; Lloyd, J.R.; Macià, N.; Ray, B.; Romaszko, L.; Sebag, M.; et al. A brief review of the ChaLearn AutoML challenge: Any-time any-dataset learning without human intervention. *Workshop Autom. Mach. Learn.* **2016**, *64*, 21–30.
22. Bustillo, A.; Reis, R.; Machado, A.R.; Pimenov, D.Y. Improving the accuracy of machine-learning models with data from machine test repetitions. *J. Intell. Manuf.* **2020**, 1–19. [CrossRef]
23. Koutsoyiannis, D. Time's arrow in stochastic characterization and simulation of atmospheric and hydrological processes. *Hydrol. Sci. J.* **2019**, *64*, 1013–1037. [CrossRef]
24. Shamshirband, S.; Mosavi, A.; Rabczuk, T.; Nabipour, N.; Chau, K.W. Prediction of significant wave height; comparison between nested grid numerical model, and machine learning models of artificial neural networks, extreme learning and support vector machines. *Eng. Appl. Comput. Fluid Mech.* **2020**, *14*, 805–817. [CrossRef]
25. Alizamir, M.; Kim, S.; Zounemat-Kermani, M.; Heddam, S.; Kim, N.W.; Singh, V.P. Kernel Extreme Learning Machine: An Efficient Model for Estimating Daily Dew Point Temperature Using Weather Data. *Water* **2020**, *12*, 2600. [CrossRef]

26. Yaseen, Z.M.; Sulaiman, S.O.; Deo, R.C.; Chau, K.W. An enhanced extreme learning machine model for river flow forecasting: State-of-the-art, practical applications in water resource engineering area and future research direction. *J. Hydrol.* **2019**, *569*, 387–408. [CrossRef]

27. Abbaa, S.I.; Elkiranb, G.; Nouranic, V. Improving novel extreme learning machine using PCA algorithms for multi-parametric modeling of the municipal wastewater treatment plant. *Desalin. Water Treat.* **2021**, *215*, 414–426. [CrossRef]

28. Ghazvinei, P.T.; Hassanpour Darvishi, H.; Mosavi, A.; Yusof, K.B.W.; Alizamir, M.; Shamshirband, S.; Chau, K.W. Sugarcane growth prediction based on meteorological parameters using extreme learning machine and artificial neural network. *Eng. Appl. Comput. Fluid Mech.* **2018**, *12*, 738–749. [CrossRef]

29. Bonakdari, H.; Ebtehaj, I. A comparative study of extreme learning machines and support vector machines in prediction of sediment transport in open channels. *Int. J. Eng.* **2016**, *29*, 1499–1506.

30. Ebtehaj, I.; Bonakdari, H.; Moradi, F.; Gharabaghi, B.; Khozani, Z.S. An integrated framework of Extreme Learning Machines for predicting scour at pile groups in clear water condition. *Coast. Eng.* **2018**, *135*, 1–15. [CrossRef]

31. Huang, G.B.; Zhu, Q.Y.; Siew, C.K. Extreme learning machine: Theory and applications. *Neurocomputing* **2006**, *70*, 489–501. [CrossRef]

32. Feng, Z.K.; Niu, W.J.; Tang, Z.Y.; Xu, Y.; Zhang, H.R. Evolutionary artificial intelligence model via cooperation search algorithm and extreme learning machine for multiple scales nonstationary hydrological time series prediction. *J. Hydrol.* **2021**, *595*, 126062. [CrossRef]

33. UN General Assembly. Resolution Adopted by the General Assembly on 25 September 2015. 2015. Available online: http://www.un.org/ga/search/view_doc.asp?symbol=A/RES/70/1&Lang=E (accessed on 31 March 2020).

34. Rebelatto, B.G.; Salvia, A.L.; Reginatto, G.; Daneli, R.C.; Brandli, L.L. Energy efficiency actions at a Brazilian university and their contribution to sustainable development Goal 7. *Int. J. Sustain. Higher Educ.* **2019**, *20*, 842–855. [CrossRef]

35. Dimitriadis, P.; Koutsoyiannis, D.; Iliopoulou, T.; Papanicolaou, P. A global-scale investigation of stochastic similarities in marginal distribution and dependence structure of key hydrological-cycle processes. *Hydrology* **2021**, *8*, 59. [CrossRef]

36. Sözen, A.; Arcaklioğlu, E.; Özalp, M. Estimation of solar potential in Turkey by artificial neural networks using meteorological and geographical data. *Energy Convers. Manag.* **2004**, *45*, 3033–3052. [CrossRef]

37. Dorvlo, A.S.; Jervase, J.A.; Al-Lawati, A. Solar radiation estimation using artificial neural networks. *Appl. Energy* **2002**, *71*, 307–319. [CrossRef]

38. Kim, S.; Seo, Y.; Rezaie-Balf, M.; Kisi, O.; Ghorbani, M.A.; Singh, V.P. Evaluation of daily solar radiation flux using soft computing approaches based on different meteorological information: Peninsula vs continent. *Theor. Appl. Climatol.* **2019**, *137*, 693–712. [CrossRef]

39. Shook, K.; Pomeroy, J. Synthesis of incoming shortwave radiation for hydrological simulation. *Hydrol. Res.* **2011**, *42*, 433–446. [CrossRef]

40. Gitelson, A.A.; Viña, A.; Masek, J.G.; Verma, S.B.; Suyker, A.E. Synoptic monitoring of gross primary productivity of maize using Landsat data. *IEEE Geosci. Remote Sens.* **2008**, *5*, 133–137. [CrossRef]

41. Che, Y.; Zhang, M.; Li, Z.; Wei, Y.; Nan, Z.; Li, H.; Wang, S.; Su, B. Energy balance model of mass balance and its sensitivity to meteorological variability on Urumqi River Glacier No. 1 in the Chinese Tien Shan. *Sci. Rep. UK* **2019**, *9*, 1–13. [CrossRef]

42. Hamlet, A.F.; Lettenmaier, D.P. Effects of climate change on hydrology and water resources in the Columbia River Basin 1. *J. Am. Water Resour. Assoc.* **1999**, *35*, 1597–1623. [CrossRef]

43. Pellicciotti, F.; Brock, B.; Strasser, U.; Burlando, P.; Funk, M.; Corripio, J. An enhanced temperature-index glacier melt model including the shortwave radiation balance: Development and testing for Haut Glacier d'Arolla, Switzerland. *J. Glaciol.* **2005**, *51*, 573–587. [CrossRef]

44. Thornton, P.E.; Thornton, M.M.; Mayer, B.W.; Wei, Y.; Devarakonda, R.; Vose, R.S.; Cook, R.B. *Daymet: Daily Surface Weather Data on a 1-km Grid for North America*; Version 3; ORNL DAAC: Oak Ridge, TN, USA, 2016. [CrossRef]

45. Gorelick, N.; Hancher, M.; Dixon, M.; Ilyushchenko, S.; Thau, D.; Moore, R. Google Earth Engine: Planetary-scale geospatial analysis for everyone. *Remote Sens. Environ.* **2017**, *202*, 18–27. [CrossRef]

46. Sazib, N.; Mladenova, I.; Bolten, J. Leveraging the google earth engine for drought assessment using global soil moisture data. *Remote Sens.* **2018**, *10*, 1265. [CrossRef] [PubMed]

47. Soltani, K.; Ebtehaj, I.; Amiri, A.; Azari, A.; Gharabaghi, B.; Bonakdari, H. Mapping the spatial and temporal variability of flood susceptibility using remotely sensed normalized difference vegetation index and the forecasted changes in the future. *Sci. Total Environ.* **2021**, *770*, 145288. [CrossRef] [PubMed]

48. Bonakdari, H.; Ebtehaj, I. Verification of equation for non-deposition sediment transport in flood water canals. In Proceedings of the 7th International Conference on Fluvial Hydraulics, RIVER FLOW, Lausanne, Switzerland, 1 January 2014; pp. 1527–1533.

49. Bonakdari, H.; Ebtehaj, I.; Samui, P.; Gharabaghi, B. Lake Water-Level fluctuations forecasting using Minimax Probability Machine Regression, Relevance Vector Machine, Gaussian Process Regression, and Extreme Learning Machine. *Water Resour. Manag.* **2019**, *33*, 3965–3984. [CrossRef]

50. Bonakdari, H.; Moradi, F.; Ebtehaj, I.; Gharabaghi, B.; Sattar, A.A.; Azimi, A.H.; Radecki-Pawlik, A. A Non-Tuned Machine learning technique for Abutment Scour Depth in Clear Water Condition. *Water* **2020**, *12*, 301. [CrossRef]

51. Ebtehaj, I.; Bonakdari, H.; Shamshirband, S. Extreme learning machine assessment for estimating sediment transport in open channels. *Eng. Comput.* **2016**, *32*, 691–704. [CrossRef]

52. Rao, C.R.; Mitra, S.K. *Generalized Inverse of Matrices and Its Applications*; John Wiley & Sons Inc.: New York, NY, USA, 1971.
53. Azimi, H.; Bonakdari, H.; Ebtehaj, I. Gene expression programming-based approach for predicting the roller length of a hydraulic jump on a rough bed. *ISH J. Hydraul. Eng.* **2019**, 1–11. [CrossRef]
54. Ebtehaj, I.; Bonakdari, H.; Gharabaghi, B. A reliable linear method for modeling lake level fluctuations. *J. Hydrol.* **2019**, *570*, 236–250. [CrossRef]
55. Moeeni, H.; Bonakdari, H.; Ebtehaj, I. Integrated SARIMA with neuro-fuzzy systems and neural networks for monthly inflow prediction. *Water Resour. Manag.* **2017**, *31*, 2141–2156. [CrossRef]
56. Willmott, C.J.; Matsuura, K. Advantages of the mean absolute error (MAE) over the root mean square error (RMSE) in assessing average model performance. *Clim. Res.* **2005**, *30*, 79–82. [CrossRef]
57. Chai, T.; Draxler, R.R. Root mean square error (RMSE) or mean absolute error (MAE)—Arguments against avoiding RMSE in the literature. *Geosci. Model Dev.* **2014**, *7*, 1247–1250. [CrossRef]
58. Shcherbakov, M.V.; Brebels, A.; Shcherbakova, N.L.; Tyukov, A.P.; Janovsky, T.A.; Kamaev, V.A.E. A survey of forecast error measures. *World Appl. Sci. J.* **2013**, *24*, 171–176. [CrossRef]
59. De Myttenaere, A.; Golden, B.; Le Grand, B.; Rossi, F. Mean absolute percentage error for regression models. *Neurocomputing* **2016**, *192*, 38–48. [CrossRef]

Review

Prediction Interval Estimation Methods for Artificial Neural Network (ANN)-Based Modeling of the Hydro-Climatic Processes, a Review

Vahid Nourani [1,2,*], Nardin Jabbarian Paknezhad [1] and Hitoshi Tanaka [3]

[1] Center of Excellence in Hydroinformatics and Faculty of Civil Engineering, University of Tabriz, Tabriz 51368, Iran; n.jabbarian@tabrizu.ac.ir
[2] Faculty of Civil and Environmental Engineering, Near East University, N. Cyprus, via Mersin 10, Nicosia 99138, Turkey
[3] Department of Civil Engineering, Tohoku University, 6-6-06 Aoba, Sendai 980-8579, Japan; hitoshi.tanaka.b7@tohoku.ac.jp
* Correspondence: vnourani@yahoo.com or nourani@tabrizu.ac.ir; Tel.: +98-914-403-0332

Abstract: Despite the wide applications of artificial neural networks (ANNs) in modeling hydro-climatic processes, quantification of the ANNs' performance is a significant matter. Sustainable management of water resources requires information about the amount of uncertainty involved in the modeling results, which is a guide for proper decision making. Therefore, in recent years, uncertainty analysis of ANN modeling has attracted noticeable attention. Prediction intervals (PIs) are one of the prevalent tools for uncertainty quantification. This review paper has focused on the different techniques of PI development in the field of hydrology and climatology modeling. The implementation of each method was discussed, and their pros and cons were investigated. In addition, some suggestions are provided for future studies. This review paper was prepared via PRISMA (preferred reporting items for systematic reviews and meta-analyses) methodology.

Keywords: artificial neural network; uncertainty; sustainability; prediction intervals

Citation: Nourani, V.; Paknezhad, N.J.; Tanaka, H. Prediction Interval Estimation Methods for Artificial Neural Network (ANN)-Based Modeling of the Hydro-Climatic Processes, a Review. *Sustainability* **2021**, *13*, 1633. https://doi.org/10.3390/su13041633

Academic Editor: Saeed Chehreh Chelgani
Received: 9 January 2021
Accepted: 1 February 2021
Published: 3 February 2021

Publisher's Note: MDPI stays neutral with regard to jurisdictional claims in published maps and institutional affiliations.

1. Introduction

Sustainable water resources management includes designing and managing various aspects such as ecology, environment and hydrology integrity in the present and future [1]. Sustainable management requires adequate information about the state of water resources. Thus, appropriate modeling to investigate the situation in present and future times is necessary. On the other hand, reliability of the modeling is an important issue that directly influences the management and decision-making of the problems. Therefore, the uncertainty involved in modeling should be carefully considered so as to achieve more realistic decisions.

Recently, the artificial neural network (ANN) as a prevalent modeling method has been used for identification of the complicated non-linear relationship of inputs and output (e.g., see, [2–8]). The relationship between the hydrological phenomena is a complicated issue and hot topic in hydrological studies, due to spatial and temporal changes of factors that influence the process. Therefore, many hydrological models with various degrees of complexity have been used for the simulation of such a stochastic process [9]. In spite of the numerous applications of the ANN, it has been indicated in many previous studies that ANN models are inherently stochastic, as identical results would be difficult to be reproduced on different occasions [10]. Classic applications of ANN include some imperfections; e.g., the weights of the ANN are randomly assigned, which leads to a long training time; ANN behavior is unexpected and there is not a specified way for determination of the best structure. These features are a deficiency of ANNs, which can have a negative effect on the reliability of the modeling.

Generally, point prediction of the ANN has been considered in most of the studies, but the reliability of the point prediction decreases when the level of uncertainty is high. In the point prediction method, only a point is directly predicted, which is the unknown true targeted value, so its application is questionable. The point prediction is not able to give information about the uncertainty of the modeling, and it is not able to describe the prediction accuracy [11]. Furthermore, via the point prediction, only a prediction error can be obtained, and the probability for correct predictions remains unknown, which can make decision making more difficult. Most of the models associated with the water management and modeling are in the form of point prediction, so various sources of uncertainties have not been investigated, which may affect the modeling outcome. Uncertainty includes model uncertainty, input uncertainty and parameter uncertainty. Prediction uncertainty sources can be errors in measuring, lack of knowledge of constants, sparse and noisy input data, and model approximation errors (e.g., due to imperfections in the model formulation) [12,13], or the target values are affected by some probabilistic events [14]. The importance of determining the total model uncertainty is the same as model output and it has efficient impact on decision making. Without investigating the resources of the uncertainty, the assessment of the ANN modeling quality is impossible.

Prediction interval (PI) is a powerful measure of uncertainties associated with prediction to inform decision makers [15]. PIs lead to the proper arrangement of the future trends and plans, appropriate risk management and an increment of the benefits of modeling. PI presents a bound that captures the observed values by measuring the indication of accuracy called the confidence level ((1-a)%) [16–18]. As PIs contain more sources of uncertainties compared to similar tools such as the confidence interval (CI), they are superior and more practical in order to help decision makers distinguish the best and the worst of the modeling scenarios. Wider PIs present more uncertainty, so more awareness is needed in decisions, and narrower PIs show more confidence in decisions. There are some methods to quantify uncertainty, such as ensemble of the ANN, sensitivity analysis and the self-organizing map. Moreover, some techniques are applied to compute the PIs, such as delta, Bayesian, Monte Carlo, bootstrap and lower upper bound estimation (LUBE). The delta method basic concept is about analyzing the ANNs regression models and application of Taylor's series [19,20]. The procedure of this technique is based on the homogeneity of the noise and its normal distribution. As most of the natural phenomena are heterogeneous, this method's reliability may be questionable. The Bayesian method is based on the consideration of a pre-defined probability distribution of the ANN's parameters, instead of a single value, so the output will also have distributions conditional on the observed training set [21]. The Bayesian technique includes great computations; moreover, to construct the PIs, calculating the Hessian matrix is needed. The Monte Carlo method constructs the PIs based on allocation of the ranges and probability distribution of each variable. These classic methods contain some assumptions about the data distribution [22]. The bootstrap method is a simple and frequently used technique to calculate the PIs [23,24]. This method is based on resampling and training different ANNs. It does not need any assumption about the data distribution but consists of high computational costs for large datasets. Implementation of this method is easy, and it is independent of massive calculation. The main disadvantage of this method is its computational cost for large datasets. The LUBE technique [25] is a non-parametric method, independent of information about the data or error distributions. Kasiviswanathan and Sudheer [22] have reviewed some techniques to quantify the uncertainty of the ANN models in hydrology. Thirty-six research articles associated with uncertainty analysis of the ANN-based stream flow and flood prediction from the years 2002 to 2015 were reviewed, and it is concluded that in order to distinguish the best procedure to encompass different sources of uncertainty, more investigations are needed. The applied methodologies in papers from well-known international journals about uncertainty analysis in hydrology from the year 2002 to quantify uncertainty are tabulated in Table 1.

Table 1. Papers from well-known international journals about uncertainty analysis in hydrology.

Method	Author	Name of the Journal
Ensemble ANN	Cannon and Whitfield [26]	*Journal of Hydrology*
	Jeong and Kim [27]	*Hydrological Processes*
	Fleming, Bourdin, Campbell, Stull and Gardner [28]	*Water Resources Research*
	Kan, Yao, Li, Li, Yu, Liu, Ding, He and Liang [29]	*Stochastic Environment Research and Risk Assessment*
	Kim and Seo [30]	*Journal of Hydro-environment Research*
Sensitivity analysis	Kim and Kim [31]	*The Journal of the American Water Resources Association*
Self-organizing map	Yang and Chen [32]	*Hydrological Processes*
Bootstrap	Srivastav, Sudheer and Chaubey [33]	*Water Resources Research*
	Boucher, Perreault and Anctil [34]	*Hydrology and Earth System Sciences*
	Sharma and Tiwari [35]	*Journal of Hydrology*
	Kant, Suman, Giri, Tiwari, Chatterjee, Nayak and Kumar [36]	*Neural Computing and Applications*
Bayesian	Zhang, Liang, Srinivasan and Van Liew [37]	*Water Resources Research*
	Khan and Coulibaly [38]	*Journal of Hydrometeorology*
	Zhang, Liang, Yu and Zong [39]	*Journal of Hydrology*
	Zhang and Zhao [40]	*Journal of Hydrology*
	Humphrey, Gibbs, Dandy and Maier [41]	*Journal of Hydrology*
Markov Chain Monte Carlo	Shen, Zeng, Liang, Li, Tan, Li and Li [42]	*Water Resources Research*
GLUBE	Tongal and Booij [43]	*Stochastic Environmental Research and Risk Assessment*

The application of PIs to quantify uncertainty has increased and attracted significant attention in the last decade. Therefore, it is worthy to evaluate different PIs development methods in order to identify the best performance of the methods and assess each method suitability.

The lack of review papers investigating PIs development methods in the field of hydrology caused the preparation of the current review paper. The number of published papers regarding PIs construction methods is depicted in Figure 1. The major objective of this review paper is to categorize and enumerate the PIs construction methods and their applications in hydro-climatic studies. Moreover, some suggestions for future works in order to develop and improve the PIs applications are presented. The reviewed sources are mostly included by the Scopus abstract and citation database (www.scopus.com (accessed on 26 January 2021)). Elsevier's Scopus is the most frequently used research engine, and it is updated earlier than the Web of Science on which the papers may be updated lately. In addition, as authors can load any paper onto Google Scholar, some information may not be reliable. The search terms were ("ANN"; "uncertainty") and ("ANN"; "Prediction interval"), respectively, for the uncertainty analysis and PIs. The search operator was "and". Then, the appropriate papers in the fields of hydrology and climatology were selected by abstract reviewing. Moreover, only journal articles published in English were considered, as most of the papers in Scopus research engine are in English. The initial number of obtained papers about the uncertainty assessment of hydrological and hydro-climatological studies was 36 papers; 18 papers from well-known international journals from the year 2002 are tabulated in Table 1. In addition, 69 papers associated with PIs construction of the ANNs were investigated, then 17 papers were selected according to an abstract review and relation to hydrology and climatology. Papers from the years 2002 to 2020 were presented as selected papers.

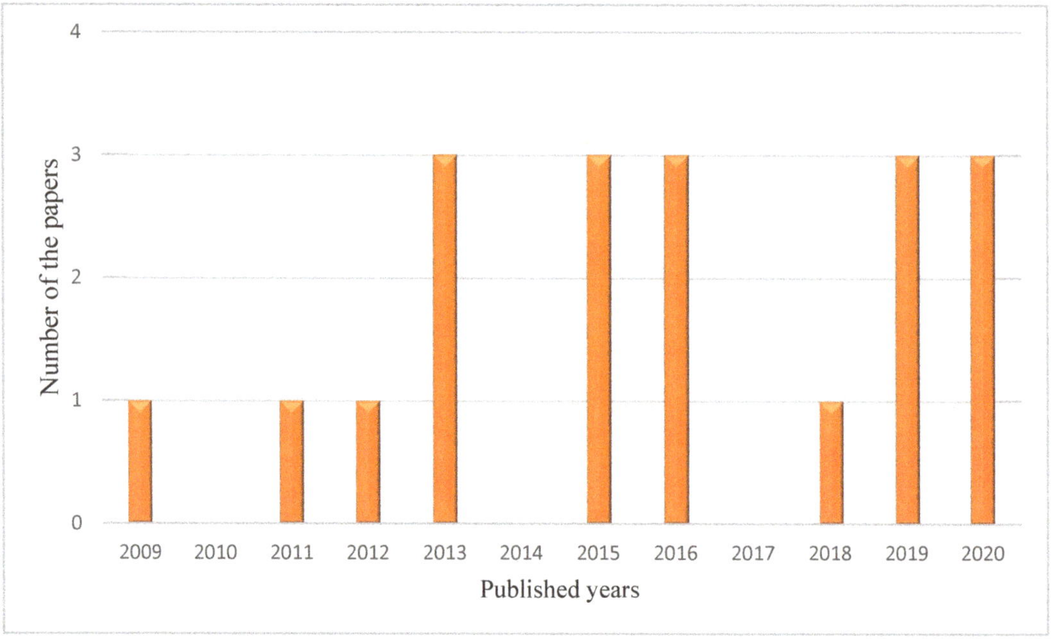

Figure 1. Number of published papers regarding prediction intervals (PIs) construction (indexed in Scopus) with respect to year of publication.

Some methodologies concerning standards of literature reviews and the way of reporting and structuring of them are RAMESES (realist and meta-narrative evidence syntheses: evolving standards), PRISMA (preferred reporting items for systematic reviews and meta-analyses) and PSALSAR (research protocol, appraisal, synthesis and analysis, reporting results). RAMSES could be an appropriate choice for systematic narrative reviews. PRISMA was developed for systematic literature reviews and meta-analyses [44]. PRISMA consists of a 27-item evaluation checklist and a specific flowchart to follow [45]. Moreover, PRISMA protocols (PRISMA-P) checklist and the Explanation and Elaboration [46] document could lead to the improvement of a more complete and reliable review report [47]. The PSALSAR method consists of six basic steps [48], while the common systematic literature review methods include four steps, which are search, appraisal, synthesis and analysis (SALSA). The PSALSAR method contains research protocol and reporting results at the first and last steps. This review paper attempts to follow most of the checklist's items of the PRISMA method, since this method is used as a basis for reporting systematic reviews for most research types. Some examples of systematic literature reviews are [49,50]. The systematic research was conducted on September 1, 2020, and it was updated on December 20, 2020 and also on January, 20, 2021 for preparing a revision. The list of reviewed papers is tabulated in Table 2.

In the following, Section 2 presents the base concepts of PIs and measurement criteria of PIs, Section 3 describes the different PIs construction methods, Section 4 compares different pros and cons of the methods and finally Section 5 recommends some suggestions for future studies.

Table 2. Number of the papers that applied PI to quantify the uncertainty of the artificial neural networks (ANN).

Keywords	Number of Cites	Author	PI Construction Method	Number of Papers
ANN; Bayesian; Bootstrap; PI; Uncertainty	27	Kasiviswanathan and Sudheer [51]	Bayesian	1
Uncertainty analysis; PIs;Lake level; ANN; ANFIS; Bootstrapping	129	Talebizadeh and Moridnejad [52]	Bootstrap	7
ANN; Bootstrap technique; Hydrological processes; Non-linear function; Taylor series	72	Kasiviswanathan and Sudheer [53]		
Uncertainty; Flood forecasting; Bootstrap; ANNs; Ensemble	60	Kumar, Tiwari, Chatterjee and Mishra [54]		
Water quality forecasting; Wavelet neural network; Bootstrap; Uncertainty; Data missing; Data filling; Songhua River	33	Wang, Zheng, Zhao, Jiang, Wang, Guo and Wang [55]		
ANN; Bayesian; Bootstrap; PI; Uncertainty	27	Kasiviswanathan and Sudheer [51]		
ANNs; ensemble simulation; input uncertainty; prediction uncertainty; rainfall–runoff modeling	82	Kasiviswanathan, He, Sudheer and Tay [56]		
General circulation models;Downscaling; PIs; ANN	9	Nourani, Paknezhad, Sharghi and Khosravi [57]		
	111	Shrestha, Kayastha and Solomatine [58]	Monte Carlo	3
groundwater; artificial intelligence; hydrologic model; groundwater level prediction; machine learning;artificial neural network	7	Seifi, Ehteram, Singh and Mosavi [59]		
ANNs; Bayesian uncertainty; fuzzy logic; kriging; uncertainty analysis	5	Tapoglou, Varouchakis, Trichakis and Karatzas [60]		
Ensemble Optimization; PI;Rainfall runoff models	68	Kasiviswanathan, Cibin, Sudheer and Chaubey [61]	LUBE	9
MOFIPS; PSO; Prediction interval; LUBE; Neural networks; Streamflow prediction	63	Taormina and Chau [62]		
PI; Symmetry; ANN; Uncertainty; Flood forecasting; Shuffled complex evolution	16	Zhang, Zhou, Ye, Zeng and Chen [63]		
ANN; Bayesian; Bootstrap; PI; Uncertainty	27	Kasiviswanathan and Sudheer [51]		
ANN; ensemble simulation; input uncertainty; prediction uncertainty; rainfall–runoff modeling	8	Kasiviswanathan, Sudheer and He [64]		
General circulation models; DownscalingPI; ANN	9	Nourani, Paknezhad, Sharghi and Khosravi [57]		
Evaporation; Neural network; Prediction interval; Uncertainty quantifying; Wavelet de-noising; Jitterd data	1	Nourani, Sayyah-Fard, Alami and Sharghi [65]		
Uncertainty analysis; Hybrid double feedforward neural network; Sediment load estimation; Lower upper bound estimation	27	Chen and Chau [66]		
ANN; Crop simulation; Reservoir operation; Optimization; Uncertainty	0	Kasiviswanathan, Sudheer, Soundharajan and Adeloye [67]		

2. PIs Concepts

An interval includes the upper and lower limits that capture indeterminate future value with a prescribed probability [68]. This limit and interval respectively are known as prediction limit and PI (see Figure 2).

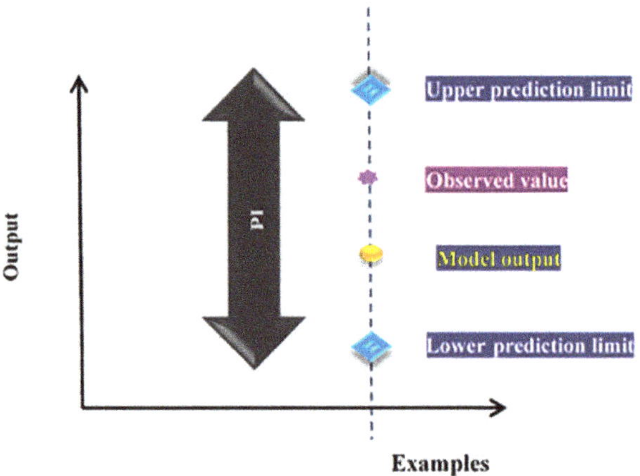

Figure 2. PI description [68].

PI and CI are different measures, and it is important to distinguish them. The CI corresponds to the accuracy of the estimation of the true regression, while PI corresponds to the accuracy of the estimation concerning the observed target value. Actually, PI is more applicable than CI, since it is associated with the accuracy of observed target prediction, whereas CI presents the accuracy of true regression estimation. In order to investigate the differences between the two measures, the following description should be considered. For the estimation of the unknown function $f(x_i; \theta)$ presenting the true underlying model, where θ is actual parameter set, for N data samples $\{(x_i, t_i)\}_{i=1}^{N}$ we have:

$$t_i = f(x_i; \theta) + e_i \tag{1}$$

where x_i and t_i are, respectively, input, observed target and model error. So, the objective is to estimate the true model $f(x_i; \theta)$. The approximate model $f(x_i; \hat{\theta})$ is the mean of the distribution of the targets, where the estimated parameters $\hat{\theta}$ are determined using machine learning methods. To quantify the uncertainty, two aspects should be considered. One of the aspects is that CI that indicates the accuracy of the estimation of the true model. CI is measured by the distribution of the quantity $f(x_i; \theta) - f(x_i; \hat{\theta})$. The second aspect is that PI indicates the accuracy of the prediction of the target. Therefore, Equation (2) can present the relation between PI and CI as:

$$t_i - f(x_i; \hat{\theta}) = [f(x_i; \theta) - f(x_i; \hat{\theta})] + e_i \tag{2}$$

It can be concluded from Equation (2) that PIs are wider than CIs, where PI contains CI and covers more sources of uncertainty [54].

3. PIs Assessment Measures

The most commonly used measures for quantifying PIs construction are PI coverage probability (PICP) and mean PI width (MPIW). The coverage measure corresponds to the encompassments of the obtained bounds. The wider PIs increase the PICP. The PICP is calculated as [24]:

$$\text{PICP} = \frac{1}{N} \sum_{i=1}^{N} c_i c_i = \begin{cases} 1, & x_i \in [L_i, U_i] \\ 0, & x_i \notin [L_i, U_i] \end{cases} \tag{3}$$

where N is the number of samples, and L_i and U_i are the lower and upper bounds of the ith PI, respectively. The second measure is used to evaluate the width of PIs. Normalized MPIW (NMPIW) shows the normalized width as [24]:

$$\text{NMPIW} = \frac{1}{NR} \sum_{i=1}^{N} L(X_i) - U(X_i) \tag{4}$$

where R is the range of the observed values. NMPIW, as a dimensionless criterion, indicates the mean width of PIs. Each of these criteria separately cannot lead to a clear judgment due to their inverse relationship. Therefore, the combinational coverage width-based criterion (CWC) containing both criteria can be used to evaluate the estimated PIs as [24]:

$$\text{CWC} = \text{NMPIW}(1 + \gamma(\text{PICP})e^{-\eta(\text{PICP} - \mu)}\gamma = \begin{cases} 0, & \text{PICP} \geq \mu \\ 1, & \text{PICP} < \mu \end{cases} \tag{5}$$

η and μ are fixed parameters, which determine the PIs with the lower value of the PICP. μ represents the confidence level of the PIs. η magnifies variation of the PICP and μ. Different η values should be examined to determine the most appropriate η value via a trial–error process. The coverage and width criteria are the most commonly used measures to evaluate the PIs' quality, however some other statistical measures are also applied to evaluate the calculated PIs. For example, in order to evaluate the constructed PIs via the Monte Carlo method, mean and standard deviation of the Nash–Sutcliffe model efficiency for each run are calculated to measure the PIs' coverage [58].

4. PIs Construction Methods

There are some methods to calculate the PIs. The Bayesian and Monte Carlo are the traditional methods. The other most frequently used method is the bootstrap technique; besides LUBE, it is one of the reliable methods. In addition, there are some other methods, such as mean variance estimation [11] and first order uncertainty analysis (FOUA) [53], but as they are not so prevalent, in the following sub sections the most common methods and their applications in hydro-climatic studies are described.

4.1. Bayesian Method

The Bayesian method was introduced by MacKay [21]. Training of the classic ANN is based on minimizing the error function, which leads to obtaining the optimum weights. Whereas, the Bayesian method attempts to train the ANN for the posterior probability distribution of weights from assumed prior probability distribution using Bayes' theorem. In the Bayesian method, ANN training is performed based on the regularized cost function as:

$$E(\omega) = \rho E_\omega + \beta E_D \tag{6}$$

where E_D is the sum of squared error and E_ω is the sum of squares of the network weights. ρ and β are used to determine training goals. The concept of this method is based on consideration of the set of ANN parameters, ω as a random set of variables with presumed distributions.

The Bayes' rule is applied to update the density function of the weights as [14]:

$$P(\omega|D, \rho, \beta, M) = \frac{P(D|\omega, \beta, M)\, P(\omega|\rho, M)}{P(D|\rho, \beta, M)} \tag{7}$$

where M and D are the NN model and the training dataset. $P(D|\omega, \beta, M)$ and $P(\omega|\rho, M)$ are the likelihood function of data occurrence and the prior density of parameters, respectively. Representing our knowledge, $P(D|\rho, \beta, M)$ is a normalization factor enforcing that the total probability is 1. Assuming that noises are normally distributed and $P(D|\omega, \beta, M)$ and $P(\omega|\rho, M)$ have normal distributions, it can be concluded that [14]:

$$P(D|\omega, \beta, M) = \frac{1}{Z_D(\beta)} e^{-\beta E_D} \tag{8}$$

$$P(\omega|\rho, M) = \frac{1}{Z_\omega(\rho)} e^{-\rho E_\omega} \tag{9}$$

where $Z_D(\beta) = \left(\frac{\pi}{\beta}\right)^{n/2}$ and $Z_\omega(\rho) = \left(\frac{\pi}{\rho}\right)^{p/2}$. n and p are the number of training samples and NN parameters, respectively. By substituting Equations (8) and (9) into (7), Equation (10) is obtained as [14]:

$$P(\omega|D, \rho, \beta, M) = \frac{1}{Z_F(\beta, \rho)} e^{-(\rho E_\omega + \beta E_D)} \tag{10}$$

The ANN is trained via maximization of the posterior probability $P(\omega|D, \rho, \beta, M)$, which is based on the minimizing Equation (6). By taking derivatives with respect to the logarithm of (10) and setting it equals to zero, the optimal values for β and ρ are obtained [14]:

$$\beta^{MP} = \frac{\gamma}{E_D(\omega^{MP})} \tag{11}$$

$$\rho^{MP} = \frac{n - \gamma}{E_\omega(\omega^{MP})} \tag{12}$$

where $\gamma = p - 2\rho^{MP} \text{tr}\left(H^{MP}\right)^{-1}$ is the so-called effective number of ANN parameters, and p is the total number of ANN model parameters. ω^{MP} are the most probable values of the ANN parameters. H^{MP} (Equation (13)) is the Hessian matrix of $E(\omega)$ as [14]:

$$H^{MP} = \rho \nabla^2 E_\omega + \beta \nabla^2 E_D \tag{13}$$

The approximation of the Hessian matrix is generally performed using the Levenberg–Marquardt optimization algorithm. Application of this technique for the training process results in ANNs having the variance as [14]:

$$\sigma_i^2 = \sigma_D^2 + \sigma_{\omega^{MP}}^2 \frac{1}{\beta} + \nabla_{\omega^{MP}}^T \hat{y}_i \left(H^{MP}\right)^{-1} \nabla_{\omega^{MP}} \hat{y}_i \tag{14}$$

The uncertainties corresponding to the data and parameters, respectively, are quantified via the term in the right and left sides of Equation (15). Finally, PI can be calculated considering the total variance of the ith future sample as [14]:

$$PI = \hat{y}_i \pm z^{1-\frac{\alpha}{2}} \left(\frac{1}{\beta} + \nabla_{\omega^{MP}}^T \hat{y}_i \left(H^{MP}\right)^{-1} \nabla_{\omega^{MP}} \hat{y}_i\right)^{\frac{1}{2}} \tag{15}$$

where $z^{1-\frac{\alpha}{2}}$ is the $1-(\alpha/2)$ quantile of the normal distribution function with zero mean and unit variance. Additionally, $\nabla_{\omega^{MP}}^T \hat{y}_i$ is the gradient of the ANN output with respect to its parameters' set of ω^{MP}. The Bayesian method for PI construction has a strong mathematical foundation. This method requires calculation of the Hessian matrix, which needs a significant amount of time, but it should be considered that the computational load is lower in the process of constructing PIs because of only calculating the gradient of Neural Network (NN) output.

Some studies applied the Bayesian method in order to construct the PIs. Kasiviswanathan and Sudheer [51] used the bootstrap and Bayesian techniques to assess the uncertainty of the flood forecasting models of the ANN. The method application was based on the assumption that the model structure is deterministic, therefore, only the parameter of uncertainty was assessed in this study. It was concluded that model implementation is acceptable when the ensemble mean is considered. It was concluded that the bootstrap method is simple and easy in the case of the implementation as compared to the Bayesian method, but comparison of the obtained results showed that the Bayesian method led to narrower PIs and lower variance in parameter convergence.

4.2. Monte Carlo Method

The performance of the Monte Carlo method is based on alteration of the model inputs, parameters or structure of their ensemble. The number of iterations depends on the required level of reliability and is a problem-dependent task. More repetition leads to more reliable results, but higher computational cost should also be examined. If the model structure and the input data are assumed to be certain, Equation (16) can be presented as [58]:

$$\hat{y}_{t,i} = M(x, \theta_i); \; t = 1, 2, \ldots, n; \; i = 1, 2, \ldots, s \tag{16}$$

where θ_i is the set of parameters sampled for the ith run of the Monte Carlo simulation, $\hat{y}_{t,i}$ is the model output of the ith time step for the ith run, n is the number of time steps and s is the number of simulations. The statistical properties (such as moments and quantiles) of the model output for each time step t are estimated from the realizations $\hat{y}_{t,i}$. In order to quantile the uncertainty, the following equation can be expressed [58]:

$$P(\hat{y}_t < \hat{Q}(p)) = \sum_{i=1}^{s} w_i | \hat{y}_{t,i} < \hat{Q}(p) \tag{17}$$

where, $\hat{y}_t$ is the model output at time step t, $\hat{y}_{t,i}$ is the value of model output at time t simulated by the model $M(x, \theta_i)$ in ith simulation, $\hat{Q}(p)$ is pth [0,1] quantile, w_i is the weight given to the model output in ith at simulation. Quantiles obtained in this way are conditioned on the inputs to the model, the model structure and the weight vector w_i. The computation of the model's PI with confidence level α $(0 < \alpha < 1)$ is achieved though estimation of the $\frac{1-\alpha}{2} * 100\%$ and $\frac{1+\alpha}{2} * 100\%$ via the $\hat{y}_{t,i}$. The lower prediction limit PL^L and the upper prediction limit PL^U are calculated as [58]:

$$\hat{Q}(p) = PL^L \text{ where } p = (1 - \alpha)/2 \tag{18}$$

$$\hat{Q}(p) = PL^U \text{ where } p = (1 + \alpha)/2 \tag{19}$$

Then, the PI is derived considering the output of the calibrated (optimal) model ($\bar{y}$) as:

$$PI^L = \bar{y} - PL^L, \; PI^U = PL^L - \bar{y} \tag{20}$$

where PI^L and PI^U are the interval of the obtained results as lower and upper bounds, respectively. It should define the purpose of the work and its significance.

Shrestha, Kayastha and Solomatine [58] applied the Monte Carlo method to assess the parametric uncertainty of the analysis of hydrological models of rainfall runoff using ANN. It was shown that the Monte Carlo method could be used to determine the other sources of uncertainty, such as input, structure or their combination. Tapoglou, Varouchakis, Trichakis and Karatzas [60] applied the Monte Carlo technique to investigate the uncertainty associated with modeling of the hydraulic head in an aquifer via the ANN. The model was performed 300 times by various training sets, and initial random values and the training results constituted a sensitivity analysis of the ANN training to the kriging part of the algorithm. This study concluded that error intervals for the train and test data of the ANN and kriging PIs were narrow, considering the complexity of the study area. Application of the Bayesian kriging methodology was assessed, and it was concluded that the difference between the predicted values and the results of simulation of the actual data was low. This method led to consistent and reliable performance in different conditions. It was assessed that this method is appropriate to simulate groundwater-level, chiefly in cases with complicated behavior and unknown geological data.

Seifi, Ehteram, Singh and Mosavi [59] attempted to evaluate the uncertainty of groundwater level modeling via the hybrid ANN modeling and some other black box models and six meta-heuristic optimization methods, such as the grasshopper algorithm, cat swarm, weed algorithm, genetic algorithm, krill algorithm and particle swarm optimization, and it was mentioned that hybrid methods led to better performance and accuracy than sole methods.

4.3. Bootstrap Method

In this technique, several ANNs (B ANNs) are trained with randomly selected subsets [22] (see Figure 3). The randomly selected samples from total data are used for training each of networks. This technique is based on ensembling some ANNs, which could lead to lower estimation errors with regard to a single ANN. This method is independent of any complicated calculation using the non-linear operator or function. A model as f_{ANN} (x) is fitted to each of the generated bootstrap sub-sets, and the bootstrapping estimate is calculated as the average and variance of each model as:

$$\hat{y}_{boot}(x) = \frac{1}{B} \sum_{b=1}^{B} f_{ANN}^{b}(x) \tag{21}$$

$$\hat{\sigma}_{boot}^{2}(x) = \frac{1}{B-1} \sum_{b=1}^{B} \left(f_{ANN}^{b}(x) - \hat{y}_{boot}(x) \right)^{2} \tag{22}$$

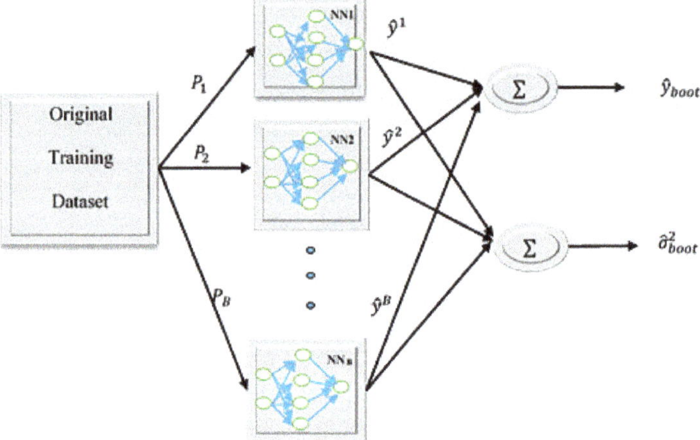

Figure 3. Schematic of the bootstrap method [14]. P_B stands for training data sub-sets.

For constructing the PIs [l.u], values of the observed $X = (x_1, x_2, \ldots, x_3)$ with normal distribution probability of P and according to Equations (21) and (22), P (l < X < u) are as:

$$P(l < X < u) = P\left(\frac{l - \hat{y}_{boot}}{\hat{\sigma}_{boot}} < \frac{X - \hat{y}_{boot}}{\hat{\sigma}_{boot}} < \frac{u - \hat{y}_{boot}}{\hat{\sigma}_{boot}} \right) \tag{23}$$

where $Z = \frac{X - \hat{y}_{boot}}{\hat{\sigma}_{boot}}$, the standard score of X, is distributed as standard normal [69], hence:

$$\frac{l - \hat{y}_{boot}}{\hat{\sigma}_{boot}} = -z \; ; \tag{24}$$

or $l = \hat{y}_{boot} - z\hat{\sigma}_{boot} \; ; \; u = \hat{y}_{boot} + z\hat{\sigma}_{boot}$.

Table 3 presents the corresponding values of Z and PIs.

Table 3. Values of Z for different PIs [70].

PI	z
75%	1.15
90%	1.64
95%	1.96
99%	2.58

Talebizadeh and Moridnejad [52] applied the bootstrap method to assess the uncertainty arising from measurement error and also the uncertainty of the ANNs' output for forecasting the lake level fluctuations. It was stressed that PIs' estimation provided beneficial information for decision making and designing. Kasiviswanathan and Sudheer [53] combined the bootstrap and the FOUA method to investigate the parametric and predictive uncertainty of the rainfall-runoff ANN-based modeling to forecast river flow. It was concluded that the FOUA method could compute the sensitivity coefficients that are the first order partial derivative of the model output and parameters of modeling. In this method, the computational burden and time of simulation for uncertainty analysis are reduced due to the usage of the statistical parameters such as mean and variance of the ANN weight vectors and biases. The parameter variability was determined via the bootstrap method. The obtained results for uncertainty analysis were quantified via the coverage and width criteria. It was concluded that the results for training and verifying data sets matched each other. Moreover, uncertainty associated with various domains of flow (low, medium and high) was assessed to identify the effect of the magnitude of flow on uncertainty; the results indicated that the uncertainty level changed with different flow regimes, proportionally. The overall results, considering both width and coverage criteria, show that the FOUA method led to a better quantification of the prediction uncertainty compared to the bootstrap method. Wang, Zheng, Zhao, Jiang, Wang, Guo and Wang [55] used the bootstrap method to calculate the PIs of water quality modeling via the wavelet-ANN approach. The uncertainty of the model structure and data noise was investigated, and it was shown that the application of the wavelet data pre-processing could lead to more accurate results. Kumar, Tiwari, Chatterjee and Mishra [39] used the combination of the bootstrap method and wavelet-ANN to quantify the uncertainty associated with the reservoir inflow forecasting. Moreover, multiple linear regression model implementation was compared to the bootstrap method and it was concluded that the performance of the bootstrap method is more reliable. It was also mentioned that PIs' estimation could provide more useful information in operational inflow forecasting. Kasiviswanathan, He, Sudheer and Tay [56] used the bootstrap technique for the quantification of the uncertainty associated with modeling streamflow and flood management. The coverage and width criteria were applied for quantification of the model's performance uncertainty. Results indicated that the bootstrap method is the proper method for streamflow forecasting and flood management. Moreover, it was mentioned that there are some limitations associated with forecasting high flow due to the lower samples and dependence of the ANN on the number of samples. Thus, it was suggested that one use hybrid modeling and integrate the data-driven models with physically-based/conceptual models and/or empirical relationships between high flows and influencing factors for the enhancement of the accuracy of the model for modeling high flow regimes.

4.4. LUBE Method

The LUBE method is based on training an ANN with two outputs for developing the PIs in one level (see Figure 4). Two outputs present the upper and lower bounds of the PIs. The proposed ANN is trained based on minimizing the defined cost function. Cost function contains both coverage and width criteria. This method is independent of special information about the PI bound. Previous studies assumed that PIs developed via this method are superior to the other PIs construction methods. In addition, its computational expense is insignificant [24]. Unlike traditional methods, LUBE method performance is independent of point prediction and it is a non-parametric technique. LUBE method application does not depend on parametric distribution of data. It is fast and simple [71].

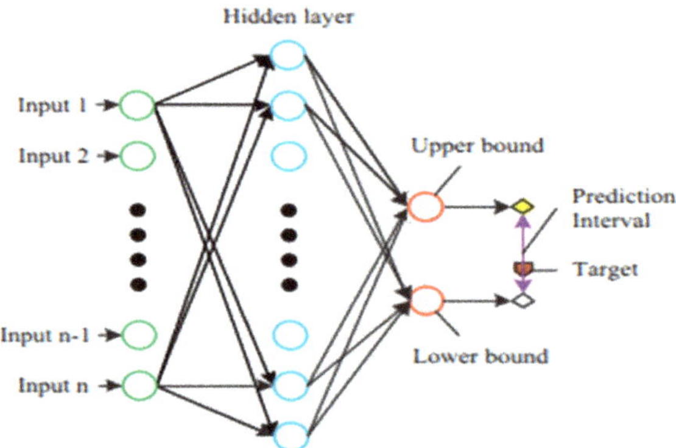

Figure 4. The structure of the ANN in the lower upper bound estimation (LUBE) method.

Kasiviswanathan, Cibin, Sudheer and Chaubey [61] attempted to develop the PIs of the ANN-based rainfall-runoff modeling by generation of the ensemble predictions (similar to the LUBE method). PIs were developed at two levels. At the first level, optimum ANN parameters were obtained via the genetic algorithm. In the second step, the ensemble of the models was created by optimization of the verity of the ANN parameters. PIs were calculated by minimizing the residual variance of the ensemble mean and maximization of the covering targets. Moreover, at the same time, the minimization of the PIs' width was taken into account. It was stated that by consideration of the ensemble mean value as the output of the model, the peak flow could be predicted more precisely compared to the classic point prediction of the ANN. Taormina and Chau [62] examined the LUBE method for construction of the PIs at different confidence levels for the 6 hours ahead streamflow discharges forecasting. Particle swarm optimization was used to minimize the CWC cost function. It was shown that the obtained results depend on the used particle swarm optimization paradigm. They claimed that the multi-objective framework led to more appropriate results than single-objective swarm optimization. It was also concluded that the applied algorithm to develop the model could have a remarkable effect on the PIs quality. Zhang, Zhou, Ye, Zeng and Chen [63] applied the LUBE method to construct the PIs of flood forecasting. This study proposed a PI symmetry index and objective function to evaluate the coverage, width and symmetry of PIs. To optimize the proposed objective function, the shuffled complex evolution algorithm was applied. The mean of the bounds was used to present deterministic forecasting. Kasiviswanathan and Sudheer [51] applied the PI method (similar to the LUBE method) for quantification of the ANN modeling uncertainty. The coverage and width criteria were used to quantify the PIs. This paper compared other PI construction methods results, such as the Bayesian and bootstrap methods, and concluded that the PI method was more reliable. In addition, it was presented that the PI method could successfully capture the peak points. Kasiviswanathan, Sudheer and He [64] assessed the uncertainty associated with input and parameters for the ANN-based rainfall-runoff modeling. A two-stage optimization method was applied to estimate the PIs. Chen and Chau [66] used the LUBE method to construct the PIs of sediment load modeling via the hybrid double feedforward NN. It was demonstrated that PIs led to appropriate results in the 90% and 95% CLs. The proposed method could generate reliable PIs. It was discussed that application of the hybrid double feedforward NN could improve performance of the PIs construction in the classification of the low, medium and high sediment loads, and coverage probability about 100% for low and medium sediment loads, but its performance was weak for modeling high sediment loads. Moreover, it was concluded that the LUBE method was efficient in quantifying the uncertainty of data-driven models. Nourani, Paknezhad, Sharghi and Khosravi [57] used the LUBE method to construct the PIs associated with the ANN-based downscaling of the general circulation models. In this study, the LUBE method was applied by generating multiple sets of weights to develop narrow PIs with high coverage probability. It was indicated that the LUBE method could be successfully used to compute the PIs of ANN-based downscaling with reliable performance. Nourani, Sayyah-Fard, Alami and Sharghi [65] quantified the uncertainty of the ANN-based evaporation modeling via the LUBE method. It was claimed that the LUBE method could construct PIs with an appropriate level of reliability; however, data pre-processing methods could affect the uncertainty. This study applied simulated annealing optimization algorithms to construct PIs with higher coverage and lower width. It was mentioned that this method could overcome the problem of trapping in local minima. Kasiviswanathan, Sudheer, Soundharajan and Adeloye [67] applied upper lower bound and mean of forecasting to evaluate uncertainty in inflow modeling via the ANN for optimizing the reservoir operation and decision making. An integrated simulation–optimization was applied, which led to minimizing the error. In Figure 5, the procedure of PIs construction is depicted.

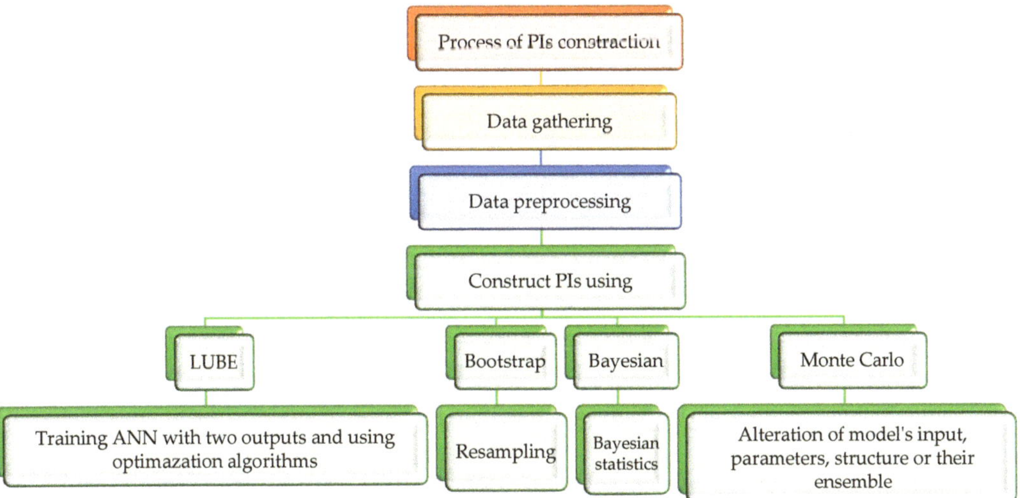

Figure 5. The PIs construction procedure.

5. Comparison of the PIs Construction Methods

Various methods of PIs construction with different levels of complexity, computational burden, difficulty in implementation, reliability and required times have been developed. Undoubtedly, it is impossible to claim that a particular technique is superior to the others, but each method has its own advantages. Therefore, in this section, the advantage and disadvantages of the prevalent methods are investigated. A brief comparison of each method is tabulated in Table 4.

Table 4. The brief comparison of the proposed methods.

Method		Run Time	Number of Required Networks	Computational Burden	Reliability	Reference
Bayesian	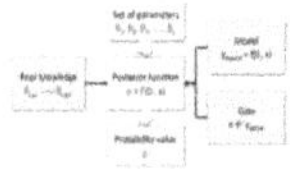 Conceptual representation of the Bayesian model for inference of parameters.	high	One	high	high	[21]
Monte Carlo	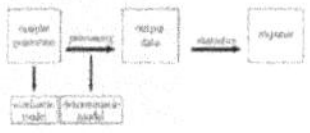 General overview of the Monte Carlo algorithm.	high	More than 1	medium	medium	[58]
Bootstrap	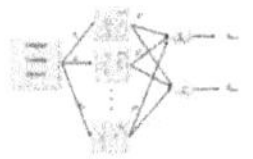 Schematic of the bootstrap method, PB stands for training data sub-sets.	high	More than 1	low	medium	[14]
LUBE	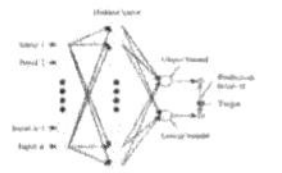 The structure of the ANN in the LUBE method	low	One	low	high	[24]

Khosravi, Nahavandi, Creighton and Atiya [14] compared the different methods of PI construction for data from various domains. It was assumed that the delta method obtains the highest quality of the PIs, but its repeatability (performance of the method in worst cases) is not acceptable. Moreover, the constructed PIs via the delta technique consist of fixed PI width. The Bayesian method is the most acceptable method in the case of reproducibility (various iterations of the method lead to similar results). Moreover, the stability of developed PIs is the other advantage of this method. The bootstrap method is favorable in the case of variability (the response of PIs to the level of uncertainty associated with data). However, the obtained PIs via the bootstrap method may lead to low quality in comparison to the delta and Bayesian methods. The variance of the outputs may be overestimated, which may cause wider PIs. In addition, it has been demonstrated that by increasing the number of training iterations, the bootstrap method might not definitely improve the results. It has also been indicated that each method has its pros and cons, and implementation of different methods by consideration of various criteria may lead to different outcomes. Kasiviswanathan and Sudheer [51] compared different PIs construction methods and showed that parameters coverage and peak flow prediction are high for the PI method and low for the bootstrap method. Fulfillment of statistical and probabilistic assumptions is low for the bootstrap method, but it is high for the Bayesian method.

The difficulty of the implementation of the bootstrap, Bayesian and LUBE methods are, respectively, medium, high and low.

6. Gaps and Suggestions for Future Studies

The following issues are suggested for future studies to fill the existing gaps in already performed studies.

i. Most of the PIs construction methods applied coverage and width criteria and their combination, but there is not any criterion in order to present information about the probability and reliability of the lower and upper bounds of the constructed PIs. Therefore, future studies may develop and present some criteria about this issue, for example, proximity of the target to the upper or lower bounds.

ii. The reviewing of multiple studies showed that there is not any study that uses the delta method for PIs construction. Although it may have high level of computational cost, according to its reliable performance in other fields as expressed in Khosravi, Nahavandi, Creighton and Atiya [14], it is proposed that one apply this method in hydrological studies as well.

iii. The performance of the LUBE method, as the most robust method of PI construction, can be improved in different aspects in order to obtain more reliable PIs. The implementation could be augmented by combining with the ANN structure selection techniques.

iv. Adaptation between point prediction and PIs has not been examined, yet significantly presented. Therefore, it can be recommended that future works analyze the adaptation and correlation between point prediction and PIs construction. Moreover, some criteria can be defined to capture simultaneously.

v. It is recommended that one apply the presented methods to assess the uncertainty associated with the improved version of the ANN, such as emotional ANN [72] and to investigate the effects of hormonal parameters in the reliability of the models.

vi. The LUBE method performance is based on the cost functions. Some studies used multi-objective optimization cost function in which the coverage and width criteria were considered as cost function simultaneously. In contrast, some other studies used coverage and width combination criteria as the cost function, in which some parameters should be determined by trial and error. Therefore, future studies can compare the implementation of multi-objective and single-objective optimization methods. Moreover, future studies can propose the appropriate value of CWC parameters or propose a method for better and faster determination of the parameters' values.

vii. As there are few studies attempting to investigate other artificial intelligence methods such as ANFIS, it is recommended that one construct the PIs of those methods too.

7. Conclusions

This study investigated multiple PIs construction methods applied in hydro-climatic studies. It concluded that the bootstrap method has been used in the majority of the studies as it is simple and can be applied easily. Moreover, the LUBE method has gained noticeable attention recently in hydrological studies due to its superiority in implementation and reliability compared to other methods. Nevertheless, there are few applications of the Bayesian or delta method in the development of PIs in the hydrological issues.

Author Contributions: All authors have contributed equally to all sections of this manuscript. All authors have read and agreed to the published version of the manuscript.

Funding: This research received no external funding.

Institutional Review Board Statement: Not applicable.

Informed Consent Statement: Not applicable.

Data Availability Statement: Data will be available upon request.

Acknowledgments: This work has been supported by University of Tabriz, International and Academic Cooperation Directorate, in the framework of TabrizU-300 program.

References

1. Loucks, D.P. Sustainable Water Resources Management. *Water Int.* **2000**, *25*, 3–10. [CrossRef]
2. Nourani, V.; Kalantari, O. Integrated Artificial Neural Network for Spatiotemporal Modeling of Rainfall–Runoff–Sediment Processes. *Environ. Eng. Sci.* **2010**, *27*, 411–422. [CrossRef]
3. Nourani, V.; Kisi, O.; Komasi, M. Two hybrid Artificial Intelligence approaches for modeling rainfall–runoff process. *J. Hydrol.* **2011**, *402*, 41–59. [CrossRef]
4. Sharghi, E.; Nourani, V.; Najafi, H.; Molajou, A. Emotional ANN (EANN) and Wavelet-ANN (WANN) Approaches for Markovian and Seasonal Based Modeling of Rainfall-Runoff Process. *Water Resour. Manag.* **2018**, *32*, 3441–3456. [CrossRef]
5. Sharifi, S.S.; Delirhasannia, R.; Nourani, V.; Sadraddini, A.A.; Ghorbani, A. Using artificial neural networks (ANNs) and adaptive neuro-fuzzy inference system (ANFIS) for modeling and sensitivity analysis of effective rainfall. *Recent Adv. Contin. Mech. Hydrol. Ecol.* **2013**, *3*, 133–139.
6. Latifoglu, L. Evaluating Stream Flow Forecasting Performance Using Adaptive Network Based Fuzzy Logic Inference System, Artificial Neural Networks with Feature Selection. *EPSTEM* **2020**, *11*, 125–130.
7. Klomjit, J.; Ngaopitakkul, A. Comparison of Artificial Intelligence Methods for Fault Classification of the 115-kV Hybrid Transmission System. *Appl. Sci.* **2020**, *10*, 3967. [CrossRef]
8. Niu, W.-J.; Feng, Z.-K.; Feng, B.-F.; Xu, Y.-S.; Min, Y.-W. Parallel computing and swarm intelligence based artificial intelligence model for multi-step-ahead hydrological time series prediction. *Sustain. Cities Soc.* **2021**, *66*, 102686. [CrossRef]
9. Nourani, V.; Komasi, M.; Mano, A. A Multivariate ANN-Wavelet Approach for Rainfall–Runoff Modeling. *Water Resour. Manag.* **2009**, *23*, 2877–2894. [CrossRef]
10. Elshorbagy, A.; Corzo, G.; Srinivasulu, S.; Solomatine, D.P. Experimental investigation of the predictive capabilities of data driven modeling techniques in hydrology—Part 1: Concepts and methodology. *Hydrol. Earth Syst. Sci.* **2010**, *14*, 1931–1941. [CrossRef]
11. Huang, F.; Huang, J.; Jiang, S.; Zhou, C.-B. Landslide displacement prediction based on multivariate chaotic model and extreme learning machine. *Eng. Geol.* **2017**, *218*, 173–186. [CrossRef]
12. Refsgaard, J.C.; Van Der Sluijs, J.P.; Brown, J.; Van Der Keur, P. A framework for dealing with uncertainty due to model structure error. *Adv. Water Resour.* **2006**, *29*, 1586–1597. [CrossRef]
13. Zio, E.; Aven, T. Uncertainties in smart grids behavior and modeling: What are the risks and vulnerabilities? How to analyze them? *Energy Policy* **2011**, *39*, 6308–6320. [CrossRef]
14. Khosravi, A.; Nahavandi, S.; Creighton, D.; Atiya, A.F. Comprehensive Review of Neural Network-Based Prediction Intervals and New Advances. *IEEE Trans. Neural Netw.* **2011**, *22*, 1341–1356. [CrossRef]
15. Quan, H.; Srinivasan, D.; Khosravi, A. Incorporating Wind Power Forecast Uncertainties into Stochastic Unit Commitment Using Neural Network-Based Prediction Intervals. *IEEE Trans. Neural Netw. Learn. Syst.* **2015**, *26*, 2123–2135. [CrossRef]
16. Chatfield, C. Calculating interval forecasts. *J. Bus. Econ. Stat.* **1993**, *11*, 121–135.
17. Ma, J.; Niu, X.; Tang, H.; Wang, Y.; Wen, T.; Zhang, J. Displacement Prediction of a Complex Landslide in the Three Gorges Reservoir Area (China) Using a Hybrid Computational Intelligence Approach. *Complexity* **2020**, *2020*, 1–15. [CrossRef]
18. Shuai, Y.; Notton, G.; Duchaud, J.-L.; Almorox, J.; Yaseen, Z.M. Solar irradiation prediction intervals based on Box–Cox transformation and univariate representation of periodic autoregressive model. *Renew. Energy Focus* **2020**, *33*, 43–53. [CrossRef]
19. Chryssolouris, G.; Lee, M.; Ramsey, A. Confidence interval prediction for neural network models. *IEEE Trans. Neural Netw.* **1996**, *7*, 229–232. [CrossRef]
20. Momotaz, B.; Dohi, T. Prediction Interval of Cumulative Number of Software Faults Using Multilayer Perceptron. *Flex. Gen. Uncertain. Optim.* **2015**, *619*, 43–58. [CrossRef]
21. Mackay, D.J.C. A Practical Bayesian Framework for Backpropagation Networks. *Neural Comput.* **1992**, *4*, 448–472. [CrossRef]
22. Kasiviswanathan, K.S.; Sudheer, K. Methods used for quantifying the prediction uncertainty of artificial neural network based hydrologic models. *Stoch. Environ. Res. Risk Assess.* **2017**, *31*, 1659–1670. [CrossRef]
23. Efron, B.; Tibshirani, R.J. *An Introduction to the Bootstrap*; CRC Press: Boca Raton, FL, USA, 1994.
24. Lu, J.; Ding, J.; Dai, X.; Chai, T. Ensemble Stochastic Configuration Networks for Estimating Prediction Intervals: A Simultaneous Robust Training Algorithm and Its Application. *IEEE Trans. Neural Netw. Learn. Syst.* **2020**, *31*, 5426–5440. [CrossRef]
25. Khosravi, A.; Nahavandi, S.; Creighton, D.; Atiya, A.F. Lower Upper Bound Estimation Method for Construction of Neural Network-Based Prediction Intervals. *IEEE Trans. Neural Netw.* **2010**, *22*, 337–346. [CrossRef]
26. Cannon, A.J.; Whitfield, P.H. Downscaling recent streamflow conditions in British Columbia, Canada using ensemble neural network models. *J. Hydrol.* **2002**, *259*, 136–151. [CrossRef]
27. Jeong, D.-I.; Kim, Y. Rainfall-runoff models using artificial neural networks for ensemble streamflow prediction. *Hydrol. Process.* **2005**, *19*, 3819–3835. [CrossRef]
28. Fleming, S.W.; Bourdin, D.R.; Campbell, D.; Stull, R.B.; Gardner, T. Development and Operational Testing of a Super-Ensemble Artificial Intelligence Flood-Forecast Model for a Pacific Northwest River. *JAWRA J. Am. Water Resour. Assoc.* **2014**, *51*, 502–512. [CrossRef]

29. Kan, G.; Yao, C.; Li, Q.; Li, Z.; Yu, Z.; Liu, C.; Ding, L.; He, X.; Liang, K. Improving event-based rainfall-runoff simulation using an ensemble artificial neural network based hybrid data-driven model. *Stoch. Environ. Res. Risk Assess.* **2015**, *29*, 1345–1370. [CrossRef]
30. Kim, S.E.; Seo, I.W. Artificial Neural Network ensemble modeling with conjunctive data clustering for water quality prediction in rivers. *HydroResearch* **2015**, *9*, 325–339. [CrossRef]
31. Kim, S.; Kim, H.S. Uncertainty Reduction of the Flood Stage Forecasting Using Neural Networks Model1. *JAWRA J. Am. Water Resour. Assoc.* **2008**, *44*, 148–165. [CrossRef]
32. Yang, C.-C.; Chen, C.-S. Application of integrated back-propagation network and self organizing map for flood forecasting. *Hydrol. Process.* **2009**, *23*, 1313–1323. [CrossRef]
33. Srivastav, R.; Sudheer, K.; Chaubey, I. A simplified approach to quantifying predictive and parametric uncertainty in artificial neural network hydrologic models. *Water Resour. Res.* **2007**, *43*. [CrossRef]
34. Boucher, M.; Perreault, L.; Anctil, F. Tools for the assessment of hydrological ensemble forecasts obtained by neural networks. *J. Hydroinform.* **2009**, *11*, 297–307. [CrossRef]
35. Sharma, S.K.; Tiwari, K. Bootstrap based artificial neural network (BANN) analysis for hierarchical prediction of monthly runoff in Upper Damodar Valley Catchment. *J. Hydrol.* **2009**, *374*, 209–222. [CrossRef]
36. Kant, A.; Suman, P.K.; Giri, B.K.; Tiwari, M.K.; Chatterjee, C.; Nayak, P.C.; Kumar, S. Comparison of multi-objective evolutionary neural network, adaptive neuro-fuzzy inference system and bootstrap-based neural network for flood forecasting. *Neural Comput. Appl.* **2013**, *23*, 231–246. [CrossRef]
37. Zhang, X.; Liang, F.; Srinivasan, R.; Van Liew, M. Estimating uncertainty of streamflow simulation using Bayesian neural networks. *Water Resour. Res.* **2009**, *45*. [CrossRef]
38. Khan, M.S.; Coulibaly, P. Assessing Hydrologic Impact of Climate Change with Uncertainty Estimates: Bayesian Neural Network Approach. *J. Hydrometeorol.* **2010**, *11*, 482–495. [CrossRef]
39. Zhang, X.; Liang, F.; Yu, B.; Zong, Z. Explicitly integrating parameter, input, and structure uncertainties into Bayesian Neural Networks for probabilistic hydrologic forecasting. *J. Hydrol.* **2011**, *409*, 696–709. [CrossRef]
40. Zhang, X.; Zhao, K. Bayesian Neural Networks for Uncertainty Analysis of Hydrologic Modeling: A Comparison of Two Schemes. *Water Resour. Manag.* **2012**, *26*, 2365–2382. [CrossRef]
41. Humphrey, G.B.; Gibbs, M.S.; Dandy, G.; Maier, H.R. A hybrid approach to monthly streamflow forecasting: Integrating hydrological model outputs into a Bayesian artificial neural network. *J. Hydrol.* **2016**, *540*, 623–640. [CrossRef]
42. Shen, S.; Zeng, G.; Liang, J.; Li, X.; Tan, Y.; Li, Z.; Li, J. Markov Chain Monte Carlo Approach for Parameter Uncertainty Quantification and Its Impact on Groundwater Mass Transport Modeling: Influence of Prior Distribution. *Environ. Eng. Sci.* **2014**, *31*, 487–495. [CrossRef]
43. Tongal, H.; Booij, M.J. Quantification of parametric uncertainty of ANN models with GLUE method for different streamflow dynamics. *Stoch. Environ. Res. Risk Assess.* **2017**, *31*, 993–1010. [CrossRef]
44. Snyder, H. Literature review as a research methodology: An overview and guidelines. *J. Bus. Res.* **2019**, *104*, 333–339. [CrossRef]
45. Torres-Carrión, P.V.; González-González, C.S.; Aciar, S.; Rodríguez-Morales, G. Methodology for Systematic Literature Review Applied to Engineering and Education. In Proceedings of the 2018 IEEE Global Engineering Education Conference (EDUCON), Tenerife, Spain, 17–20 April 2018; pp. 1364–1373.
46. Shamseer, L.; Moher, D.; Clarke, M.; Ghersi, D.; Liberati, A.; Petticrew, M.; Shekelle, P.; Stewart, L.A. Preferred reporting items for systematic review and meta-analysis protocols (PRISMA-P) 2015: Elaboration and explanation. *BMJ* **2015**, *349*, 1–25. [CrossRef]
47. Moher, D.; Shamseer, L.; Clarke, M.; Ghersi, D.; Liberati, A.; Petticrew, M.; Shekelle, P.; Stewart, L.A. Preferred reporting items for systematic review and meta-analysis protocols (PRISMA-P) 2015 statement. *Syst. Rev.* **2015**, *4*, 1. [CrossRef]
48. Mengist, W.; Soromessa, T.; Legese, G. Method for conducting systematic literature review and meta-analysis for environmental science research. *MethodsX* **2020**, *7*, 100777. [CrossRef]
49. Hasan, H.H.; Razali, S.F.M.; Muhammad, N.; Ahmad, A. Research Trends of Hydrological Drought: A Systematic Review. *Water* **2019**, *11*, 2252. [CrossRef]
50. Harrison-Atlas, D.; Theobald, D.M.; Goldstein, J.H. A systematic review of approaches to quantify hydrologic ecosystem services to inform decision-making. *Int. J. Biodivers. Sci. Ecosyst. Serv. Manag.* **2016**, *12*, 160–171. [CrossRef]
51. Kasiviswanathan, K.S.; Sudheer, K.P. Comparison of methods used for quantifying prediction interval in artificial neural network hydrologic models. *Model. Earth Syst. Environ.* **2016**, *2*, 1–11. [CrossRef]
52. Talebizadeh, M.; Moridnejad, A. Uncertainty analysis for the forecast of lake level fluctuations using ensembles of ANN and ANFIS models. *Expert Syst. Appl.* **2011**, *38*, 4126–4135. [CrossRef]
53. Kasiviswanathan, K.S.; Sudheer, K. Quantification of the predictive uncertainty of artificial neural network based river flow forecast models. *Stoch. Environ. Res. Risk Assess.* **2012**, *27*, 137–146. [CrossRef]
54. Kumar, S.; Tiwari, M.K.; Chatterjee, C.; Mishra, A. Reservoir Inflow Forecasting Using Ensemble Models Based on Neural Networks, Wavelet Analysis and Bootstrap Method. *Water Resour. Manag.* **2015**, *29*, 4863–4883. [CrossRef]
55. Wang, Y.; Zheng, T.; Zhao, Y.; Jiang, J.; Wang, Y.; Guo, L.; Wang, P. Monthly water quality forecasting and uncertainty assessment via bootstrapped wavelet neural networks under missing data for Harbin, China. *Environ. Sci. Pollut. Res.* **2013**, *20*, 8909–8923. [CrossRef]
56. Kasiviswanathan, K.; He, J.; Sudheer, K.; Tay, J.-H. Potential application of wavelet neural network ensemble to forecast streamflow for flood management. *J. Hydrol.* **2016**, *536*, 161–173. [CrossRef]

57. Nourani, V.; Paknezhad, N.J.; Sharghi, E.; Khosravi, A. Estimation of prediction interval in ANN-based multi-GCMs downscaling of hydro-climatologic parameters. *J. Hydrol.* **2019**, *579*, 124226. [CrossRef]
58. Shrestha, D.L.; Kayastha, N.; Solomatine, D.P. A novel approach to parameter uncertainty analysis of hydrological models using neural networks. *Hydrol. Earth Syst. Sci.* **2009**, *13*, 1235–1248. [CrossRef]
59. Seifi, A.; Ehteram, M.; Singh, V.P.; Mosavi, A. Modeling and Uncertainty Analysis of Groundwater Level Using Six Evolutionary Optimization Algorithms Hybridized with ANFIS, SVM, and ANN. *Sustainability* **2020**, *12*, 4023. [CrossRef]
60. Tapoglou, E.; Varouchakis, E.A.; Trichakis, I.; Karatzas, G.P. Hydraulic head uncertainty estimations of a complex artificial intelligence model using multiple methodologies. *J. Hydroinform.* **2020**, *22*, 205–218. [CrossRef]
61. Kasiviswanathan, K.; Cibin, R.; Sudheer, K.; Chaubey, I. Constructing prediction interval for artificial neural network rainfall runoff models based on ensemble simulations. *J. Hydrol.* **2013**, *499*, 275–288. [CrossRef]
62. Taormina, R.; Chau, K.-W. ANN-based interval forecasting of streamflow discharges using the LUBE method and MOFIPS. *Eng. Appl. Artif. Intell.* **2015**, *45*, 429–440. [CrossRef]
63. Zhang, H.; Zhou, J.; Ye, L.; Zeng, X.; Chen, Y. Lower Upper Bound Estimation Method Considering Symmetry for Construction of Prediction Intervals in Flood Forecasting. *Water Resour. Manag.* **2015**, *29*, 5505–5519. [CrossRef]
64. Kasiviswanathan, K.; Sudheer, K.; He, J. Probabilistic and ensemble simulation approaches for input uncertainty quantification of artificial neural network hydrological models. *Hydrol. Sci. J.* **2018**, *63*, 101–113. [CrossRef]
65. Nourani, V.; Sayyah-Fard, M.; Aalami, M.T.; Sharghi, E. Data pre-processing effect on ANN-based prediction intervals construction of the evaporation process at different climate regions in Iran. *J. Hydrol.* **2020**, *588*, 125078. [CrossRef]
66. Chen, X.-Y.; Chau, K.-W. Uncertainty Analysis on Hybrid Double Feedforward Neural Network Model for Sediment Load Estimation with LUBE Method. *Water Resour. Manag.* **2019**, *33*, 3563–3577. [CrossRef]
67. Kasiviswanathan, K.S.; Sudheer, K.P.; Soundharajan, B.-S.; Adeloye, A.J. Implications of uncertainty in inflow forecasting on reservoir operation for irrigation. *Paddy Water Environ.* **2020**, 1–13. [CrossRef]
68. Shrestha, D.L.; Solomatine, D.P. Machine learning approaches for estimation of prediction interval for the model output. *Neural Netw.* **2006**, *19*, 225–235. [CrossRef]
69. Shrivastava, N.A.; Khosravi, A.; Panigrahi, B. Prediction Interval Estimation of Electricity Prices Using PSO-Tuned Support Vector Machines. *IEEE Trans. Ind. Inform.* **2015**, *11*, 322–331. [CrossRef]
70. Grant, E.L.; Leavenworth, R.S. *Statistical Quality and Control*; McGraw-Hill: New York, NY, USA, 1972.
71. Quan, H.; Srinivasan, D.; Khosravi, A. Particle swarm optimization for construction of neural network-based prediction intervals. *Neurocomputing* **2014**, *127*, 172–180. [CrossRef]
72. Nourani, V. An Emotional ANN (EANN) approach to modeling rainfall-runoff process. *J. Hydrol.* **2017**, *544*, 267–277. [CrossRef]

 *sustainability*

Review

Review of Nitrogen Compounds Prediction in Water Bodies Using Artificial Neural Networks and Other Models

Pavitra Kumar [1], Sai Hin Lai [1], Jee Khai Wong [2,3], Nuruol Syuhadaa Mohd [1], Md Rowshon Kamal [4], Haitham Abdulmohsin Afan [5], Ali Najah Ahmed [6], Mohsen Sherif [7,8], Ahmed Sefelnasr [8] and Ahmed El-Shafie [1,*]

[1]   Department of Civil Engineering, Faculty of Engineering, University of Malaya, Kuala Lumpur 50603, Malaysia; pavitrakumar27@gmail.com (P.K.); laish@um.edu.my (S.H.L.); n_syuhadaa@um.edu.my (N.S.M.)

[2]   Department of Civil Engineering, College of Engineering, University Tenaga Nasional (UNITEN), Jalan Ikram-UNITEN, Kajang 43000, Selangor, Malaysia; wongjk@uniten.edu.my

[3]   Institute for Sustainable Energy (ISE), University Tenaga Nasional (UNITEN), Kajang 43000, Selangor, Malaysia

[4]   Department of Biological and Agricultural Engineering, Faculty of Engineering, University Putra Malaysia, Selangor 43400, Malaysia; rowshon@upm.edu.my

[5]   Department of Civil Engineering, Al-Maaref University College, Ramadi 31001, Iraq; haitham.afan@gmail.com

[6]   Institute for Energy Infrastructure (IEI), University Tenaga Nasional (UNITEN), Kajang 43000, Selangor, Malaysia; mahfoodh@uniten.edu.my

[7]   Civil and Environmental Eng. Dept., College of Engineering, United Arab Emirates University, Al Ain 15551, UAE; msherif@uaeu.ac.ae

[8]   National Water Center, United Arab Emirate University, Al Ain P.O. Box 15551, UAE; ahmed.sefelnasr@uaeu.ac.ae

*   Correspondence: elshafie@um.edu.my

Received: 22 April 2020; Accepted: 22 May 2020; Published: 26 May 2020

Abstract: The prediction of nitrogen not only assists in monitoring the nitrogen concentration in streams but also helps in optimizing the usage of fertilizers in agricultural fields. A precise prediction model guarantees the delivering of better-quality water for human use, as the operations of various water treatment plants depend on the concentration of nitrogen in streams. Considering the stochastic nature and the various hydrological variables upon which nitrogen concentration depends, a predictive model should be efficient enough to account for all the complexities of nature in the prediction of nitrogen concentration. For two decades, artificial neural networks (ANNs) and other models (such as autoregressive integrated moving average (ARIMA) model, hybrid model, etc.), used for predicting different complex hydrological parameters, have proved efficient and accurate up to a certain extent. In this review paper, such prediction models, created for predicting nitrogen concentration, are critically analyzed, comparing their accuracy and input variables. Moreover, future research works aiming to predict nitrogen using advanced techniques and more reliable and appropriate input variables are also discussed.

Keywords: nitrogen compound; nitrogen prediction; prediction models; neural network

---

1. Introduction

Human activities have provoked serious effects on the nutrient cycle, ecological functioning of streams, and water quality [1–3]. Presently, agriculture production consummately depends on the amount of fertilizers and pesticides used. Fertilizers mainly contain nitrogen compared with other chemicals. Crops require nitrogen for their growth and for the production of fruits or grains. Some agricultural specialists have also recommended using the fertilizers that carry a higher percentage of nitrogen [4]. However, only 40–70% of nitrogen compounds applied as fertilizers are absorbed by the crops. The remaining nitrogen compounds either percolate downward with water to join groundwater or flow along with the runoff water to join the streams [5,6]. In both cases, the nitrogen concentration in water escalates, which can affect human health [7–9]. If pesticides and fertilizers are added to the fields at a high rate, there is more chance for nitrate to percolate to the aquifer, increasing the nitrate level in groundwater [10–12]. In warmer countries, the loss of total nitrogen is more, as mineralization rate is probably higher due to the higher temperature; thus, the percolation of total nitrogen is increased [13].

The major proportion of the surplus nitrogen is transported by the runoff water to the streams, and consequently, nitrogen compounds such as ammonia-nitrogen, nitrite, and nitrate, are escalated in the streams. A surfeit of nitrogen in streams seems to be deleterious for both human beings and aquatic lives. In water bodies, it may lead to the magnification of aquatic plants and algae, which can result in the depletion of dissolved oxygen and hinder the contact of water with air and light. The presence of such excess nitrogen in drinking water reduces the amount of oxygen transported in the blood [5]. Mostly, treatment plants are not designed for the full removal of nitrogen compounds from river water. In China, sewage treatment systems remove total nitrogen by 40–70% [14,15]. In Malaysia, sewage treatment plants are not designed for ammonia removal [16]. Recently, several water treatment plants have been forced to shut down when, after testing the samples, it was found that ammonia-nitrogen pollution has crossed the acceptable limit in different rivers in Malaysia. The abrupt closure of the water treatment plant affects the water supply to the consumers; thus, adding additional pressure on the government for arranging an alternate source of water supply.

The lack of monitoring systems leads to an abrupt increase in pollution, which can result in the closure of the water treatment plants. Monitoring systems should contain a proper predictive system: which works based on the historical data; and a treatment system: that deals with the nitrogen pollutant, should be developed in treatment plants. Predictive systems could provide the daily data of pollutants and thus save the daily effort of quantifying such data in the laboratory. Moreover, predictive systems would create an alert for nitrogen surge in rivers before it actually happens. Hence, the government would have ample time to optimize various nitrogen inputs in the rivers. Different river basins require a separate predictive model, trained on historical data of the basin's parameters because a model well-trained on historical data of one particular basin, not necessarily will perform with the same accuracy on different basins. Hence, the government requires a separate predictive model for each basin. Additionally, to consider the upcoming seasonal changes, the predictive models need to be re-trained with the real-time data on a quarterly or yearly basis. Observing the increased pollution of nitrogen in rivers, this topic becomes important to be evaluated.

Artificial neural networks (ANN) models have been utilized for developing better-precision water quality predictive models [17–21]. The computational intelligence, among which ANN is one, has become a fast-evolving area [22]. The applications of ANN are not limited to water quality prediction. According to He, Oki, Sun, Komori, Kanae, Wang, Kim and Yamazaki [18], ANNs have been successfully used for reservoir operations [23–27], water resources management [28,29], and hydrological processes [30,31]. Application in water resources management includes river flow forecasting [32,33], rainfall-runoff modeling [31,34], and water quality predictions [35–38]. The present study is confined to water quality predictive systems only.

The primary objective of this study is to classify different types of ANN used for predicting nitrogen content in streams in different rivers all around the world. Furthermore, the states of different rivers in the world were also evaluated, resulting in the scope of future research work. This review paper also highlights the prediction accuracy and reliability, the parameters and methods used for prediction, and the details of ANNs of different models used for nitrogen prediction. This review paper will, surely, add some valuable points on the table for those researchers working for modeling using ANN, for those modeling for nitrogen compounds pollution and for those seeking information about nitrogen pollution level in water bodies. The articles cited in this review are those published in reputable journals.

2. Nitrogen Sources in Streams

Nitrogen is a vital element for plants, as it helps them in their growth and productivity. Nitrogen present as N_2 in the atmosphere cannot be utilized directly by plants until it is converted to its reactive compounds, such as NH_3, NH_4^+, NO_2^-, or NO_3^- [39]. This process is naturally done by bacteria present in the soil and in the root nodules of legume crops. Additionally, nitrogen compounds are provided to the soil in the form of fertilizers. Nitrate is the main constituent of fertilizers, but ammonia, ammonium, urea, and amines are also present in minor proportions. Nowadays, fertilizers contain more of a percentage of nitrogen compounds in order to boost the agricultural productivity.

In addition, the landscapes of the farmlands have been modified extensively. Farmlands are now designed to drain off the excess rainwater or irrigation water [40]. This drained water is rich in nitrogen compounds, which had been applied to the field for crop nourishment. The drained water then joins either running rivers or still water bodies such as lakes, leading to a surfeit of nitrogen entering the water system.

Sources of nitrogen to streams are not confined to agricultural fields. Industries and municipal and residential areas also contribute nitrogen compounds to streams. Comprehensively, the sources of nitrogen are classified into two:

a. Point Sources

A point source of nitrogen pollution is any single identifiable source of nitrogen pollution into rivers. Point sources include industries and municipal sewage treatment plants [15,41,42]. In urban areas, the contribution of nitrogen from point sources is dominant. Industries and municipal sewage treatment plants deliver more than 50% of the total nitrogen in rivers [39].

b. Non-point sources

Non-point sources are sources of nitrogen pollution whose specific locations of input to rivers are not defined. They mainly consist of agricultural fields and atmospheric and biological nitrogen fixation [15,41,42]. In rural areas, the contribution by non-point sources is dominant. In different regions of rural areas, different parts of non-point sources contribute major amounts of nitrogen in streams; for example, in farming regions, agricultural fields provide significant nitrogen to the streams, and in the regions of rivers surrounded by dense forests, atmospheric nitrogen deposition dominates [39].

3. Effects of Nitrogen

Nitrogen, if present in river water, causes different disorders, which are deleterious for both human and aquatic animals. Nitrogen present in streams are mainly found in three compound states: ammonia, nitrate, and nitrite. Some amounts of ammonia present in the river water get converted to nitrate depending on the dissolved oxygen concentration in the water [43]. As stated earlier, nitrate is not much deleterious, but if present in surplus amount, it starts converting into nitrite, which is very harmful even in minute concentration. The Environmental Protection Agency has set standards which state that for water which is to be distributed for public use, the maximum acceptable nitrate concentration is 10 mg/L [5,25] and that for nitrite is 1 mg/L.

There are two major effects of ammonia on the whole ecosystem: eutrophication of marine and terrestrial ecosystems [44,45] and increase in the acidity of water bodies [46]. Excessive nutrients such as nitrogen and phosphorus when present in water bodies lead to the growth of algae on the top surface of water; this process is termed as eutrophication. Excess grown algae cover the whole water surface, blocking the contact of water from sunlight and air. Additionally, the algae growth decreases the oxygen level in the water body, which affects the aquatic lives. Stream eutrophication was recognized as a major problem years ago, and the United States along with other countries commenced nutrient control measures in rivers [47,48].

Streams may get acidified due to the presence of surfeit ammonia. The most common form of ammonia, ammonium sulphate, leads to formation of a considerable amount of acid, as hydrogen ions are released during nitrification. Additionally, nitrite ions present in the streams lead to the formation of nitric acid under different situations along with sulfate ions, consequently acidifying the stream water [49]. Acidic stream water is not even suitable for reuse to satisfy human water requirements. As stated by Gündüz [50], one day, reuse of treated water would be a reality for the rural population, and this would result in serious problems such as human health issues. Compared with urban areas, agricultural areas are more susceptible to health risks by the presence of nitrate-nitrogen in groundwater [51,52].

Nitrite has been found to be more toxic than nitrate and if present in drinking water can cause human health problems such as liver damage and, in worst cases, can lead to various types of cancer [53] and two types of birth defects [54,55]. Nitrite present in surplus quantity in drinking water will eventually lower the ability of bloodstreams to carry oxygen, leading to the lack of oxygen in the body. Infants and young livestock are lamentably affected, as this causes "blue baby syndrome" [53]. The reaction of nitrites with amines either enzymatically or chemically leads to the formation of potent carcinogenic nitrosamines [53,56].

Consumption of nitrates leads to various tumors in the human body [53,57]. In the digestive system, nitrate leads to the formation of N-nitroso compounds [53,58], which are considered to be carcinogenic. Iodine uptakes can be restricted by nitrates, causing thyroid-related problems [53].

4. ANN

ANN is a black-box computational model [59] that contains interconnected network-like structures passing values to other nodes of the connections. It contains an input layer, hidden layers as required, and an output layer. It is well known for its capability of predicting the non-linear variables [60]. ANN forms the same structure as neurons in the human brain [6,20,61]. It functions like a biological neuron, receiving the input as stimulus, evaluating the stimulus, and then providing the output as the response to the stimulus. Figure 1 represents a simple example of the neural network. The inputs are fed to the nodes in the input layer, and those nodes pass the values of input data to the nodes in hidden layer 1 via interconnecting links. As the values are passed from input nodes to the following nodes, it is multiplied with the weights and then passed to the corresponding layer through a transfer function [62]. Likewise, it is passed up to the output layer, where the error is calculated using target vector. Based on this error, weights get adjusted to obtain the exact weighted combination of the input data for forecasting the target vector.

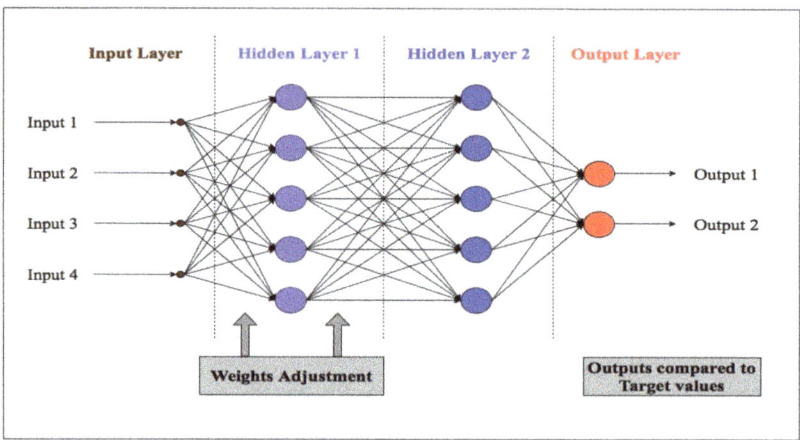

Figure 1. Basic structure of neural network.

The major advantage of application of the ANN model, over the traditional model, such as a statistical model, is that it learns itself the complexity of nature, without being explicitly transformed into mathematical form [63,64]. Statistical models have a limitation of assuming additional information to derive a sharp conclusion [65]. The major disadvantage of ANN is that it is susceptible to overfitting. Overfitting is the state in training, beyond which, training error decreases but the model starts losing its ability of generalizing the relation between input and output for the new data set i.e., the testing set data. This results in increasing the testing error and decreasing the overall performance of the model. There are several ways to prevent the model from overfitting, among which a well-known method is early-stopping; in which training process is stopped early. However, if the training is stopped too early then the model fails to learn important information. Hence, training should be stopped accordingly to learn all important information without overfitting.

Many types of ANNs feature different concepts of data processing. Each type is designed differently to obtain a more precise output with less data processing time. This is achieved by changing the network's architecture. According to Jain et al. [66], based on the network connection pattern, i.e., their architecture, ANN is classified into two categories:

a. Feed-Forward Neural Networks (FFNNs)

FFNN has the simplest network connection pattern in which data flow in the forward direction only, starting from the input layer to hidden layers, and then to the output layer. No loops are formed in the paths of the data flow. As shown in Figure 2, FFNN is classified into three subcomponents: single-layer perceptron, multilayer perceptron, and radial basis function neural network (RBFNN). Single-layer perceptron, which consists of one layer, i.e., the output layer, is the simplest form of neural network. It is mainly used for classifying the linearly separable cases that use binary targets. The connection patterns of multilayer perceptron and RBFNN are the same: an input layer, as many hidden layers as required, and an output layer. The only difference between these two is the use of the data processing function. Multilayer perceptron utilizes either threshold function or sigmoidal function [67] in each of its computational units, whereas RBFNN utilizes radial basis function as the activation function in each unit of its hidden layers. The Table 1 presents the advantages and disadvantages of different models of FFNN. These models are generally used for time series prediction, system control, and data classification.

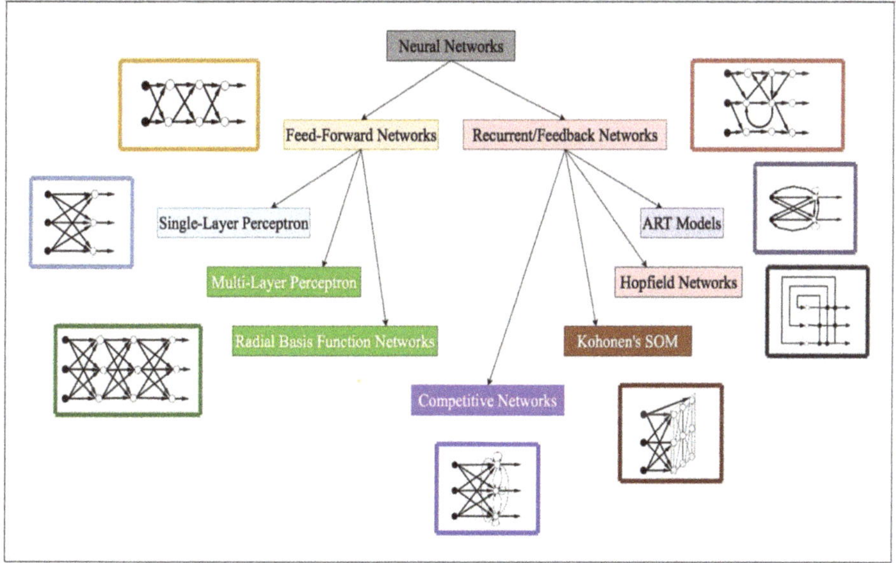

Figure 2. Classification of neural network [66].

Table 1. Advantages and disadvantages of different ANN (artificial neural network) models.

Main Type	Model Name	Advantages	Disadvantages
FFNN (Feed-Forward Neural Network)	Single Layer Perceptron	• Less computation—time • Easy to setup	• Can only be used in linearly separable data
	Multi-Layer Perceptron	• Can be used for complex problems	• Need more time for training • Can get stuck in local minima
	RBFNN	• Less susceptible to be stuck in local minima • Can tolerate with input noise	• Classification is slow, as the network have to calculate the radial basis function for each input vector during classification
Recurrent/Feedback Network	ART (Adaptive Resonance Theory) model	• Can be integrated with other models to enhance the performance	• Some ART models are inconsistent. They depend upon the order of the training data
	Hopfield Network	• No training needed	• Handles a smaller number of memories. • More number of patterns results in spurious output
	Kohonen's SOM	• Provides deterministic and reproducible results • Simplicity of computation	• Performance depends on initialization
	Competitive Network	• Groups the similar pattern based on the data correlation	• Susceptible to stability issue

b. Recurrent or Feedback Neural Networks

Recurrent or feedback neural networks experience the backward flow of data in some computational cells. The data flow is not unidirectional; loops within the cells transfer back the feedback of the errors encountered in computations, with reference to the target values. The feedback of errors helps in updating the weights of the corresponding inputs. As shown in Figure 2, feedback neural network is classified into four subcomponents: adaptive resonance theory model, Hopfield networks, Kohonen's networks, and competitive networks. Table 1 presents their advantages and disadvantages. These networks form very complex architectures, composed of a number of loops. These networks are utilized for complex computations, such as speech recognition, image processing, robotics, and process controls. This study is limited to the review of the FFNN.

5. Hybrid Model

Hybrid model is the combination of different models to solve a computational task. The need of hybridization aroused when the learning models were observed to be very efficient in some cases and inefficient in most of the cases [68]. The main aim of hybridization is to resolve the limitations of an individual model by fusion of decision making models with learning models [69]. The main advantage of a hybrid model is that it provides better results in comparison to the standalone model. The decision making model integrated in the hybrid model provides a good start with selected initial values of the internal parameters of learning models; hence, increasing the productivity of the learning model. The disadvantages of the hybrid models are: overall training process is time consuming, and complex architecture and training requires modern computational resources. Some of the examples of hybrid models are [70]:

- ANN and genetic algorithm
- ANN and fruit fly optimization algorithm
- ANN and firefly algorithm
- ANN and artificial immune systems

- ANN and particle swarm-optimization algorithm

6. Methods and Evaluation

This study is based on nitrogen compounds prediction in water bodies using ANN and other predictive models. In this study, in the section of 'Application of ANN', authors have first analyzed the sources of data collection, methods used, internal parameters of the predictive model, and then the final results of the previous research works in literature. On the basis of this analysis, authors have recommended various steps to be followed in future studies for achieving better accuracy models.

As used by [71], authors of this study have used relevant search engines such as Google Scholar and Science Direct. Additionally, the authors of [72] concluded, in their study, that Google Scholar is the most comprehensive source. While searching the relevant literature research works, the following keywords have been used: nitrogen compounds prediction, use of ANN in nitrogen prediction and nitrogen prediction in water bodies.

6.1. Nitrogen Monitoring

More than 60% of the world's rivers are affected by pollution [43], from point sources or non-point sources. Wastes generated by industrial, municipal, and agricultural activities are discharged into the rivers and pollute them [43,73]. Over time, human activities have escalated nitrogen species concentration in water bodies. Nitrate concentrations in many European rivers have surged by 5- to 10-fold since the 20th century [39]. In Malaysia, because of the excessive chemical pollution in rivers, more than one among the nine water treatment plants in Langat River basin has been closed several times between 2012 and 2015 [41]. According to Selangor Water Management Authority, Malaysia, between 2012 and 2015, the ammonia concentration level in the Langat River exceeded 7.0 mg/L, which led to the repeated closure of many water treatment plants during the period [41]. Moreover, in the Johor River basin, nearly five treatment plants were repeatedly closed between 2017 and 2019 due to the high concentration of ammonia in the Johor River [74–76].

There is no specific standard set for ammonia discharge in water bodies, but different agencies have provided separate guidelines for ammonia concentration in water bodies. "Canadian Water Quality Guidelines for the Protection of Aquatic Lives", [77] states that the guideline value for unionized ammonia discharge in freshwater is a concentration of 0.019 mg/L. The guidelines for drinking water quality (2003) published by WHO states that natural levels of ammonia in groundwater are usually below 0.2 mg/L, and this level may go up to 12 mg/L for surface waters.

For analyzing nitrate variations, Rekacewicz [76] designed a map, as shown in Figure 3, by considering all the river data at continental level, which represent the concentration of nitrate-nitrogen in streams at various locations around the world. Rekacewicz [76] compared the data of two decades and observed that rivers in North America and Europe were fairly stable, but those of south-central Asia and southeast Asia showed high nitrate concentrations.

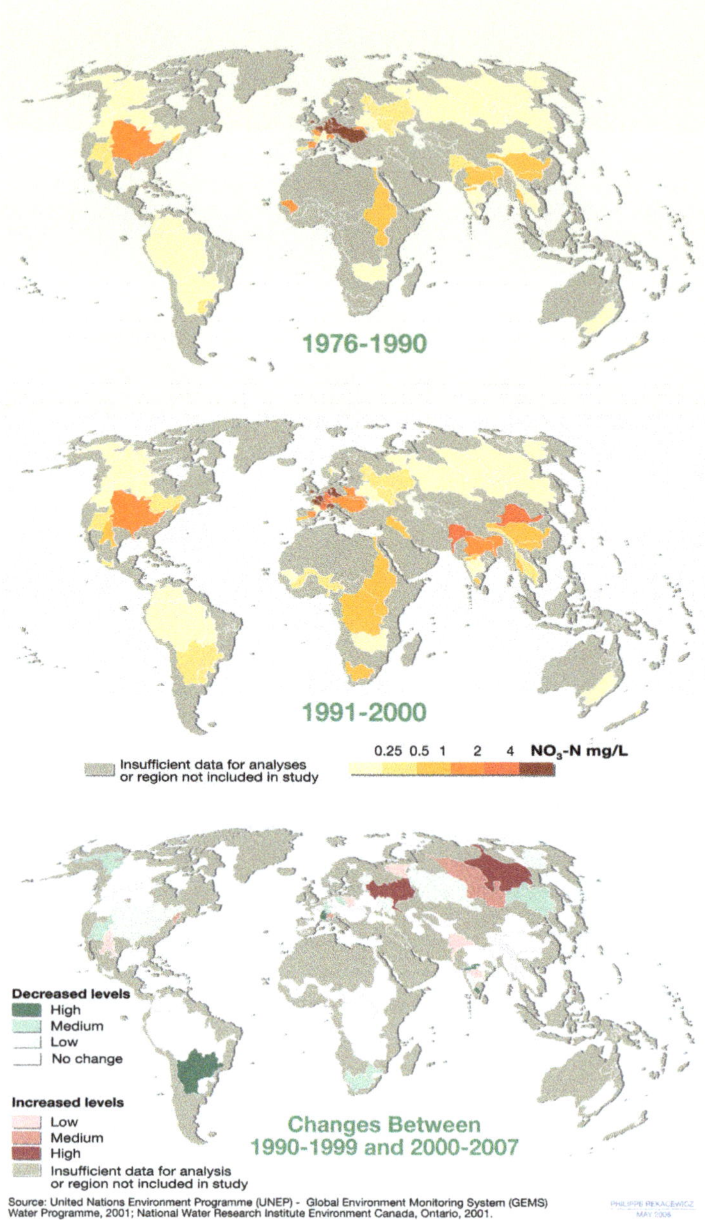

Figure 3. Comparison of nitrate-nitrogen in rivers for two decades of data [76].

Furthermore, Basheer et al. [78] studied the water quality of the Langat River in Malaysia. They utilized 10 samples from different locations to quantify different water quality parameters. Their results showed that the pH range for the Langat River was between 5.91 and 6.79. The average value of ammonia for the Langat River was measured to be 0.24 mg/L. The total ammonia-nitrogen amounts added to the Langat River from point and non-point sources were calculated to be 9.51 ton/day and 12.67 ton/day, respectively [41,79], as displayed in Figure 4.

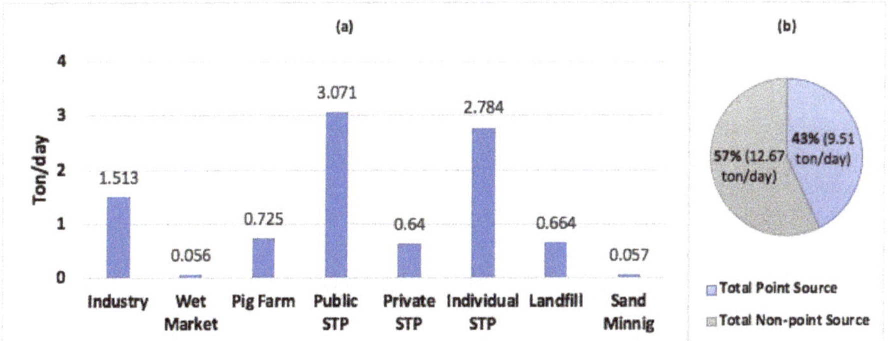

Figure 4. (a) Classification of different point sources showing their contribution of ammonia-nitrogen in the Langat River; (b) comparison of the contributions of ammonia-nitrogen from point and non-point sources [41,79].

Moreover, Zhang, Swaney, Li, Hong, Howarth and Ding [15] tried to calculate nitrogen input to the Huai River in China from anthropogenic point and non-point sources, and also the impact of nitrogen discharge on the riverine ammonia-nitrogen flux. They used the data from Yan et al. [80], which stated that the average nitrogen concentration in the sewage discharged from industries in the Changjiang River basin was 25 mg/L. From the previous studies, they could conclude that ammonia-nitrogen in the river was about 10% (or less) of the total nitrogen [15,81,82], and it could be as high as 70% in heavily polluted Asian rivers in the urban areas [15,83,84]. They used the data of Zhang et al. [85], which suggested that nitrate had become a major constituent of riverine nitrogen flux; the data was obtained from measurement in 2008, at several stations in the Huai River basin; the values of riverine nitrate concentration was found to vary between 0 and 15.7 mg/L nitrate-nitrogen, with a mean of 2.1 mg/L nitrate-nitrogen. When the authors of [15] measured the ammonia-nitrogen in the same river basin, they found that the average ammonia-nitrogen concentration varied between 0.2 and 3.3 mg/L N, with an average of 1 mg/L N, which was half of the average nitrate-nitrogen concentration measured in 2008. The calculation of nitrogen input to the Huai River showed that on average, 27200 ± 1100 kg km^{-2}y^{-1} of nitrogen was added to the river from 2003 to 2010 as the net anthropogenic nitrogen input.

6.2. Application of ANN

ANNs have been extensively used worldwide in the past as a predictive model for nitrogen prediction in streams. Table 2 lists studies on the use of ANN by various authors. Various authors had utilized different methodology, as shown in Table 3. For nitrogen prediction, ANN was utilized, for the first time, probably by Lek, Guiresse and Giraudel [20]. They used ANN to predict inorganic and total nitrogen concentration in streams using eight input parameters from the catchments along with the historical data of inorganic and total nitrogen. The input database was obtained from U.S. National Eutrophication Survey (NES); which had many variables in record but according to the scope of the research (prediction of stream nitrogen concentration), the following eight variables were included: average annual flow; animal unit density; mean annual streamflow; the percentages of forest cover, wetland, urban areas, and agriculture areas; and the percentages of the remaining area in the catchment. Sensitivity analysis showed five different types of variation in total nitrogen concentration and three different types of variation in inorganic nitrogen concentration. The sensitivity types (or contribution) for total nitrogen concentration are: (i) Increasing sigmoid contribution: wetland and animal unit density. Low values of these independent variables lead to low (minimum) value of total nitrogen; which then enhances to reach its maximum value with the independent variable. (ii) Weakly growing contribution: agricultural areas. For low values of agricultural areas, the total nitrogen is less and likewise increasing gradually. (iii) Decreasing contribution: average annual flow and percentage of remaining area. (iv) Gaussian: Urban areas. (v) Weak contribution: percentage of forest cover. For inorganic nitrogen: (i) Growing contribution: urban and agricultural areas. For low values of urban and agricultural areas, inorganic nitrogen concentration is less and then rapidly increases with these independent variables. (ii) Gaussian: percentage of wetland areas. (iii) Decreasing contribution: Percentage of forest cover, animal unit density and remaining area. Forest cover rapidly and constantly decreases the inorganic nitrogen concentration. The other two independent variables also reduce the inorganic nitrogen concentration but at low levels only. Input variables were auto-scaled by centered and reduced variables. Autoscaling reduces the chance of domination of any one particular input variable over the prediction. This input database was divided into a training and independent testing set (two thirds and one third of the total database, respectively). Using data from 927 sites from different parts of the United States, Lek, Guiresse and Giraudel [20] developed a multilayer feed-forward ANN model having 10 neurons and 1 hidden layer, with a correlation coefficient of 0.82 for total nitrogen concentration and 0.8 for inorganic nitrogen concentration. Examining the results obtained, they concluded that the urban areas produced most of the inorganic nitrogen, and animal husbandry contributed the most to the total nitrogen concentration in streams. It was assumed that fertilizers were used in less quantities as its contribution was less in stream nitrogen. Forest cover lowered the inorganic nitrogen concentration in streams and has less effect on total nitrogen concentration. Percentage of wetland areas helped in reducing the inorganic nitrogen in streams, but they increased the total nitrogen.

Table 2. A summary of studies that utilize ANN model for nitrogen prediction, including their specific area, location, and methods used.

	Authors	Specific Area	Location	Method
1	Anctil, Filion and Tournebize [61]	Streams	Melarchez, France	Stacked multilayer perceptron
2	He, Oki, Sun, Komori, Kanae, Wang, Kim and Yamazaki [18]	Streams	Japan	Feed-forward model
3	Holmberg, Forsius, Starr and Huttunen [19]	Streams	Finland	Backpropagation algorithm
4	Lek, Guiresse and Giraudel [20]	Streams	The United States	Multilayer feed-forward
5	Suen and Eheart [25]	Streams	Illinois, The United States	Backpropagation and radial basis
6	Sharma, Negi, Rudra and Yang [6]	Drainage water	Canada	Fast backpropagation and self-organizing radial basis
7	Wang et al. [86]	Groundwater	Australia	13 machine learning models
8	Zhang et al. [87]	Lake	China	ARIMA, radial basis, and hybrid
9	Markus et al. [88]	Streams	Illinois	Backpropagation, Evolutionary Polynomial Regression (EPR), and Naïve Bayes Model (NBM)
10	Amiri and Nakane [89]	Stream	Japan	Backpropagation and Multiple Linear Regression (MLR)
11	Zeleňáková et al. [90]	Streams	Slovakia	Dimensional analysis

Table 3. Details of methodology of the reviewed research work.

No.	Authors	Duration of Data	Data Pre-Processing	Internal Parameters
1	Anctil, Filion and Tournebize [61]	1975–1993 (Daily)	Standardization (linearly)	2 Inputs, 12 hidden neurons
2	He, Oki, Sun, Komori, Kanae, Wang, Kim and Yamazaki [18]	1995 (Monthly)	Sensitivity Analysis	8 Inputs, 7 hidden neurons
3	Holmberg, Forsius, Starr and Huttunen [19]	1990–2000 (Daily)	-	13 Inputs, 1 hidden layer, 7 nodes
4	Lek, Guiresse and Giraudel [20]	One year	Sensitivity Analysis, Autoscaling	8 Input, 10 hidden neurons
5	Suen and Eheart [25]	1993–2000, (Daily)	-	-
6	Sharma, Negi, Rudra and Yang [6]	1991–1994, (Daily)	Sensitivity Analysis	**Fast Backpropagation:** 8 Inputs, 20 hidden neurons, learning rate: 0.02 **RBFNN:** Tolerance 20, spread 15
7	Wang, Oldham and Hipsey [86]	2006–2014, (401 samples)	-	-
8	Zhang, Zhang and Li [87]	2006–2011, (Monthly)	-	**ARIMA:** • Nitrogen: p = 1, d = 1, q = 1 • Phosphorus: p = 2, d = 1, q = 1 **RBFNN:** • 2 hidden layers • Training width $\sigma = 0.6$
9	Markus, Hejazi, Bajcsy, Giustolisi and Savic [88]	1994–1999, (Weekly)	-	**ANN:** 4 Input, 2 hidden nodes, epochs: 100,000; performance gradient: 1E-10; goal: zero **EPR equations:** $N_{t+1} = 0.827 N_t$ $N_{t+1} = 0.659 N_t + 0.560 N_t \sqrt{Q_t}$ **NBM equations:** $N_{t+1} = f[N_t, Q_t, P_t, T_t]$ $N_{t+1} = f[N_t, Q_t, Q_{t-1}, P_t, P_{t-1}, T_t, T_{t-1}]$
10	Amiri and Nakane [89]	2001, (Monthly)	Statistical Analysis	6 Input nodes, 2 hidden nodes, 1 output nodes, 11,600 epochs
11	Zeleňáková, Čarnogurská, Šlezingr and Słyś [90]	2003–2010, (Monthly)	Sensitivity Analysis	**Dimensional analysis equations:** $\pi_1 = 0.0039 \pi_2^{13.805}$ $\pi_1 = 0.1868 \pi_2^{9.7892}$

The condition of the United States seemed to be critical in terms of nitrogen in streams, as four years after the study by Lek, Guiresse and Giraudel [20], a research work published by Suen and Eheart [25] stated that nitrate has become an important problem. They conducted a study in the Upper Sangamon River, Illinois, and pointed out the use of chemical fertilizers in agriculture to be responsible for the high nitrate concentration in streams. In their study, they developed two models, RBFNN and backpropagation neural network (BPNN), and compared the models on the basis of accuracy. The parameters used for modeling were daily highest temperature, seven-day cumulative daily rainfall, daily streamflow, and Julian date. To include the common practice of fertilizer application, Julian date was used as an input parameter to the model. They used a dataset of eight years, i.e., 1993–2000. To divide the dataset into the training set and testing set, two methods were adopted. In the first method, data from 1993 to 1996 were used as the training dataset and the remaining were used for testing. For the second method, the data of odd years (i.e., 1993, 1995, 1997, and 1999) were used for training, and those of even years were used for testing. Comparing the results obtained from the models, they concluded that the odd-even years method proved to be more accurate. The overall accuracy of the first method was obtained to be 0.784 and 0.752 for BPNN and RBFNN, respectively, and that of the second method was 0.832 for both the networks. Neural network models predicted with greater precision when tested for Boolean output considering the second method. The network signaled 1 when the nitrate concentration exceeded 10 mg/L and 0 when the nitrate concentration was below 10 mg/L. Considering Boolean output, they concluded that RBFNN had a higher accuracy (0.893) than BPNN (0.866).

In 2003, a research work published in Canada by Sharma, Negi, Rudra and Yang [6] stated that subsurface waters in Canada were being polluted by the nitrate from the fertilizers used in agricultural fields. Their experimental site was a field, of area 14 ha, located at the Greenbelt Research Farm of Agriculture and Agri-Food Canada, near Ottawa. The authors proposed a neural network model to assist in optimizing the use of fertilizers. The input database was collected from the experimental field for the period of 1991–1994, except for the temperature and precipitation data. Data of these two variables were collected at the station of Agriculture and Agri-Food Canada, located 12 km from the site. Two neural network models, fast BPNN and self-organizing RBFNN, were examined, aiming to select the superior network. Inputs to the model used were treatment (tillage or no tillage, i.e., whether the land was prepared or not), Julian day, rainfall per day, cumulative rainfall, total nitrogen applied, snowfall per day, and maximum and minimum temperature. Sensitivity analysis was performed to determine the optimum internal parameters of both the networks. The input data were divided into two sets: training and testing set. Training set consisted of eight input variables and two output, and the testing set consisted of only the unexposed inputs from the replicate plots. For fast BPNN, the parameters varied for sensitivity analysis were learning rate and number of hidden neurons. This analysis comprised of two stages: First stage was to keep the number of hidden neurons constant at 20 and vary the learning rate from 0.02 to 0.08. Analysis of the fluctuation of error on every variation led to the selection of optimum learning rate as 0.02. In the second stage, learning rate was kept constant to 0.02 and number of hidden neurons were varied from 5 to 25. Analyzing the similar way, optimum number of hidden neurons were selected as 20. Similarly, sensitivity analysis was performed for RBFNN, in two stages, by varying the tolerance and spread values from 5 to 20 and 1 to 20, respectively. The selected optimum value for tolerance and spread values were 20 and 15, respectively. Using these parameter values, both the models were further trained. Comparing the results of both networks, the authors concluded that the self-organizing RBFNN, with a correlation coefficient of 0.8079 for conventional tillage and 0.6911 for no tillage, outperformed the fast BPNN, with a correlation coefficient of 0.8017 for conventional tillage and 0.6635 for no tillage, for nitrate-nitrogen concentration prediction in drainage water.

Holmberg, Forsius, Starr and Huttunen [19], predicted the future data of total organic carbon, total nitrogen, and total phosphorus in streams, considering the climate change effect and utilizing the data of three streams (Kelopuro, Hietapuro and Valkea-Kotinen) located in two catchments of the same name (Hietajärvi) in Finland. They developed a BPNN model employing the database of 13 input variables: month of data sampling, mean temperatures of 3 and 10 preceding days, runoff of sampling day, maximum and minimum runoffs of 3 preceding days, days of peak flow, days of low flow, catchment area, fractions of lake area and peatland area with respect to catchment area, catchment latitude, and elevation. This database was collected from the catchment, except for the daily temperature and precipitation, which was collected from the nearby Finnish Meteorological Institute weather station, Lammi, from 1990 to 2000. Samples of these variables were divided into two sets: training set and testing set. The samples were allocated into these sets by random choosing, provided it was ensured that the highest and lowest 10-percentile data were included in both the sets. While training, they were to test all the possible set of models with the available inputs, hence, they varied the number of inputs from 2 to 16, fixing the number of hidden layer to 1 and the neurons in the hidden layer were set as the integer part of (1 + number of inputs)/2. Training 10 sessions for each combination, resultant models were analyzed on the basis of their efficiency. The model resulted the best efficiency with 13 input variables and 1 hidden layer with 7 nodes, having the values of flux efficiencies of total organic carbon, total nitrogen, and total phosphorus as 0.94, 0.92, and 0.90, respectively. Using this model, they forecasted the total nitrogen data until 2050. They stated that if there is a low change in climate, then the total nitrogen flux will be near the value in 2005, but for a scenario of high change in climate, the nitrogen flux will increase by 26%, with respect of the value in 2005.

Similar conditions have been stimulated in Melarchez, a catchment near Paris, France, where Anctil, Filion and Tournebize [61] investigated an agricultural catchment area to develop a neural network model for predicting the nitrate-nitrogen flux. Considering the soil moisture at different depths as the input parameter, the authors analyzed its effect on the nitrate-nitrogen flux. They developed a stacked multilayer perceptron model focusing mainly on the selection of best performing model among the list of models developed, based on different combinations of input variables and neurons in hidden layers. Fifty models were trained for each combination of inputs and neurons in hidden layers. Neurons in hidden layers were varied from 2 to 20. Every issue was tested discretely to make the final decision on the basis of the model accuracy. They had 12 different options for the input parameter: same-day stream flow, previous-day stream flow, increment in the flow from the previous day, same-day precipitation, previous-day precipitation, same-day historical mean flux, increment in the historical mean flux from the previous day, same-day 10 cm-, 20 cm-, 40 cm-, 80 cm-, and 120 cm-depth soil moisture indices. These input variables were collected from the gauge station for the period of 1975 to 1993. Since the important step, in pre-processing of data, is standardization [91], all the input variables were ensured to be on the same scale by standardizing them linearly such that their standard deviation as 1 and mean as 0. After optimizing, the final model had 2 input parameters (same-day stream flow and same-day 80 cm-depth soil moisture index), 12 neurons in hidden layers, and Levenberg-Marquardt with Bayesian regulation as the calibration procedure, which performed well with an efficiency index of 0.888. The utilization of soil moisture content at different depths revealed that the soil moisture also had an effect on nitrate nitrogen flux generated from the agricultural field.

Since a large number of input variables are available to decide for the neural network, these inputs should be chosen using sensitivity analysis [92]. Numerous authors have provided models with different sets of input parameters, which according to them, were suitable for their models (Table 4). He, Oki, Sun, Komori, Kanae, Wang, Kim and Yamazaki [18] investigated 59 river basins all over Japan and developed an FFNN to predict the monthly total nitrogen concentrations in streams. They had to choose the most important independent input variables from a set of 16 input variables: the area of each basin, amount of fertilizer applied in each basin, average temperature, precipitation, sunshine duration and river discharge of each basin, ratio of paddy area, farmland area, forest area, bare land area, urban area, road area, river area, lake area, seashore area, and other land areas in the total basin area. This input database was collected from different sources. The land use variables were collected from Ministry of Land, Infrastructure, Transport and Tourism (MLIT land use database), a digital database in Japan. Total nitrogen concentration was collected from MLIT water information system. Sunshine duration, precipitation and temperature data were obtained from Automated Meteorological Data Acquisition System. The input data were divided into three subsets: Training, overfitting test and validation subsets. Among the data of 59 river basins, 40 river basin data were used for training and overfitting test (80% and 20%, respectively). The remaining 19 river basin data were never exposed to the network for training and were used for validation only. FFNN was trained with backpropagation algorithm with different combinations of input variables and internal parameters: input variables were varied from 7 to 9, number of hidden layers was fixed to 1 with number of neurons in it fixed to 7 and 8. Analyzing the results of all the trained network on the basis of coefficient of regression, the authors found that the model with 8 input variables (river discharge, average temperature and precipitation of each basin, amount of fertilizer applied in each basin, the proportions of forest land area, urban land area, road area, and other areas in the total basin area) and one hidden layer with seven nodes provided the best accuracy with R^2 for training as 0.96, R^2 for validation as 0.84, and R^2 for overfitting as 0.90.

Table 4. A summary of studies that utilize ANN model for nitrogen prediction, including their input variables, prediction variables, and accuracy.

No.	Authors	Input Variables	Prediction Variables	Accuracy
1	Anctil, Filion and Tournebize [61]	• Same-day stream flow • Same-day 80 cm-depth soil moisture index	Nitrate-nitrogen flux	Efficiency index = 0.888
2	He, Oki, Sun, Komori, Kanae, Wang, Kim and Yamazaki [18]	• River discharge • Average temperature and precipitation of each basin • Amount of fertilizer applied in each basin • Proportions of forest land area, urban land area, road area, and other areas in the total basin area	Monthly total nitrogen concentrations	$R^2_{training} = 0.96$ $R^2_{Validation} = 0.84$ $R^2_{Overfitting} = 0.9$
3	Holmberg, Forsius, Starr and Huttunen [19]	• Month of data sampling • Mean temperatures of 3 and 10 preceding days • Runoff of sampling day • Maximum and minimum runoffs of 3 preceding days • Days of peak flow, days of low flow • Catchment area • Fractions of lake area and peatland area with respect to catchment area • Catchment latitude and elevation	• Total organic carbon • Total nitrogen • Total phosphorus	**Flux efficiency:** • Total organic carbon = 0.94 • Total nitrogen = 0.92 • Total phosphorus = 0.90
4	Lek, Guiresse and Giraudel [20]	• Average annual flow • Animal unit density • Mean annual streamflow • Percentage of forest cover, wetland, urban, agriculture and the percentage of remaining area in the catchment	Inorganic and total nitrogen concentration	**Correlation coefficient:** Total nitrogen = 0.82 Inorganic nitrogen = 0.8

Table 4. *Cont.*

No.	Authors	Input Variables	Prediction Variables	Accuracy
5	Suen and Eheart [25]	• Daily highest temperature • Seven-day cumulative daily rainfall • Daily streamflow • Julian date	Nitrate concentration	Overall accuracy: • **Method one:** • BPNN = 0.784 • RBFNN = 0.752 • **Method two:** • BPNN = 0.832 • RBFNN = 0.832 • **Boolean output (Method two)** • BPNN = 0.866 • RBFNN = 0.893
6	Sharma, Negi, Rudra and Yang [6]	• Treatment • Julian day • Rainfall per day • Cumulative rainfall • Total nitrogen applied • Snowfall per day • Maximum and minimum temperature	Nitrate concentration	**Correlation coefficient** • **RBFNN** • Tillage = 0.8079 • No tillage = 0.6911 • **BPNN** • Tillage = 0.8017 • No tillage = 0.6635

Sustainability **2020**, *12*, 4359

Table 4. *Cont.*

No.	Authors	Input Variables	Prediction Variables	Accuracy
7	Wang, Oldham and Hipsey [86]	• **Scenario 1** • Nutrients (dissolved organic nitrogen (DON), total nitrogen, NH_4^+, NO_x^-) • Landscape (vegetation, land use, and soil) • Hydrological conditions (surface water subarea, groundwater subarea, and catchment area) • Sampling condition (temperature, sample depth, and sampling date, pH) • **Scenario 2** • Total nitrogen • All other non-nutrient data	DON	R^2 of best models: • **Scenario 1** • Cubist = 0.897 • Bagged multivariate adaptive regression spline (Bagged mars) = 0.882 • Random forest (RF) = 0.856 • **Scenario 2** • Cubist = 0.849 • Bagged mars = 0.887 • RF = 0.858
8	Zhang, Zhang and Li [87]	Monthly data for total nitrogen	• Monthly total nitrogen • Monthly total phosphorus	**Mean absolute percentage error:** • **Nitrogen** • ARIMA = 18.194% • RBFNN = 34.633% • Hybrid = 7.017% • **Phosphorus** • ARIMA = 27.299% • RBFNN = 126.957% • Hybrid = 14.528%

Table 4. *Cont.*

No.	Authors	Input Variables	Prediction Variables	Accuracy
9	Markus, Hejazi, Bajcsy, Giustolisi and Savic [88]	• Observed weekly river discharge • Precipitation • Air temperature • Nitrate-nitrogen concentration	• Weekly nitrate-nitrogen	**Root mean square error (RMSE) for ANN:** • Training = 0.787 mg/L • Testing = 0.935 mg/L **RMSE for EPR:** • Training = 0.991 mg/L • Testing = 1.010 mg/L **Critical success index for NBM:** • NBM1: • Training = 0.214 • Testing = 0.200 • NBM2: • Training = 0.286 • Testing = 0.188
10	Amiri and Nakane [89]	• Percentage land use • Urban • Forest • Agriculture • Grassland • Water body • Population density	Total nitrogen	**R^2 Value:** • BPNN = 0.94 • MLR = 0.85
11	Zeleňáková, Čarnogurská, Šlezingr and Słyś [90]	• Stream discharge • Catchment area • Stream velocity • Temperature of air and water • Concentration of pollutant	Nitrogen and phosphorus concentration	**Average Uncertainty:** • Nitrogen = 31.33% • Phosphorus = 32.30%

In addition to ANN, other machine learning methods can also be used to predict nonlinear environmental variables. Wang, Oldham and Hipsey [86] compared 13 machine learning models, including ANN, on the basis of precision in the prediction of DON (dissolved organic nitrogen) in groundwater in urban areas in southwestern Australia. These 13 machine learning models are classified into five different groups: (1) tree-based and rule-based model (generalized busted model (GBM), RF (Random Forest), conditional inference random forest (cforest), and cubist); (2) kernel-based machine learning model (Gaussian process with radial basis function kernel (GPR), Gaussian process with linear kernel (GPL), support vector machine with radial basis function kernel (SVMR), and support vector machine with linear kernel (SVML)); (3) generalized stepwise linear regression models (bagged mars, multivariate adaptive regression spline (mars), and generalized linear model with stepwise feature selection (GLM)); (4) instance-based model (k-nearest neighbors (KNNs)); and (5) ANNs. Using 401 groundwater samples (60% for training and 40% for testing), the models were examined based on two scenarios: (1) to train the models with all the data available such as nutrients (DON, total nitrogen, NH_4^+, and NO_x^-), landscape (vegetation, land use, and soil), hydrological conditions (surface water subarea, groundwater subarea, and catchment area), and sampling conditions (temperature, sample depth, sampling date, and pH); (2) to train the models with only total nitrogen and all other non-nutrient data. Database of nutrients were obtained from the Western Australian Department of Water for the period of 2006-2014. ArcGIS spatial mapping feature provided the data of soil type, land use and vegetation type. These models were analyzed on the basis of their RMSE and R^2 values and compared with the manually calculated DON (DONcal) (Figure 5). Analysis of all the results revealed that scenario 1 produced lower errors in models than scenario 2, stating that nutrients can improve the performance of models. Among the 13 tested models, 3 models showed higher R^2 value. For scenarios 1 and 2, the cubist model had R^2 values of 0.897 and 0.849; bagged mars, 0.882 and 0.887; random forest, 0.856 and 0.858; and ANN, about 0.72 and 0.65, respectively.

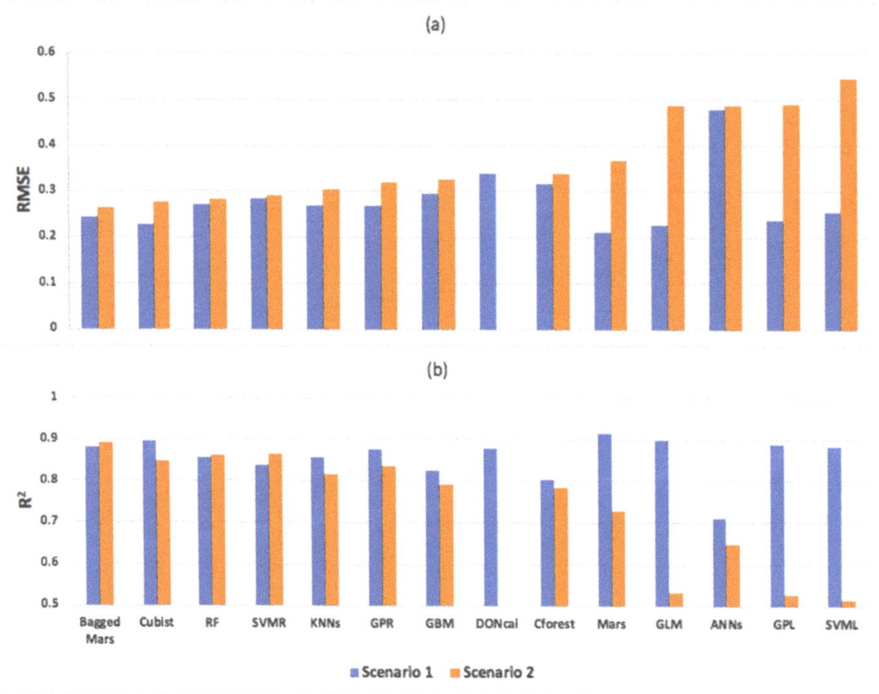

Figure 5. Comparison of 13 different models results with the DONcal.

Zhang, Zhang and Li [87] compared ARIMA model, RBFNN model, and hybrid ARIMA-RBFNN model based on the analysis and prediction of water quality in Chagan Lake, China. Database of water quality was collected from "The Second Songhua River Diversion Project Record" from the Chinese Academy of Science. The water quality parameters utilized for analysis were monthly total nitrogen and total phosphorus for the period of 2006–2011. The parameters of ARIMA model for total nitrogen were p = 1, d = 1 and q = 1 and for total phosphorus were p = 2, d = 1 and q = 1. Water quality data from 2006 to 2010 were used for training and the trained model was used for prediction of water quality data of 2011. The width of training, σ, was 0.6 for RBFNN model with 2 nodes in hidden layers. ARIMA-predicted values were linearly super-positioned with RBFNN-derived ARIMA residual prediction values to generate the hybrid ARIMA-RBFNN model. These models were analyzed on the basis of their RMSE and mean absolute percentage error. Results showed that RBFNN model had bad prediction results for total phosphorus; though, this model had learned the pattern of total nitrogen, but the predicted values were not satisfactory. Although ARIMA model did not have high prediction accuracy, it had successfully learned various trends for both total nitrogen and total phosphorus. Analyzing the results obtained, the mean absolute percentage error for the monthly total nitrogen was 18,194%, 34,633%, and 7017% for ARIMA, RBFNN, and hybrid ARIMA-RBFNN, respectively, and the mean absolute percentage error for the monthly total phosphorus was 27,299%, 126,957%, and 14,528% for ARIMA, RBFNN, and hybrid ARIMA-RBFNN, respectively. Following the results, it was stated that hybrid models had more capacity in predicting nonlinear variables.

Markus, Hejazi, Bajcsy, Giustolisi and Savic [88] developed three models—BPNN, EPR and NBM—for predicting weekly nitrate-nitrogen in a small agricultural watershed in Illinois. For the ANN part, the authors utilized observed weekly river discharge, precipitation, air temperature, and nitrate-nitrogen concentration as input variables. The study used the historical data of nitrate-nitrogen concentration and was collected from the Upper Sangamon River near Decatur for the period of 1994-1999. Employing half of the data for training and the other half for testing, they predicted the weekly data of nitrate-nitrogen in streams. The input selection was performed on the basis of trial and error with two sets of variables and their time lags. The first set consisted of four variables: N_t, Q_t, T_t, P_t; and the second set consisted of four variables and three time lags N_t, Q_t, T_t, P_t, Q_{t-1}, T_{t-1}, P_{t-1}. The first set predicted better results and hence was used for ANN modeling. ERP model has the capability of selecting the input subset, hence it is fed with the larger input set, the second set. In case of NBM, both the sets were used for modeling. For modeling in the ANN part, the internal parameters selected were: epochs: 100,000; performance gradient: 1E-10; goal: 0; number of hidden nodes: 1, 2, 3, 4 and 5; input variables: 4 (air temperature, discharge, nitrate-N concentration and precipitation) and output variable: 1 (next week nitrate-N concentration). The results indicated that the ANN with 2 nodes showed more accurate results in terms of RMSE as 0.787 mg/L and 0.935 mg/L for training and testing, respectively. For EPR, two models (EPR1 and EPR2) were generated which had their equations as: $N_{t+1} = 0.827N_t$ and $N_{t+1} = 0.659N_t + 0.560N_t \sqrt{Q_t}$, respectively. The RMSE obtained for EPR1 was 1.092 mg/L for training and 1.170 mg/L for testing. The RMSE obtained for the EPR2 was more accurate: 0.991 mg/L and 1.010 mg/L for training and testing, respectively. The NBM model utilized two categories: high and low values for variables. Each variable, except for nitrate-N concentration, had its categories divided by the average values as threshold. For nitrate-N concentration, the separation point was the emergency cutoff level (8.5 mg/L). NBM1 and NBM2 were the two models tested with the equations as: $N_{t+1} = f[N_t, Q_t, P_t, T_t]$ and $N_{t+1} = f[N_t, Q_t, Q_{t-1}, P_t, P_{t-1}, T_t, T_{t-1}]$, respectively. The results of these models indicated that, for low concentration, NBM1 had accurately predicted 79 of 80 concentrations, but for high concentrations, the prediction rate was 2 of 9. For NBM2, the predicted high flows (10) were somewhat similar to the observed ones (9). However, the false alarm rate for NBM2 was higher (7) than NBM1 (1). The critical success index for NBM1 was obtained as 0.214 and 0.200 for training and testing, respectively, and that for NBM2 was 0.286 and 0.188 for training and testing, respectively. The authors concluded that none of these models can be considered superior based on this analysis criteria, hence, suggesting a multi-tool approach. In their previous study,

Markus et al. [93] compared the ANN model and linear regression model to calculate the uncertainty in forecasting the weekly nitrate-nitrogen in the Sangamon River, Illinois. They stated that the ANN model was more accurate than the linear regression model. The ANN model surpassed the linear regression model by 3.30% and 4.42% of RMSE in testing and training phases, respectively.

Amiri and Nakane [89] compared BPNN and MLR on the basis of the total nitrogen prediction in streams. The study was conducted in the Chugoku district of Japan, which contains 21 river basins. Total nitrogen database, for year 2001, was collected from prefecture offices from Okayama, Shimane, Hiroshima, Tottori and Yamaguchi. Six input variables were used for the prediction, which included five variables for land cover percentage (urban area, forest area, agriculture area, grassland, and water body) and the last variable for population density. The total nitrogen was predicted by utilizing 60% of the data for training, 25% for controlling, and the remaining 15% for testing. BPNN consisted of six input nodes for the corresponding six input variables, one hidden layer and one node in output layer for total nitrogen prediction. The optimum number of nodes in hidden layer were selected by varying the nodes from 0 to 13 and training the network 5 times for each variation and evaluating them on the basis of correlation coefficient. The selected optimum BPNN had the following internal parameters: input nodes: 6, hidden layer: 1, hidden layer node: 2, output node: 1, epochs: 11, 600. MLR model had the same inputs as for the BPNN. For MLR modeling, a normality test was conducted for total nitrogen and land cover data using Sharpio-Wilk test having p-value less than 0.05. Models were analyzed on the basis of regression statistics and coefficient of the model (if the resultant was normally distributed). Final regression model was developed by using backward approach. The goodness of fit of the models was evaluated by regression of observed versus predicted and scatter plot. Comparison of the results for both the models showed that the backpropagation model ($R^2 = 0.94$) predicted the results more precisely than the multiple regression model ($R^2 = 0.85$)

Zeleňáková, Čarnogurská, Šlezingr and Słyś [90] predicted nitrogen and phosphorus concentrations in river Laborec in Slovakia, employing dimensional analysis method. They used Buckingham theorem to develop a prediction model utilizing important variables such as stream discharge, area of catchment, stream velocity, temperatures of air and water, and pollutant concentration. The equation established for nitrogen concentration was: $\pi_1 = 0.0039\pi_2^{13.805}$ and for phosphorus was: $\pi_1 = 0.1868\pi_2^{9.7892}$. These models were tested for the data of eight years (2003–2010); which was collected from Slovak Hydrometeorological Institute and Slovakian Water Management Company in Košice. Sensitivity analysis of the model stated that air and water temperature have major influence on the prediction of concentration of nitrogen and phosphorus. Velocity and flow of water have less influence and the catchment area has no influence on the prediction. By exploring the results of the model, it was found that the model equations calculated the prediction values with an average uncertainty of 31.33% for nitrogen and 32.30% for phosphorus.

7. Recommendation for Future Works

The precision of the predictive ANN model relies on many factors such as the amount of input data provided to the model for training and testing, relevant input variables, and different types of ANN methods used in the model. Based on the reviewed research works, we suggest some techniques to improve the accuracy of the nitrogen predicting model and also to account for a large range of inputs.

a)   Being the first step of modeling, the training is the most important part of the modeling procedure. Various kinds of important information are provided to the model during training. The model learns different patterns in the input data. Weights are updated during training [94]. Providing ample data for training can lead to better precision of the model. Input data is divided into three sets: training, testing and validation sets [95], and sometimes divided into two sets: training and testing set, depending on the model. Training set is used for updating the weights and biases of the model. Validation set is used for preventing the model from overfitting. While training, if the validation accuracy is decreasing, then the model seems to be overfitting and the training should be stopped. Testing set is used for testing the output of the model in order to confirm the accuracy of the model. These sets are divided on certain percentage of input data, either provided by user or divided, by default, by the model. By default, ANN modeling software uses 70% of the input data as the training data, which may be less for getting higher accuracy, 15% for validation and the remaining 15% for testing. In order to increase the accuracy of the model, we suggest using a higher percentage of data for training, i.e., about 80% to 90%. The remaining is to be divided equally for validation and testing. While dividing the input data into the training, validation and testing set, it should be ensured that these sets are statistically similar. In order to increase the learning capacity of the model, it should be ensured that the model is exposed to the maximum and minimum values of the inputs while training.

b) The accuracy of the AI model also depends on the types of inputs provided to the model [96]. Since there are many input variables upon which the nitrogen in streams depends, we suggest considering all the relevant inputs and then performing a sensitivity analysis to select the highly sensitive input variables for the prediction. Some of the relevant inputs are daily average rainfall data, daily average river discharge, daily average water temperature, historical data of nitrogen in streams, land use pattern, Julian day, amount of fertilizer applied in the catchment area, and the amount of nitrogen per day added from point sources. Using many input variables leads to the increase in the complexity of the network, which often effects the results of the network. To avoid this complexity, the user should avoid selecting the inter-dependent variables, for example: if the runoff data is included in the input data then the precipitation data can be avoided because runoff is dependent on precipitation and has the same pattern as that of precipitation.

c) ANN is divided into different types, which are utilized for modeling hydrological parameters having different complexity levels. For creating a model involving a huge set of input variables, we suggest creating a hybrid model, which has higher accuracy. The ANN model has to be clipped with other models to create a hybrid model, and hence, it improves the accuracy of the resultant model. Zhang, Zhang and Li [87] utilized a hybrid model (ARIMA and RBFNN) to predict the monthly total nitrogen, and the mean absolute percentage error was reduced to 7.017%. However, in this case, they used only historical monthly data as input to the hybrid model; hence, a hybrid model with a wide range of relevant stochastic input variables will attain increased accuracy.

8. Conclusions

This research paper reviews the previous uses of ANN for the prediction of nitrogen compounds in streams. The efforts that have been made in past decades to predict the nitrogen compounds with greater accuracy are also demonstrated in this work. The current condition of rivers in terms of nitrogen compound concentration is discussed. The major non-point source of nitrogen in the streams is the fertilizer applied in agricultural fields. Excess nitrogen concentration in streams leads to human health issues. The operations of many water treatment plants depend on the concentration of nitrogen in the river. In the past two decades, ANNs have shown greater reliability in predicting the nitrogen compounds and have also helped in optimizing the sources of nitrogen input to the streams. The analysis of the literature reveals that published papers on the prediction of nitrogen compounds using hybrid models are limited. This study suggests the usage of a hybrid model along with the set of suggested relevant input variables and training procedures.

Author Contributions: Conceptualization, A.E.-S. and H.A.A.; Methodology, P.K. and A.E.-S.; Formal Analysis, S.H.L.; Investigation, P.K.; Resources, A.E.-S.; Data Curation, S.H.L.; Writing—Original Draft Preparation, P.K.; Writing—Review & Editing, J.K.W., M.S., and A.S.; Visualization, M.S., A.S., and A.N.A.; Supervision, A.E.-S., S.H.L., N.S.M. and M.R.K.; Project Administration, A.E.-S.; Funding Acquisition, A.E.-S. All authors have read and agreed to the published version of the manuscript.

Funding: This research was funded by University of Malaya Research Grant (UMRG), grant number RP025A-18SUS.

Acknowledgments: The authors appreciate so much the facilities support by the Civil Engineering Department, Faculty of Engineering, University of Malaya, Malaysia.

Conflicts of Interest: We declare no conflicts of interest with any person or institute.

References

1. Maloney, K.O.; Weller, D.E. Anthropogenic disturbance and streams: Land use and land-use change affect stream ecosystems via multiple pathways. *Freshw. Biol.* **2011**, *56*, 611–626. [CrossRef]

2. Kilonzo, F.; Masese, F.O.; Van Griensven, A.; Bauwens, W.; Obando, J.; Lens, P.N.L. Spatial–temporal variability in water quality and macro-invertebrate assemblages in the Upper Mara River basin, Kenya. *Phys. Chem. EarthParts A/B/C* **2014**, *67–69*, 93–104. [CrossRef]

3. Jacobs, S.R.; Breuer, L.; Butterbach-Bahl, K.; Pelster, D.E.; Rufino, M.C. Land use affects total dissolved nitrogen and nitrate concentrations in tropical montane streams in Kenya. *Sci. Total Env.* **2017**, *603–604*, 519–532. [CrossRef] [PubMed]

4. Hessong, A. The Composition of Fertilizers. Available online: http://homeguides.sfgate.com/composition-fertilizers-48898.html (accessed on 26 June 2019).

5. Salehi, F.; Prasher, S.O.; Amin, S.; Madani, A.; Jebelli, S.J.; Ramaswamy, H.S.; Tan, C.; Drury, C.F. Prediction of annual nitrate-n losses in drain outflows with artificial neural networks. *Am. Soc. Agric. Eng.* **2000**, *43*, 1137–1143. [CrossRef]

6. Sharma, V.; Negi, S.C.; Rudra, R.P.; Yang, S. Neural networks for predicting nitrate-nitrogen in drainage water. *Agric. Water Manag.* **2003**, *63*, 169–183. [CrossRef]

7. Fewtrell, L. Drinking-water nitrate, methemoglobinemia, and global burden of disease: A discussion. *Environ. Health Perspect* **2004**, *112*, 1371–1374. [CrossRef]

8. Gallo, E.L.; Meixner, T.; Aoubid, H.; Lohse, K.A.; Brooks, P.D. Combined impact of catchment size, land cover, and precipitation on streamflow and total dissolved nitrogen: A global comparative analysis. *Glob. Biogeochem. Cycles* **2015**, *29*, 1109–1121. [CrossRef]

9. Ward, M.H.; deKok, T.M.; Levallois, P.; Brender, J.; Gulis, G.; Nolan, B.T.; VanDerslice, J.; International Society for Environmental. Workgroup report: Drinking-water nitrate and health–recent findings and research needs. *Environ. Health Perspect* **2005**, *113*, 1607–1614. [CrossRef]

10. Reddy, A.G.S.; Niranjan Kumar, K.; Subba Rao, D.; Sambashiva Rao, S. Assessment of nitrate contamination due to groundwater pollution in north eastern part of Anantapur District, A.P. India. *Environ. Monit. Assess.* **2009**, *148*, 463–476. [CrossRef]

11. Hamed, Y.; Awad, S.; Ben Sâad, A. Nitrate contamination in groundwater in the Sidi Aïch–Gafsa oases region, Southern Tunisia. *Environ. Earth Sci.* **2013**, *70*, 2335–2348. [CrossRef]

12. Rahmati, O.; Samani, A.N.; Mahmoodi, N.; Mahdavi, M. Assessment of the Contribution of N-Fertilizers to Nitrate Pollution of Groundwater in Western Iran (Case Study: Ghorveh–Dehgelan Aquifer). *Water Qual. Expo. Health* **2015**, *7*, 143–151. [CrossRef]

13. Räike, A.; Pietiläinen, O.P.; Rekolainen, S.; Kauppila, P.; Pitkänen, H.; Niemi, J.; Raateland, A.; Vuorenmaa, J. Trends of phosphorus, nitrogen and chlorophyll a concentrations in Finnish rivers and lakes in 1975–2000. *Sci. Total Environ.* **2003**, *310*, 47–59. [CrossRef]

14. Qiu, Y.; Shi, H.-C.; He, M. Nitrogen and Phosphorous Removal in Municipal Wastewater Treatment Plants in China: A Review. *Int. J. Chem. Eng.* **2010**, *2010*, 1–10. [CrossRef]

15. Zhang, W.S.; Swaney, D.P.; Li, X.Y.; Hong, B.; Howarth, R.W.; Ding, S.H. Anthropogenic point-source and non-point-source nitrogen inputs into Huai River basin and their impacts on riverine ammonia–nitrogen flux. *Biogeosciences* **2015**, *12*, 4275–4289. [CrossRef]

16. Indah Water, M. Ammonia. Available online: https://www.iwk.com.my/do-you-know/ammonia (accessed on 6 July 2019).

17. Fogelman, S.; Blumenstein, M.; Zhao, H. Estimation of chemical oxygen demand by ultraviolet spectroscopic profiling and artificial neural networks. *Neural Comput. Appl.* **2005**, *15*, 197–203. [CrossRef]

18. He, B.; Oki, T.; Sun, F.; Komori, D.; Kanae, S.; Wang, Y.; Kim, H.; Yamazaki, D. Estimating monthly total nitrogen concentration in streams by using artificial neural network. *J. Env. Manag.* **2011**, *92*, 172–177. [CrossRef]

19. Holmberg, M.; Forsius, M.; Starr, M.; Huttunen, M. An application of artificial neural networks to carbon, nitrogen and phosphorus concentrations in three boreal streams and impacts of climate change. *Ecol. Model.* **2006**, *195*, 51–60. [CrossRef]

20. Lek, S.; Guiresse, M.; Giraudel, J.-L. Predicting stream nitrogen concentration from watershed features using neural networks. *Water Resour. Res.* **1999**, *33*, 3469–3478. [CrossRef]

21. Sarangi, A.; Bhattacharya, A.K. Comparison of Artificial Neural Network and regression models for sediment loss prediction from Banha watershed in India. *Agric. Water Manag.* **2005**, *78*, 195–208. [CrossRef]

22. Ehtram, M.; Karami, H.; Mousavi, S.-F.; El-Shafie, A.; Amini, Z. Optimizing Dam and Reservoirs Operation Based Model Utilizing Shark Algorithm Approach. *Knowl. -Based Syst.* **2017**. [CrossRef]

23. Aguilera, P.A.; Frenich, A.G.; Torres, J.A.; Castro, H.; Vidal, J.L.M.; Canton, M. Application of the kohonen neural network in coastal water management: Methodological development for the assessment and prediction of water quality. *Water Resources* **2001**, *35*, 4053–4062. [CrossRef]

24. Chang, L.-C.; Chang, F.-J. Intelligent control for modelling of real-time reservoir operation. *Hydrol. Process.* **2001**, *15*, 1621–1634. [CrossRef]

25. Suen, J.-P.; Eheart, J.W. Evaluation of Neural Networks for Modeling Nitrate Concentrations in Rivers. *J. Water Resour. Plan. Manag. ASCE* **2003**, *129*, 505–510. [CrossRef]

26. Zaheer, I.; Bai, C.-G. Application of artificial neural network for water quality management. *Lowl. Technol. Int.* **2003**, *5*, 10–15.

27. Tayfur, G.; Swiatek, D.; Wita, A.; Singh, V.P. Case Study: Finite Element Method and Artificial Neural Network Models for Flow through Jeziorsko Earthfill Dam in Poland. *J. Hydraul. Eng.* **2005**, *131*, 431–440. [CrossRef]

28. Mazvimavi, D.; Meijerink, A.M.J.; Savenije, H.H.G.; Stein, A. Prediction of flow characteristics using multiple regression and neural networks: A case study in Zimbabwe. *Phys. Chem. EarthParts A/B/C* **2005**, *30*, 639–647. [CrossRef]

29. He, B.; Takase, K. Application of the Artificial Neural Network Method to Estimate the Missing Hydrologic Data. *J. Jpn. Soc. Hydrol. Water Resour.* **2006**, *19*, 249–257. [CrossRef]

30. Cigizoglu, H.K.; Alp, M. Rainfall-Runoff Modelling Using Three Neural Network Methods. *ICAISC* **2004**, 166–171.

31. Riad, S.; Mania, J.; Bouchaou, L.; Najjar, Y. Rainfall-runoff model usingan artificial neural network approach. *Math. Comput. Model.* **2004**, *40*, 839–846. [CrossRef]

32. Shamseldin, A.Y.; Nasr, A.E.; O'Connor, K.M. Comparison of different forms of the Multi-layer Feed-Forward Neural Network method used for river flow forecasting. *Hydrol. Earth Syst. Sci. Discuss.* **2002**, *6*, 671–684. [CrossRef]

33. Teschl, R.; Randeu, W.L. A neural network model for short term river flow prediction. *Nat. Hazards Earth Syst. Sci.* **2006**, *6*, 629–635. [CrossRef]

34. Wu, C.L.; Chau, K.W. Rainfall–runoff modeling using artificial neural network coupled with singular spectrum analysis. *J. Hydrol.* **2011**, *399*, 394–409. [CrossRef]

35. Gazzaz, N.M.; Yusoff, M.K.; Aris, A.Z.; Juahir, H.; Ramli, M.F. Artificial neural network modeling of the water quality index for Kinta River (Malaysia) using water quality variables as predictors. *Mar. Pollut Bull.* **2012**, *64*, 2409–2420. [CrossRef] [PubMed]

36. Khalil, B.; Ouarda, T.B.M.J.; St-Hilaire, A. Estimation of water quality characteristics at ungauged sites using artificial neural networks and canonical correlation analysis. *J. Hydrol.* **2011**, *405*, 277–287. [CrossRef]

37. Palani, S.; Liong, S.Y.; Tkalich, P. An ANN application for water quality forecasting. *Mar. Pollut Bull.* **2008**, *56*, 1586–1597. [CrossRef]

38. Singh, K.P.; Basant, A.; Malik, A.; Jain, G. Artificial neural network modeling of the river water quality—A case study. *Ecol. Model.* **2009**, *220*, 888–895. [CrossRef]

39. Suo, W.Q.; Dong-Bao, S.; Wei-Ping, H.; Yu-Zhong, L.; Xu-Rong, M.; Yan-Qing, Z. Human activities and nitrogen in waters. *Acta Ecol. Sin.* **2012**, *32*, 174–179. [CrossRef]

40. USGS. Nitrogen and Water. Available online: https://www.usgs.gov/special-topic/water-science-school/science/nitrogen-and-water?qt-science_center_objects=0#qt-science_center_objects (accessed on 26 June 2019).

41. Farid, A.M.; Lubna, A.; Choo, T.G.; Rahim, M.C.; Mazlin, M. A Review on the Chemical Pollution of Langat River, Malaysia. *Asian J. Water Environ. Pollut.* **2016**, *13*, 9–15. [CrossRef]

42. Yi, Q.; Chen, Q.; Hu, L.; Shi, W. Tracking nitrogen sources, transformation and transport at a basin scale with complex plain river networks. *Environ. Sci. Technol.* **2017**. [CrossRef]

43. Nuruzzaman, M.; Mamun, A.A.; Salleh, M.N.B. Determining ammonia nitrogen decay rate of Malaysian river water in a laboratory flume. *Int. J. Environ. Sci. Technol.* **2017**, *15*, 1249–1256. [CrossRef]

44. Rabalais, N.N.; Turner, R.E.; Scavia, D. Beyond Science into Policy: Gulf of Mexico Hypoxia and the Mississippi River. *Bioscience* **2002**, *52*, 129–142. [CrossRef]

45. Rabalais, N.N.; Turner, R.E. Oxygen depletion in the gulf of mexico adjacent to the mississippi river. *Past Present Water Column Anoxia* **2006**, 225–245.

46. Hessen, D.O.; Hindar, A.; Holtan, G. The Significance of Nitrogen Runoff for Eutrophication of Freshwater and Marine Recipients. *R. Swed. Acad. Sci.* **1997**, *26*, 312–320.

47. Dodds, W.; Smith, V. Nitrogen, phosphorus, and eutrophication in streams. *Inland Waters* **2016**, *6*, 155–164. [CrossRef]

48. Dodds, W.K.K.; Welch, E.B. Establishing nutrient criteria in streams. *J. N. Am. Benthol. Soc.* **2000**, *19*, 186–196. [CrossRef]

49. Murdoch, P.S.; Stoddard, J.L. The Role of Nitrate in the Acidification of Streams in the Catskill Mountains of New York. *Water Resour. Res.* **1992**, *28*, 2707–2720. [CrossRef]

50. Gündüz, O. Water Quality Perspectives in a Changing World. *Water Qual. Expo. Health* **2015**, *7*, 1–3. [CrossRef]

51. Su, X.; Wang, H.; Zhang, Y. Health Risk Assessment of Nitrate Contamination in Groundwater: A Case Study of an Agricultural Area in Northeast China. *Water Resour. Manag.* **2013**, *27*, 3025–3034. [CrossRef]

52. He, S.; Wu, J. Hydrogeochemical Characteristics, Groundwater Quality, and Health Risks from Hexavalent Chromium and Nitrate in Groundwater of Huanhe Formation in Wuqi County, Northwest China. *Expo. Health* **2019**, *11*, 125–137. [CrossRef]

53. Hossain, F.; Chang, N.-B.; Wanielista, M.; Xuan, Z.; Daranpob, A. Nitrification and Denitrification in a Passive On-site Wastewater Treatment System with a Recirculation Filtration Tank. *Water Qual. Expo. Health* **2010**, *2*, 31–46. [CrossRef]

54. Gulis, G.; Czompolyova, M.; R Cerhan, J. An Ecologic Study of Nitrate in Municipal Drinking Water and Cancer Incidence in Trnava District, Slovakia. *Environ. Res.* **2002**, *88*, 182–187. [CrossRef] [PubMed]

55. Chen, J.; Wu, H.; Qian, H.; Gao, Y. Assessing Nitrate and Fluoride Contaminants in Drinking Water and Their Health Risk of Rural Residents Living in a Semiarid Region of Northwest China. *Expo. Health* **2017**, *9*, 183–195. [CrossRef]

56. Sawyer, C.N.; McCarty, P.L.; Parkin, G.F. *Chemistry for Environmental Engineering and Science*, 5th ed.; McGraw Hill: New York, NY, USA, 2003; p. 667.

57. Aslan, S.; Turkman, A. Biological denitrification of drinking water using various natural organic solid substrates. *Water Sci. Technol. A J. Int. Assoc. Water Pollut. Res.* **2003**, *48*, 489–495. [CrossRef]

58. Della Rocca, C.; Belgiorno, V.; Meric, S. Cotton-supported heterotrophic denitrification of nitrate-rich drinking water with a sand filtration post-treatment. *Water SA* **2005**, *31*, 229–236. [CrossRef]

59. Akrami, S.A.; El-Shafie, A.; Jaafar, O. Improving Rainfall Forecasting Efficiency Using Modified Adaptive Neuro-Fuzzy Inference System (MANFIS). *Water Resour Manag.* **2013**. [CrossRef]

60. Farzad, F.; El-Shafie, A.H. Performance Enhancement of Rainfall Pattern – Water Level Prediction Model Utilizing Self-Organizing-Map Clustering Method. *Water Resour Manag.* **2016**. [CrossRef]

61. Anctil, F.; Filion, M.; Tournebize, J. A neural network experiment on the simulation of daily nitrate-nitrogen and suspended sediment fluxes from a small agricultural catchment. *Ecol. Model.* **2009**, *220*, 879–887. [CrossRef]

62. El-Shafie, A.H.; El-Shafie, A.; Mazoghi, H.G.E.; Shehata, A.; Taha, M.R. Artificial neural network technique for rainfall forecasting applied to Alexandria, Egypt. *Int. J. Phys. Sci.* **2011**, *6*, 1306–1316. [CrossRef]

63. Raju, M.M.; Srivastava, R.K.; Bisht, D.C.S.; Sharma, H.C.; Kumar, A. Development of Artificial Neural-Network-Based Models for the Simulation of Spring Discharge. *Adv. Artif. Intell.* **2011**, *2011*, 1–11. [CrossRef]

64. Shafie, A.H.E.; El-Shafie, A.; Almukhtar, A.; Taha, M.R.; Mazoghi, H.G.E.; Shehata, A. Radial basis function neural networks for reliably forecasting rainfall. *J. Water Clim. Chang.* **2012**. [CrossRef]

65. Xie, Y. Values and Limitations of Statistical Models. *Res. Soc. Strat. Mobil* **2011**, *29*, 343–349. [CrossRef] [PubMed]

66. Jain, A.K.; Mao, J.; Mohiuddin, K.M. Artificial Neural Networks: A Tutorial. *IEEE* **1996**, 31–44. [CrossRef]

67. El-Shafie, A.; Noureldin, A.; Taha, M.; Hussain, A.; Mukhlisin, M. Dynamic versus static neural network model for rainfall forecasting at Klang River Basin, Malaysia. *Hydrol. Earth Syst. Sci.* **2012**, 1151–1169. [CrossRef]

68. Voyant, C.; Muselli, M.; Paoli, C.; Nivet, M.-L. Numerical weather prediction (NWP) and hybrid ARMA/ANN model to predict global radiation. *Energy* **2012**, 341–355. [CrossRef]

69. Grosan, C.; Abraham, A. *Intelligent Systems A Modern Approach*; Springer: Berlin/Heidelberg, Germany, 2011; Volume 17.

70. Fallah, S.N.; Deo, R.C.; Shojafar, M.; Conti, M.; Shamshirband, S. Computational Intelligence Approaches for Energy Load Forecasting in Smart Energy Management Grids: State of the Art, Future Challenges, and Research Directions. *Energies* **2018**, *11*, 596. [CrossRef]

71. Fiyadh, S.S.; AlSaadi, M.A.; Jaafar, W.Z.; AlOmar, M.K.; Fayaed, S.S.; Mohd, N.S.; Hin, L.S.; El-Shafie, A. Review on heavy metal adsorption processes by carbon nanotubes. *J. Clean. Prod.* **2019**, 783–793. [CrossRef]

72. Martín-Martín, A.; Thelwall, M.; Orduna-Malea, E.; López-Cózar, E.D. Google Scholar, Microsoft Academic, Scopus, Dimensions, Web of Science, and OpenCitations' COCI: A multidisciplinary comparison of coverage via citations. 2020.

73. Kannel, P.R.; Lee, S.; Lee, Y.S.; Kanel, S.R.; Pelletier, G.J. Application of automated QUAL2Kw for water quality modeling and management in the Bagmati River, Nepal. *Ecol. Model.* **2007**, *202*, 503–517. [CrossRef]

74. The Star. Five Water Treatment Plants Shut down due to Ammonia Pollution Fully Operational. Available online: https://www.thestar.com.my/news/nation/2019/04/06/five-water-treatment-plants-shut-down-due-to-ammonia-pollution-fully-operational/ (accessed on 26 June 2019).

75. New Straits Times. Another Johor Water Treatment Plant Shuts down over Ammonia Pollution. Available online: https://www.nst.com.my/news/nation/2017/11/304914/update-another-johor-water-treatment-plant-shuts-down-over-ammonia (accessed on 26 June 2019).

76. Rekacewicz, P. Nitrate Levels: Concentrations at River Mouths. Available online: http://www.grida.no/resources/5650 (accessed on 26 June 2019).

77. Canadian Council of Ministers of the Environment. *Canadian Water Quality Guidelines for the Protection of Aquatic Life*; Ammonia: Regina, SK, Canada, 2010.

78. Basheer, A.O.; Hanafiah, M.M.; J. Abdulhasan, M. A Study on Water Quality from Langat River, Selangor. *Acta Sci. Malays.* **2017**, *1*, 01–04. [CrossRef]

79. Juahir, H.; Zain, S.M.; Yusoff, M.K.; Hanidza, T.I.; Armi, A.S.; Toriman, M.E.; Mokhtar, M. Spatial water quality assessment of Langat River Basin (Malaysia) using environmetric techniques. *Env. Monit Assess.* **2011**, *173*, 625–641. [CrossRef]

80. Yan, W.; Mayorga, E.; Li, X.; Seitzinger, S.P.; Bouwman, A.F. Increasing anthropogenic nitrogen inputs and riverine DIN exports from the Changjiang River basin under changing human pressures. *Glob. Biogeochem. Cycles* **2010**, *24*. [CrossRef]

81. Singh, K.P.; Malik, A.; Sinha, S. Water quality assessment and apportionment of pollution sources of Gomti river (India) using multivariate statistical techniques—A case study. *Anal. Chim. Acta* **2005**, *538*, 355–374. [CrossRef]

82. Li, S.; Cheng, X.; Xu, Z.; Han, H.; Zhang, Q. Spatial and temporal patterns of the water quality in the Danjiangkou Reservoir, China. *Hydrol. Sci. J.* **2009**, *54*, 124–134. [CrossRef]

83. Pernet-Coudrier, B.; Qi, W.; Liu, H.; Muller, B.; Berg, M. Sources and pathways of nutrients in the semi-arid region of Beijing-Tianjin, China. *Env. Sci Technol* **2012**, *46*, 5294–5301. [CrossRef] [PubMed]

84. Li, W.; Li, X.; Su, J.; Zhao, H. Sources and mass fluxes of the main contaminants in a heavily polluted and modified river of the North China Plain. *Env. Sci. Pollut. Res. Int.* **2014**, *21*, 5678–5688. [CrossRef] [PubMed]

85. Zhang, L.; Song, X.; Xia, J.; Yuan, R.; Zhang, Y.; Liu, X.; Han, D. Major element chemistry of the Huai River basin, China. *Appl. Geochem.* **2011**, *26*, 293–300. [CrossRef]

86. Wang, B.; Oldham, C.; Hipsey, M.R. Comparison of Machine Learning Techniques and Variables for Groundwater Dissolved Organic Nitrogen Prediction in an Urban Area. *Procedia Eng.* **2016**, *154*, 1176–1184. [CrossRef]

87. Zhang, L.; Zhang, G.X.; Li, R.R. Water Quality Analysis and Prediction Using Hybrid Time Series and Neural Network Models. *J. Agr. Sci. Tech.* **2016**, *18*, 975–983.

88. Markus, M.; Hejazi, M.I.; Bajcsy, P.; Giustolisi, O.; Savic, D.A. Prediction of weekly nitrate-N fluctuations in a small agricultural watershed in Illinois. *J. Hydroinform.* **2010**, *12*, 251–261. [CrossRef]
89. Amiri, B.J.; Nakane, K. Comparative prediction of stream water total nitrogen from land cover using artificial neural network and multiple linear regression approaches. *Pol. J. Environ. Stud.* **2009**, *18*, 151–160.
90. Zeleňáková, M.; Čarnogurská, M.; Šlezingr, M.; Słyś, D. Model based on dimensional analysis for prediction of nitrogen and phosphorus concentration in the River Laborec. *Hydrol. Earth Syst. Sci. Discuss.* **2012**, *9*, 5611–5634. [CrossRef]
91. Akrami, S.A.; El-Shafie, A.; Naseri, M.; Santos, C.A.G. Rainfall data analyzing using moving average (MA) model and wavelet multi-resolution intelligent model for noise evaluation to improve the forecasting accuracy. *Neural Comput Applic* **2014**, *25*, 1853–1861. [CrossRef]
92. May, D.B.; Sivakumar, M. Prediction of urban stormwater quality using artificial neural networks. *Environ. Model. Softw.* **2009**, *24*, 296–302. [CrossRef]
93. Markus, M.; Tsai, C.W.-S.; Demissie, M. Uncertainty of Weekly Nitrate-Nitrogen Forecasts Using Artificial Neural Networks. *J. Environ. Eng.* **2003**, *129*, 267–274. [CrossRef]
94. Najah, A.; El-Shafie, A.; Karim, O.A.; Jaafar, O. Integrated versus isolated scenario for prediction dissolved oxygen at progression of water quality monitoring stations. *Hydrol. Earth Syst. Sci.* **2011**, *15*, 2693–2708. [CrossRef]
95. Ahmed, A.N.; El-Shafie, A.; Karim, O.A.; El-Shafie, A. An augmented wavelet de-noising technique with neuro-fuzzy inference system for water quality prediction. *Int. J. Innov. Comput. Inf. Control.* **2012**, *8*.
96. Maier, H.R.; Dandy, G.C. Neural networks for the prediction and forecasting of water resources variables: A review of modelling issues and applications. *Environ. Model. Softw.* **2000**, *15*, 101–124. [CrossRef]

MDPI
St. Alban-Anlage 66
4052 Basel
Switzerland
Tel. +41 61 683 77 34
Fax +41 61 302 89 18
www.mdpi.com

Sustainability Editorial Office
E-mail: sustainability@mdpi.com
www.mdpi.com/journal/sustainability